Lecture Notes in Mathematics

Edited by A. Dold and B. Eckmann

T0212364

537

Set Theory and Hierarchy Theory
A Memorial Tribute to Andrzej Mostowski

Bierutowice, Poland 1975

Edited by W. Marek, M. Srebrny and A. Zarach

Springer-Verlag
Berlin · Heidelberg · New York 1976

Editors
Wiktor Marek
Institute of Mathematics
University of Warsaw
PKiN IX p.
00-901 Warszawa/Poland

Marian Srebrny
Mathematical Institute
Polish Academy of Sciences
P.O. Box 137
00-950 Warszawa/Poland

Andrzej Zarach
Institute of Mathematics
Technical University
ul. Wybrzeże Wyspiańskiego 27
50-370 Wrocław/Poland

Library of Congress Cataloging in Publication Data

Conference on Set Theory and Hierarchy Theory, 2d, Bierutowice, Poland, 1975.
Set theory and hierarchy theory, Bierutowice, Poland, 1975.

(Lecture notes in mathematics ; 537)
1. Set theory--Congresses. 2. Model theory--Congresses. 3. Recursive functions--Congresses. 4. Mostowski, Andrzej--Bibliography. I. Marek, Wiktor. II. Srebrny, Marian, 1947- III. Zarach, Andrzej, 1948- IV. Title. V. Series: Lecture notes in mathematics (Berlin) ; 537.
QA3.L28 vol. 537 [QA248] 510'.8s [511'.32]
76-26534

AMS Subject Classifications (1970): 02F27, 02F29, 02F35, 02H05, 02H13, 02H15, 02K05, 02K10, 02K30.

ISBN 3-540-07856-8 Springer-Verlag Berlin · Heidelberg · New York
ISBN 0-387-07856-8 Springer-Verlag New York · Heidelberg · Berlin

Printed in Germany

Printing and binding: Beltz Offsetdruck, Hemsbach/Bergstr

IN MEMORIAM PROFESSOR ANDRZEJ MOSTOWSKI

ANDRZEJ MOSTOWSKI

November 1, 1913 - August 22, 1975

FOREWORD

This volume constitutes the Proceedings of the Second Conference on Set Theory and Hierarchy Theory held at the mountain resort of Bierutowice in Poland, September 17-28, 1975.
This conference was organized and sponsored by the Institute of Mathematics of Technical University of Wrocław.
The programme committee comprised Andrzej Mostowski (Warsaw), Czesław Ryll-Nardzewski (Wrocław), Wiktor Marek (Warsaw), Leszek Pacholski (Wrocław), Bogdan Węglorz (Wrocław) and the organizing committee comprised Marian Srebrny (Warsaw), Jan Waszkiewicz (Wrocław), Agnieszka Wojciechowska (Wrocław), Andrzej Zarach (Wrocław) - chairman.
The conference was attended by 71 registered participants, of these 41 were from Poland.

In view of the sudden death of Professor Mostowski we decided to dedicate this volume to his memory and to include a brief curriculum vitae, full bibliography of his works and an unpublished paper by him. Professor Mostowski contributed this paper to the similar conference held in 1974. Unfortunately, it has not been published yet. We wish to thank Mrs. M.Mostowska for her kind permission to include it to these Proceedings. The contributed papers are closely connected with Professor Mostowski's main topics of interest. During the conference a special session devoted to the memory of Professor Mostowski was held. We believe that these Proceedings constitute the participants' best tribute to him.

On behalf of the organizing committee of the conference we wish to thank the Technical University of Wrocław and all the people who by their help contributed to the success of the conference and its good scientific and friendly atmosphere. Special thanks are due to A. Wojciechowska and J.Waszkiewicz for their help in organizing the conference.We gratefuly acknowledge the assistance of Professor

G.H. Müller in suggesting the present form of the volume and for all his assitance in various stages of its preparation. We acknowledge also kind assistance of those who read and refereed the contributed papers (A.H.Lachlan, G.Wilmers and our colleagues from Warsaw and Wrocław). Finally, we owe special gratitude to Miss T.Bochynek who had enough patience to expertly type all the manuscripts.

March, 1976

Wiktor Marek

Marian Srebrny

Andrzej Zarach

C O N T E N T S

LIST OF REGISTERED PARTICIPANTS

Bohuslav Balcar	Praha
John Bell	London
Ian Bergstra	Utrecht
Konrad Bieliński	Kielce
Piotr Borowik	Częstochowa
Lev Bukovský	Košice
Wojciech Buszkowski	Poznań
Douglas Cenzer	Gainesville
Jaroslav Chudáček	Praha
Jacek Cichoń	Wrocław
John Crossley	Clayton
Bernd Dahn	Berlin
Keith Devlin	Bonn
Małgorzata Dubiel	Warszawa
Arleta Dyluś	Wrocław
Anna Fedyszak	Warszawa
Ulrich Felgner	Heidelberg
Adam Figura	Wrocław
Siegfried Gottwald	Leipzig
Edward Grzegorek	Wrocław
Andrzej Grzegorczyk	Warszawa
Wojciech Guzicki	Warszawa
Petr Hájek	Praha
Tomáš Havránek	Praha
Czesława Jakubowicz	Wrocław
Leena-Marija Jauko	Helsinki
Laurence Kirby	Manchester

Henryk Kotlarski	Kielce
Stanisław Krajewski	Warszawa
Adam Krawczyk	Warszawa
Michał Krynicki	Warszawa
Wiesław Kułaga	Wrocław
Alistair Lachlan	Vancouver
Anna Lemańska	Warszawa
Anna Lin	Warszawa
Francis Louveaux	Paris
Francis Loventhal	Brussel
Wiktor Marek	Warszawa
Mariano Martinez	Warszawa
Tom Mc Laughlin	Lubbock
Roman Murawski	Poznań
Leszek Pacholski	Wrocław
Cecylia Rauszer	Warszawa
Stanisław Roguski	Wrocław
Michał Ryszkiewicz	Warszawa
Czesław Ryll-Nardzewski	Wrocław
Detlef Seese	Berlin
Felice Serano	Warszawa
Kostas Skandalis	Wrocław
Alan Slomson	Leeds
Anton Sochor	Praha
Krzysztof Sokołowski	Warszawa
Marian Srebrny	Warszawa
Petr Štěpánek	Praha
Zbigniew Szczepaniak	Wrocław
Rastislav Telgársky	Wrocław
Jerzy Tomasik	Wrocław
John Truss	Oxford

Peter Tuschik	Berlin
Jouko Väänänen	Helsinki
Zygmunt Vetulani	Poznań
Petr Vopěnka	Praha
Anita Wasilewska	Warszawa
Jan Waszkiewicz	Wrocław
Martin Weese	Berlin
Jędrzej Wierzejewski	Wrocław
Kazimierz Wiśniewski	Gdańsk
Agnieszka Wojciechowska	Wrocław
Włodzimierz Zadrożny	Wrocław
Andrzej Zarach	Wrocław
Paweł Zbierski	Warszawa

CURRICULUM VITAE OF A.MOSTOWSKI

Professor Andrzej Mostowski was born on November 1st, 1913 in the city of Lwów. He studied mathematics at Warsaw University from 1931-1936. After receiving his master's degree he spent one year in Vienna and Zürich. During that time he worked both with Kurt Gödel and Paul Bernays. Mostowski's Doctoral Dissertation (written under supervision of Alfred Tarski) was devoted to the interrelationships of various definitions of the notion of infinite set. His thesis was defended in 1938 and Mostowski then started to work in the Warsaw Hydro-Meteorological Institute. When the Nazis occupied Poland he had to work as an accountant in a tile factory, at the same time taught in the Underground Warsaw University. After the war Mostowski defended in Kraków, 1945 his thesis for the degree of habilitation. The theme of this work was the axiom of choice for finite sets. In 1946 he settled in Warsaw as assistant professor of Mathematics in Warsaw University. In 1947 he became associate professor, and full professor in 1951. In 1952 he was the dean of the Faculty of Mathematics and Physics. From 1948 to 1968 he was the head of the Division of Foundations of Mathematics at the Mathematical Institute of the Polish Academy of Sciences. In 1956 he was elected to associate membership of the Polish Academy of Sciences and in 1966 to full membership. He received a Polish State Prize in 1953 and in 1966 and a Jurzykowski Foundation Prize in 1972. In 1972 Mostowski was elected President of the Section of Logic, Methodology and Philosophy of Science of the International Union of History and Philosophy of Science.

Professor Andrzej Mostowski died in Vancouver, B.C., on August 22nd, 1975.

Professor Mostowski obtained his most important scientific results in Recursion Theory, Foundations of Set Theory and in Model Theory. Many of his results have become classical now. He was a coauthor (with K.Kuratowski) of a monograph on Set Theory and the author of a monograph on Foundations of Set Theory. Throughout his whole life he devoted much time to pedagogical activities, in particular both in Foundations of Mathematics and in Algebra. Numerous mathematicians in various parts of the world (including the editors of these Proceedings) owe to Mostowski their mathematical education.

In the history of Foundations of Mathematics after the second world war Mostowski distinguished himself as one of those who set the trends in this branch of science. Owing to his deep knowledge of other branches of mathematics Professor Mostowski was able to extend considerably the scope of the Foundations of Mathematics. His activity helped to overcome the harmful isolation of the Foundations of Mathematics from other branches of modern mathematics.

BIBLIOGRAPHY OF MOSTOWSKI'S WORKS

1. Abzahlbare Boole'sche Körper und ihre Anwendungen auf die allgemeine Metamathematik,
Fundamenta Mathematicae XXIX (1937) pp 34-53.

2. O niezależności definicji skończoności w systemie logiki. Appendix to Roczniki Polskiego Towarzystwa Matematycznego, volume XI (1938) pp 1-54.

3. Über gewisse universelle Relationen.
Ann. Soc. Pol. de Math. XVII (1939) pp 117-118.

4. Über den Begriff der endlichen Menge.
Sprawozdania Towarzystwa Naukowego Warszawskiego XXXI (1938) pp 13-20.

5. (+ A. Lindenbaum) Über Unabhägigkeit des Auswahlaxioms und einiger seiner Folgerungen.
Sprawozdania Towarzystwa Naukowego Warszawskiego XXXI (1939) pp 27-32.

6. (+ A. Tarski) Boole'sche Ringe mit geordneter Basis.
Fundamenta Mathematicae XXXII (1939) pp 69-86.

7. Über die Unabhängigkeit des Wohlordnungastzes von Ordnungsprinzip.
Fundamenta Mathematicae XXXII (1939) pp 201-252.

8. Bemerkungen zum Begriff der inhaltlichen Widerspruchsfreiheit.
Journal of Symbolic Logic 4 (1939) pp 113-114.

9. Remarque sur une note de W. Sierpiński.
Fundamenta Mathematicae XXXIII (1946) pp 7-8.

10. Axiom of choice for finite sets.
Fundamenta Mathematicae XXXIII (1945) pp 137-168.

11. O zdaniach nierozstrzygalnych w sformalizowanych systemach matematyki.
Kwartalnik Filozoficzny XVI (1946) pp 223-276.

12. Zarys teorii Galois.
Appendix to course book "Algebra Wyższa" by W. Sierpiński (1946) pp 371-428.

13. On definable sets of positive integers.
Fundamenta Mathematicae XXXIV (1947) pp 81-112.

14. On absolute properties of relations.
Journal of Symbolic Logic 12 (1947) pp 33-42.

15. On the principle of dependent choices.
Fundamenta Mathematicae XXXV (1948) pp 127-130.

16. Proofs of non-deducibility in intuitionistic functional calculus.
Journal of Symbolic Logic 13 (1948) pp 193-203.

17. On a set of integers not definable by means of one quantifier predicates.
Ann. Soc. Pol. de Math. XXI (1948) pp 114-119.

18. Logika Matematyczna.
Monografie Matematyczne t.XVIII (1948) Warszawa-Wrocław pp XIII + 338.

19. Un theoreme sur les nombres cos $2k\pi/n$.
Colloquium Mathematicum I (1948) pp 195-196.

20. Sur l'interpretation geometrique et topologique des notions logiques.
Actes du X-eme Congres International de Philosophie (Amsterdam 11-18 Aout 1948) Amsterdam 1949 pp 610-617.

21. An undecidable arithmetical statement.
Fundamenta Mathematicae XXXVI (1949) pp 143-164.

22. (+ A. Tarski) Arithmetical classes and types of well-ordered systems.
Bull. Amer. Math. Soc. 55 (1949) p 65 (1192).

23. (+ A. Tarski) Undecidability on the arithmetic of integers and in the theory of rings.
Journal of Symbolic Logic 14 (1949) p 76.

24. La vie et l'oeuvre de S. Dickstein.
Prace Matematyczno-Fizyczne 47 (1949) pp 7-12.

25. Kilka refleksji na temat zadań logicznych z "Matematyki". Matematyka 3 (1950) pp 6-11.

26. On the rules of proof in the pure functional calculus of the first order.
Journal of Symbolic Logic 16 (1951) pp 107-111.

27. Some impredicative definitions in the axiomatic set theory. Fundamenta Mathematicae XXXVII (1951) pp 111-124. Correction: ibid XXXVIII (1952) p 238.

28. A classification of logical systems.
Studia Philosophica 4 (1951) pp 237-274.

29. (+ K. Kuratowski) Sur un probleme de la theorie des groupes et son rapport a la topologie.
Colloquium Mathematicum II (1951) pp 212-215.

30. Groups connected with Boolean algebras.
Colloquium Mathematicum II (1951) pp 216-219.

31. On direct product of theories.
Journal of Symbolic Logic 17 (1952) pp 1-31.

32. On models of axiomatic systems.
Fundamenta Mathematicae XXXIX (1952) pp 133-158.

33. (+ K. Kuratowski) Teoria Mnogości.
Monografie Matematyczne XXVII Warszawa-Wrocław 1952 pp IX + 311.

34. Sentences undecidable in formalized arithmetic.
in: Studies in Logic and the Foundations of Mathematics Amsterdam (1952) pp VIII + 117.

35. A lemma concerning recursive functions and its applications. Bull. Acad. Pol. Sci. I (1953) pp 277- 280 (wersja rosyjska pp 275-279).

36. On a system of axioms which has no recursively enumerable arithmetic model.
Fundamenta Mathematicae XL (1953) pp 56-61.

37. O tzw. konstruktywnych prawach w matematyce.
Myśl Filozoficzna 1 (7) (1953) pp 230-241.

38. (+ M. Stark) Algebra Wyższa, cz.I.
w serii: Biblioteka Matematyczna I 3 Warszawa 1953 pp VI + 308.

39. (+ H. Rasiowa) O geometrycznej interpretacji wyrażeń logicznych.

Studia Logica I (1952) pp 274-275.

40. (+ A. Tarski i R. M. Robinson) Undecidable theories. in:Studies in Logic and the Foundations of Mathematics Amsterdam 1953 pp IX + 98.

41. (+ M. Stark) Algebra Wyższa, cz.II. w serii:Biblioteka Matematyczna III Warszawa 1954 pp VII + 173.

42. (+ M. Stark) Algebra Wyższa, cz.III. w serii:Biblioteka Matematyczna IV Warszawa 1954 pp VII + 262.

43. Współczesny stan badań nad podstawami matematyki. Prace Matematyczne 1 (1954) pp 13-55.

44. Sovremennoje sostajanie issledovanij po osnovanijam matematiki. Uspechi Mat. Nauk 9 (1954) pp 1-38.

45. Podstawy matematyki na VIII zjeździe matematyków polskich. Myśl Filozoficzna 2 (1954) pp 328-330.

46. Development and applications of the "projective" classification of sets of integers. Proceedings of the International Congress of Mathematicians Amsterdam 1954 vol.1 pp 280-288.

47. (+ other logicians) The present state of investigations of foundations of mathematics. Rozprawy Matematyczne (Dissertationes Mathematicae) IX (1955)pp 1-48 (German translation: Die Hauptreferate des 8 polonisches Mathematiker Kongres, Berlin 1954 pp 11-44).

48. A formula without recursively enumerable model. Fundamenta Mathematicae XLII (1955) pp 125-140.

49. Examples of sets definable by means of two and three quantifiers. Fundamenta Mathematicae XLII (1955) pp 259-270.

50. Contributions to the theory of definable sets of integers and functions. Fundamenta Mathematicae XLII (1955) pp 271-275.

51. (+ J. Łoś, H. Rasiowa) A proof of Herbrand's theorem. Journal des Mathematiques Pures et Appliquees (1955) pp 19-54.

52. Eine Verallgemeinerung eines Satzes von M. Deuring. Acta Scientiarum Math. Szeged XVI (1955) pp 197-201.

53. Wyznaczanie stopni niektórych liczb algebraicznych.
Prace Matematyczne 1 (1955) pp 239-252.

54. (+ M. Stark) Elementy Algebry Wyższej.
w serii:Biblioteka Matematyczna 16 I wyd. Warszawa 1955 (i dalsze wydania).

55. (+ A. Ehrenfeucht) Models of axiomatic theories admitting automorphisms.
Fundamenta Mathematicae XLIII (1956) pp 50-68.

56. Concerning a problem of H. Scholz.
Zeitschrift für Mathematische Logik und Grundlagenforschung 2 (1956) pp 210-214.

57. On models of axiomatic set theory.
Bull. Acad. Pol. Sci. (Cl III) IV (1956) pp 663-668.

58. Zamecanija k dokazatelstvam suscestvovanija standartnych modelej.
Trudy 3-go Vsesojuznogo Matematiceskogo Zjazda 1956 Tom 4 Moskva AN SSSR 1959 pp 232-236.

59. Logika matematyczna na międzynarodowym zjeździe matematyków w Amsterdamie.
Studia Logica IV (1956) pp 245-253.

60. L'oeuvre de J. Łukasiewicz dans le domaine de la logique mathematique.
Fundamenta Mathematicae XLIV (1957) pp 1-11.

61. On a generalization of quantifiers.
Fundamenta Mathematicae XLIV (1957) pp 12-36.

62. Computable sequences.
Fundamenta Mathematicae XLIV (1957) pp 37-51.

63. On recursive models of formalized arithmetic.
Bull. Acad. Pol. Sci. Ser. Sci. Math. Astron. Phys. V (1957) pp 706-718.

64. (+ A. Grzegorczyk, Cz.Ryll-Nardzewski) The classical and ω-complete arithmotic.
Journal of Symbolic Logic 23 (1958) pp 188-206.

65. On a problem of W. Kinna and K. Wagner.
Colloquium Mathematicum VI (1958) pp 207-208.

66. Quelques observations sur l'usage des methodes infinitistes dans la meta-mathematiques.
70 Colloquium du CNRS, Paris 1958.

67. (+ Cz. Ryll-Nardzewski) Representability of sets in models of axiomatic theories.
Journal of Symbolic Logic 23 (1958)p 458.

68. (+ M. Stark) Algebra Liniowa.
w serii:Biblioteka Matematyczna 19 I wyd. Warszawa 1958 (i dalsze wydania).

69. On various degrees of constructivism.
in:Constructivity in Mathematics, Studies in Logic and foundations of mathematics Amsterdam 1959 pp 178-194.

70. A class of models of second-order arithmetic.
Bull. Acad. Pol. Sci. Ser. Sci. Math. Astron. Phys. VII (1959) pp 401-404.

71. Formal system of analysis based on an infinitary rule of proof.
Proceedings of Warsaw Symposium on Infinitistic Methods. Warszawa-London 1960 pp 141-166.

72. A generalization of the incompleteness theorem.
Fundamenta Mathematicae XLIX (1961) pp 205-232.

73. An example of a non-axiomatizable many-valued logic.
Zeitschrift für Mathematische Logik und Grundlagenforschung 7 (1961) pp 72-76.

74. Concerning the problem of axiomatizability of the field of real numbers in the weak second order logic.
Essays in Fundations of Mathematics, Jerusalem 1961 pp 269-286.

75. (+ A. Grzegorczyk, Cz. Ryll-Nardzewski) Definability of sets in models of axiomatic theories.
Bull. Acad. Pol. Sci. Ser. Sci. Math. Astron. Phys. IX (1961) pp 163-167.

76. (+ A. Ehrenfeucht) A compact space of models for 1st order theories.
Bull. Acad. Pol. Sci. Ser. Sci. Math. Astron. Phys. IX (1961) pp 369-373.

77. (+ J. Łoś, H. Rasiowa) Addition au travail "A proof of Herbrand theorem".
Journal des Mathematiques Pures et Appliques 1961 pp 129-134.

78. Organizacja i prace naukowe Instytutu Matematyki Uniwersytetu Warszawskiego.

Księga pamiątkowa dla uczczenia 140-lecia Uniwersytetu Warszawskiego. (1961).

79. Axiomatizability of some many-valued predicate calculi.
Fundamenta Mathematicae L (1961-1962) pp 165-190.

80. Representability of sets in formal systems.
Proceedings of Symposium on Recursive Functions New York 1961 (1962) pp 29-48.

81. A problem in the theory of models.
Bull. Acad. Pol. Sci. Ser. Sci. Math. Astron. Phys. X (1962) pp 121-126.

82. On invariant, dual invariant and absolute properties of relations.
Rozprawy Matematyczne (Dissertationes Mathematicae) XXIX (1962) pp 1-38.

83. L'espace des modeles d'une theorie formalisee et quelques-unes des applications.
Annales de la Faculte des Sciences de L'Universite Clevmant 7 (1962) pp 107-116.

84. The Hilbert epsilon function in many valued logics.
Acta Phil. Fennica 16 (1963) pp 169-188.

85. (+ M. Stark) Introduction to Higher Algebra.
Oxford Warszawa 1963 pp 474.

86. Widerspruchsfreiheit und Unabhangigkeit der Kontinuumhypothese.
Elemente der Mathematik 19 (1964) pp 121-125.

87. Thirty years of foundational studies
Acta Phil. Fennica 17 (1965) pp 1-180.

88. Models of Zermelo-Fraenkel set-theory satisfying the axiom of constructibility.
Acta Phil. Fennica 18 (1965) pp 135-144.

89. (+ K. Kuratowski) Teoria Mnogości. wyd. II (całkowicie zmienione) .
Monografie Matematyczne XXVII Warszawa 1966 pp 1-375.

90. Modeles transitifs de le theorie des ensembles de Zermelo-Fraenkel.
Univ. de Montreal 1967 pp 170.

91. Recent results in set theory.
in:Problems in the philosophy of mathematics, Studies in Logic and the Foundations of Mathematics Amsterdam 1967 pp 82-96 i 105-108.

92. O niektórych nowych wynikach matematycznych dotyczących teorii mnogości.
Studia Logica 20 (1967) pp 99-116.

93. Tarski Alfred.
in:The Encyclopedia of Philosophy, ed.by Paul Edwards New York 1967 Vol.8 pp 77-81.

94. (+ K. Kuratowski) Set theory. (translated from polish) Amsterdam Warszawa 1967 pp 417 Studies in Logic and the Foundations of Mathematics.

95. Craig interpolation theorem in some extended systems of logic.
in:Logic, Methodology and Philosophy of Science III Studies in Logic and the Foundations of Mathematics Amsterdam 1968 pp 87-103.

96. Models of Set Theory.
C.I.M.E. (Italy) , 1968. (Lectures delivered in Varenna September 1968) pp 67-179.

97. Niesprzeczność i niezależność hipotezy continuum.
Wiadomości Matematyczne 10 (1968) pp 175-182.

98. (+ Y. Suzuki) On ω-models which are not β-models.
Fundamenta Mathematicae LXV (1969) pp 83-93.

99. Constructible sets with applications.
Studies in Logic and Foundations of Mathematics Amsterdam 1969 pp 1-269.

100. (+ K. Kuratowski) Teorija mnozestv.
Moskva 1970 pp 416.

101. Models of second order arithmetic with definable Skolem functions.
Fundamenta Mathematicae LXXV (1972) pp 223-234. + Correction ibid LXXXIV (1974) p 173.

102. A transfinite sequence of ω-models.
Journal of Symbolic Logic 37 (1972) pp 96-102.

103. Sets.
in:Scientific thought, Paris-Hague 1972 pp 1-34.

104. Matematyka a logika. Refleksje przy lekturze książki A. Grzegorczyka "Zarys arytmetyki teoretycznej" wraz z próbą recenzji.
Wiadomości Matematyczne 15 (1972) pp 79-89.

105. Partial ordering of the family of ω-models.
Logic Methodology and Philosophy of Science IV. Studies in Logic and

the Foundation of Mathematics Amsterdam 1973 pp 13-28.

106. A contribution to teratology.
Izbrannye voprosy algebry i logiki Nowosybirsk 1973 pp 184-196.

107. Konstruktivnye mnozestva i ich prilozenije.
Moskva 1973 (translated from English) pp 256.

108. Some Problems in the Axiomatic Theory of Classes.
Bolletion U.M.I. 9 (1974) pp 161-170.

109. Observations concerning elementary extensions of ω-models I.
Proceedings of AMS Symposia in Pure Mathematics 25 (1975) pp 349-355.

110. An exposition of forcing.
in:Algebra and Logic, Springer Lecture Notes 450 (1975) pp 220-282.

111. (+ W. Marek) On extendability of the models of ZF set theory to the models of KM theory of classes.
in:Springer Lecture Notes 499, Proc. of the 1974 Kiel conference pp 460-542.

112. A remark on models of the Gödel-Bernays axioms for set-theory.
Preprint Uniwersytetu Warszawskiego 1974 (to appear in the Bernays volume).

113. Travaux de W. Sierpiński sur la theorie des ensembles et ses applications. La theorie generale des ensembles.
Introduction to:Wacław Sierpiński, Oeuvres choisies, vol.II Warszawa 1975 pp 9-13.

114. (+ K. Kuratowski) Set theory. (revised edition).
Studies in Logic and the Foundations of Mathematics Amsterdam Warszawa (w druku).

115. Refleksje na temat pewnego wykładu z historii matematyki.
O nauczaniu historii nauki. pp 231-243.

116. Two remarks on the models of Morse' set theory.
(This proceedings).

TWO REMARKS ON THE MODELS OF MORSE′ SET THEORY

by A. Mostowski (Warszawa)

Let $\underline{M}$ be the axiomatic system of the class theory of Morse. We discuss below the local and global forms of the well ordering theorem, denoted respectively by WO_L and WO_G. Our first result is that each denumerable standard transitive model of $\underline{M} + WO_L$ can be extended to a model of $\underline{M} + WO_G$ with the same sets. The second result is that one can construct a complicated tree of extensions of a given standard transitive and denumerable model of $\underline{M} + WO_L$.

Similar results are known for models of second order arithmetic (see [4] and [5]). In the proofs given below we treat in details only lemmas which are characteristic for the theory M and indicate other parts of proofs rather briefly.

1. The language of the Morse's class theory is identical with the usual first order language of set theory into which we introduce the familiar distinction of variables into set - variables $x,y,z,\ldots$ and class-variables $X,Y,Z,\ldots$. The axioms of $\underline{M}$ are given e.g. in [3].

The local well ordering theorem WO_L asserts that for every set x there is a relation $r \subset x \times x$ which well orders x. The global well ordering theorem asserts the same for classes.

Felgner [2] considered the local and global forms of the axiom of choice, AC_L and AC_G, and proved that the Gödel - Bernays system GB enriched by AC_G is a conservative extension of the Zermelo Fraenkel system ZF enriched by AC_L. He proved this by showing that each denumerable model for $GB + AC_L$ can be extended to a model of $GB + AC_L$ so that no new sets appear in the extended model.

Using a similar method we shall prove the following

THEOREM 1. Each standard transitive denumarable model M of $\underline{M} + WO_L$ has an extension to a model of $\underline{M} + AC_G$ such that both models have the same sets.

Proof makes use of Chuaqui[1]. Let M be a standard transitive and denumerable model of $\underline{M} + AC_L$. We simplify the notation by using the same letter to denote the model and its universe.

Let P be a class of M partially ordered by a relation $\leq$ which also is an element of M. We assume that there is a largest element 1 of P and that the following assumptions are valid:

(i) P is a coherent notion of forcing (see [1]) ;

(ii) If $x \subset P$ and x is a set of M linearly ordered by the relation $\leq$ then the g.l.b. $\bigwedge\{p: p \in x\}$ exists ;

(iii) If $p,q \in P$ and p,q are compatible then $p \wedge q =$ = g.l.b. $\{p,q\}$ exists.

LEMMA 1. If $D = \{D_\xi\}_{\xi<\alpha} \in M$ and D is a sequence consisting of dense sections of P then the intersection $\bigcap\{D_\xi: \xi < \alpha\}$ is dense in P.

Remark. A sequence of non-void classes is coded by a class of pairs $\langle d,\xi\rangle$ such that $d \in D_\xi$.

Proof. Let $p \in P$. In order to construct an extension q of p which is in the intersection of all D_ξ 's we first define a tree of height α with the initial vertex p.

For each $\xi < \alpha$ and q in P let $k_\xi(q)$ be the set of elements r in D_ξ which satisfy $r \leq q$ and have a possibly small rank. The vertices of the tree to be constructed are divided into levels. The 0th level consists of elements of $k_0(p)$. Now assume that π is an ordinal $< \alpha$ and that levels ξ have been constructed for all $\xi < \pi$. Consider the set of all decreasing sequences $b = \{b_\xi\}_{\xi<\pi} \in M$ such that b_ξ lies on the ξ th level. For each such sequence b let $\bar{b}$ be its g.l.b. The π th level consists of all elements of $k_\pi(\bar{b})$ where b ranges over sequences of the described form.

Using AC_L we obtain a branch of the tree of length α, say g. Then the g.l.b. of the vertices lying on g is the required $q \leq p$ which belongs to all D_ξ 's .

LEMMA 2. Each set $G \subset P$ which is generic over M is strongly generic over M (see [1]).

Proof. Put $c = \{1\}$. If D is a sequence of dense sections of P as in the previous lemma and q is an element of G which belongs to all the classes D_ξ then, since $q = q \wedge 1$, we obtain $q = q \wedge p$ where $p \in c$. Thus the conditions of strong genericity are satisfied.

LEMMA 3. If G is generic over M then $M[G] \models \underline{M}$.

This follows from the main theorem of [1] and from lemma 2.

We shall now discuss the question whether each set of $M[G]$ is a set of M.

LEMMA 4. If z is a set of M then the class

$$D = \{p \in P: (q)_{P \cap z} [(p \leq q) \vee (p \text{ and } q \text{ are incompatible})]\}$$

is dense in P.

<u>Proof</u>. Let $p \in P$. We may suppose that z is well ordered and shall denote by ζ the type of this well ordering. For each non-void subset z' of z we denote by $\Gamma(z')$ the first element of z'. Now we define by transfinite induction three sequences $\{p_\xi\}$, $\{z_\xi\}$ and $\{r_\xi\}$ all of the same type $\gamma \leq \zeta$:

$$z_0 = \{s \in z \cap P : s \text{ is compatible with } p\} ; \quad r_0 = \Gamma(z_0) ;$$

$$p_0 = r_0 \wedge p ;$$

$$z_{\beta+1} = \{s \in z_\beta : s \text{ is compatible with } p_\beta \text{ and } s \neq r_\beta\} ;$$

$$r_{\beta+1} = \Gamma(z_{\beta+1}) ; \quad p_{\beta+1} = r_{\beta+1} \wedge p_\beta ;$$

$$z_\lambda = \{s \in z : (\beta)_{\beta<\lambda} (s \in z_\beta \text{ and } s \neq r_\beta)\} ; \quad r_\lambda = \Gamma(z_\lambda) ;$$

$$p_\lambda = r_\lambda \wedge \bigwedge\{p_\beta : \beta < \lambda\} \quad \text{for } \lambda \in \mathrm{Lim} .$$

Let γ be the least ordinal λ such that $z_\lambda = \emptyset$. Such an ordinal exists because each z_β is a proper subset of all preceeding z_α's. We take γ to be the type of the sequences $\{p_\xi\}$, $\{z_\xi\}$, and $\{r_\xi\}$.

We easily show that

$s \in z_\beta - z_{\beta+1} \equiv [(s = r_\beta)$ or (s is incompatible with p_β)] for $\beta < \gamma$;

$z \cap P = \bigcup\{s \in z_\beta : s$ is incompatible with $p_\beta\} \cup \{r_\beta\}_{\beta<\gamma}$;

$p_\beta \leqslant p_\alpha \leqslant p$ for $\alpha \leqslant \beta < \gamma$.

Let $p' = \bigwedge\{p_\beta : \beta < \gamma\} \wedge p$. It follows that $p' \leqslant p$ and each q in z P is either incompatible with a p_β and hence with p' or is equal to one of the r_β 's and hence $\geqslant p'$. Thus D is dense.

LEMMA 5. If G is generic in P over M then M and $M[G]$ have the same sets.

Proof. It is obvious that each set of M is a set of $M[G]$. Now, let s be a set of $M[G]$. Hence there is Y in M such that $s \in K_G(Y)$ i.e. there is $\langle p,x\rangle \in Y$ for which $p \in G$ and $s = K_G(x)$. Hence X is a set of M . Let $z = \mathrm{Dom}(TC(X))$ and let p be an element of G such that for each q in $z \cap P$ either $p \leqslant q$ or p and q are incompatible (see lemma 4).

For each set t of M we define a set t' by induction on the rank: $\emptyset' = \emptyset$, $t' = \{s' : (Eq)_P\ [q \geqslant p$ and $\langle q,s\rangle \in t]\}$. It is obvious that t' is a set of M and that $\emptyset' = K_G(\emptyset)$. We shall prove by induction that $t' = K_G(t)$ for each t in z . For suppose that this equation is true for all t whose rank is lower than the rank of t_0 where $t_0 \in z$. Using this assumption and the definitions we obtain

$$K_G(t_0) = \{K_G(t) : t \in_G t_0\} = \{t' : (E_q)_G\ \langle q,t\rangle \in t_0\} .$$

Since $\langle q,t\rangle \in t_0$ implies $q \in z$, we obtain further, in view of the definition of p , that $\langle q,t\rangle \in t_0$ implies: $q \geqslant p$ or q is incompatible with p . The second alternative cannot occur if $q \in G$ because we assumed $p \in G$. Hence

$$K_G(t_0) = \{t' : (Eq)[p \leqslant q \text{ and } \langle q,t\rangle \in t_0]\} = t_0' .$$

Applying this result to $t_0 = X$ we obtain $K_G(X) = X'$ and hence $K_G(X)$ is a set of M .

Proof of the theorem. Let P be a class of M consisting of all functions f which are sets of M and whose domains are families of non - void sets and which satisfy the formula $f(x) \in x$ for each x in $\mathrm{Dom}(f)$. Let $f \leq g$ mean that $f \supseteq g$. It is easy to show that assumptions (ii) - (iii) are satisfied; in order to obtain (i) we use the decomposition $P = P_\xi \cup P^\xi$ where $P_\xi = P \cap R_\xi$. $P^\xi = P - R_\xi$ and R_ξ is the set of all sets of ranks $< \xi$.

The model $M[G]$ contains a universal choice function $F = \bigcup G$. Using this function we obtain effectively a well ordering of $R_\xi \cap M$, whence we obtain a class of $M[G]$ which well orders the union

$$\bigcup\{R_\xi \cap M : \xi \in On_M\} .$$

Theorem 1 is thus proved.

2. A tree of extensions of a denumerable model.

Let M and M_1 be standard models of $\underline{M}$. We say that M_1 is a C-extension of M if $M_1 \supseteq M$ and M has exactly the same sets as M_1. Using lemmata proved in section 1 we shall construct now a complicated tree of C-extensions of an arbitrary standard denumerable model of $\underline{M} + WO_L$.

THEOREM 2. Let M be a standard denumerable model of $\underline{M} + WO_L$ and $\nu \in On_M$. Furthermore let $\underline{S}$ and $\underline{T}$ be families which belong to M, consist of subsets of ν, and satisfy the conditions

$$T \in \underline{T} \rightarrow \overline{\overline{T}} \geq 2 \quad , \qquad S \in \underline{S} \wedge T \in \underline{T} \rightarrow T - S \neq \emptyset .$$

Under these assumptions there is a family $\{M_\alpha\}_{\alpha<\nu}$ consisting of standard C-extensions of M such that $M_\alpha \models \underline{M} + WO_L$ and such that for each S in $\underline{S}$ there is a standard C-extension M_S of M satisfying $M_\alpha \subseteq M_S \models \underline{M} + WO_L$ for each $\alpha \in S$ whereas for no T in $\underline{T}$ there is a standard C-extension M_T satisfying $M_\alpha \subseteq M_T \models GB$ for each α in T.

<u>Proof</u>. Let $P = \bigcup\{(2^\lambda)^\nu \cap M : \lambda \in On_M\}$ and let $p \leq q$ be defined as $p \supseteq q$. If $p \in (2^\lambda)^\nu$ then we call λ the length of p and denote it by $lh(p)$. For p in P and ξ in ν we denote by p_ξ the value of p for the argument ξ and by $p_\xi(\rho)$ the value of p_ξ for the argument ρ. Thus $p_\xi \in 2^\lambda$, $p_\xi(\rho) \in \{0,1\}$. If $f \in 2^\lambda$ and $\mu > \lambda$ then f^μ denotes a function in 2^μ such that $f^\mu | \lambda = f$ and $f^\mu(\rho) = 0$ for $\rho \in \mu - \lambda$. We call f^μ the zero-extension of f.

For $x \subseteq \nu$ and any set p in M we put $p[x] = p \cap (\nu \times V_M)$ where V_M is the universal class of M. We denote by $P[x]$ the class consisting of all $p[x]$ where $p \in P$. The ordering of $P[x]$ is the restriction of $\leq$. Finally let $\check{p}[x]$ be the set $\bigcap_{\xi \in x} \{\eta < lh(p) : p_\xi(\eta) = 1\}$.

We need the following

LEMMA. Let $\alpha, \rho \in On_M$, $p \in P$, $x \in (P(\nu) - \{0\})^\alpha \cap M$ and let $D = \{D_\beta\}_{\beta<\alpha}$ be a sequence of type α which belongs to M and whose β-th element is a dense section of $P[x_\beta]$, $\beta < \alpha$. Then there exists a p' in P such that $p' \leq p$, $p'[x_\beta] \in D_\beta$ for $\beta < \alpha$, $lh(p') \geq \rho$, $lh(p')$ has a predecessor π and p' satisfies the condition:

(*) if $t \subseteq \nu$ and $t - x_\beta \neq \emptyset$ for each $\beta < \alpha$, then $\check{p}'[t] = \check{p}[t] \cup \{\pi\}$.

Proof. For $\beta < \alpha$ and $q \in P[x_\beta]$ we let $k_\beta(q)$ be the set of those elements of D_β which are $\leq q$ and have a possibly small rank. Let $\mu_\beta(q) = \sup\{lh(r) : r \in k_\beta(q)\}$.

We define now a tree Δ of height α ordered by $\leq$. The initial vertex of Δ is p. All the other vertices are divided into levels and all vertices lying on the same level have equal heights. Let us assume that levels $< \sigma$ have already been defined and that their union forms a tree Δ'_σ. We denote by Γ_σ the set of all branches of length σ of Δ'_σ and put $\bar{g} = \bigwedge_{\xi<\sigma} g_\xi$ for each g in Γ_σ. Each vertex of the σ-th level is determined by a branch g in Γ_σ and an element q of $k_\sigma(\bar{g})$. Thus we denote the elements of the σ-th level by $r_{g,q}$. We define $r_{g,q} = r$ by the equations

$$r_\xi = q_\xi^{\mu_\sigma} \quad \text{if} \quad \xi \in x_\sigma, \quad r_\xi = \bar{g}^{\,\mu_\sigma} \quad \text{if} \quad \xi \in \nu - x_\sigma .$$

Notice that

(**) if $r_\xi(\eta) = 1$ and $\xi \in \nu - x_\sigma$ then $\eta < lh(\bar{g})$ and $\bar{g}_\xi(\eta) = 1$.

The tree $\Delta = \bigcup\{\Delta'_\sigma : \sigma < \alpha\}$ is thus defined by induction and we easily see that $\Delta \in M$ and that all the elements of the σ-th level have the same lenght μ'_σ. Moreover we can show that the vertex $r = r_{g,q}$ of Δ belonging to the level σ determines uniquely the branch connecting it with the initial vertex. This branch consists of vertices obtained from r by restricting the functions r_ξ $(\xi < \nu)$ to μ_τ where τ ranges over ordinals $< \sigma$.

Let b be a branch of Δ of length α. Since $M \models WO_L$ and Δ is a set of M, there exists such a branch in M. The g.l.b. of the elements b_ξ, $\xi < \alpha$, of this branch is an element of P which we denote by $\overline{b}$; let $\lambda = lh(\overline{b})$ and $\pi = \max(\lambda, \rho)$. We define now p' by the equation

$$(***) \quad p' = \overline{b} \cup \{\langle \xi, \langle \rho, 0 \rangle \rangle : \xi < \nu \wedge \lambda \leq \rho < \pi\} \cup \{\langle \xi, \langle \pi, 1 \rangle : \xi < \nu\}.$$

It is evident that $p' \in M \cap P$, $p' \leq p$, $lh(p') \geq \rho$ and $lh(p') = \pi + 1$.

Furthermore $p'[x_\beta] \leq b_\beta[x_\beta]$ for $\beta < \alpha$; since b_β belongs to the β-th level, we have $b_\beta = r_{g,q}$ where $q \in k_\beta(\overline{g})$ and g is a branch of Δ'_β. Since this branch can be extended to a branch connecting p with b_β, we infer that $g = b \restriction \beta$ because there is just one branch connecting p with b_β. It follows that $b_\beta = r_{b\restriction\beta, q}$ where $q \in k_\beta(\bigwedge\{b_\gamma : \gamma < \beta\})$. Hence $(b_\beta)_\xi = q_\xi^{\mu_\beta}$ for $\xi \in x_\beta$ and so $(b_\beta)_\xi \leq q_\xi$ for $\xi \in x_\beta$. Since $q \in P[x_\beta]$ and $q \in D_\beta$, it follows that $b_\beta[x_\beta] \in D_\beta$. Since $p'[x_\beta] \leq b_\beta[x_\beta]$, we obtain $p'[x_\beta] \in D_\beta$.

It remains to prove (*). Let $t \subseteq \nu$ be a set such that $t - x_\beta \neq \emptyset$ for each $\beta < \alpha$. Since p' is an extension of p we evidently have $\check{p}[t] \subseteq \check{p}'[t]$. From (***) we obtain $\pi \in p'[t]$ and hence

$$\check{p}[t] \cup \{lh(p') - 1\} \subseteq \check{p}'[t].$$

Now let η belong to $\check{p}'[t] - \{\pi\}$; we have to show that $\eta \in \check{p}[t]$. If $\eta < lh(p)$, then $p_\xi(\eta) = 1$ for $\xi \in t$ because $p'_\xi(\eta) = 1$ and p' is an extension of p. Hence in this case $\eta \in \check{p}[t]$.

Now we assume that $lh(p) \leq \eta < lh(p')$. Since we assumed that $\eta < \pi$ and $p'_\xi(\eta) = 1$, we obtain $\eta < lh(\overline{b}) = \sup\{lh(b_\beta) : \beta < \alpha\}$ and hence there exists a $\beta < \alpha$ such that $\eta \leq lh(b_\beta)$. If $\eta = lh(b_\beta)$, then $lh(b_\beta)$ is smaller than $lh(\overline{b})$ and hence b_β is not the least element of the branch b. We can therefore infer that there exists a $\gamma < \alpha$ such that $\eta < lh(b_\gamma)$. The same is obviously true if $\eta < lh(b_\beta)$.

Let β be the smallest ordinal such that $\eta < lh(b_\beta)$. Hence $(b_\beta)_\xi(\eta) = 1$ for $\xi \in t$ because p' is an extension of b_β. But in view of the definition of b_β we have $b_\beta = r_{b\restriction\beta, q}$ for some q in $k_\beta(f)$ where f is the g.l.b. of all the b_γ with $\gamma < \beta$. In view of (**) the value of $(b_\beta)_\xi$ for the argument η is either

$q_\xi^{\mu_\beta}(\eta)$ or $f_\xi^{\mu_\beta}(\eta)$ according as $\xi \in x_\beta$ or $\xi \in \nu - x_\beta$. Selecting ξ in $t - x_\beta$ we obtain $f_\xi^{\mu_\beta}(\eta) = 1$. However, in view of the minimality of β we have $\eta \geqslant lh(b_\gamma)$ for each $\gamma < \beta$ and so $\eta \geqslant \sup\{lh(b_\gamma): \gamma < \beta\} = lh(f) = Dom(f_\xi)$. Hence $f_\xi^{\mu_\beta}(\eta) = 0$ because a zero extension $f_\xi^{\mu_\beta}$ of the function f_ξ takes on the value 0 for all arguments $\geqslant Dom(f_\xi)$. Thus we obtain a contradiction which proves that no η satisfying $lh(p) \leqslant \eta < \pi$ can belong to $p'[t]$. The lemma is thus proved.

<u>Proof of theorem</u> 2. Let $\{x_\beta\}_{\beta<\alpha}$ be a sequence belonging to M consisting of all the singletons $\{\xi\}$, $\xi < \nu$, and all the elements S of $\underline{S}$. Let $\{D^n\}_{n<\omega}$ be a sequence of type ω (not belonging to M) which lists all the sequences $D^n = \{D^n_\beta\}_{\beta<\alpha}$ which belong to M and have the property that for each $\beta < \alpha$ the class D^n_β is a dense section of $P[x_\beta]$. Thus each dense section of $P[x_\beta]$ which belongs to M is equal to one of the D^n_β .

Let $\{\rho_n\}_{n<\omega}$ be an increasing sequence of type ω of ordinals which is cofinal with On_M ; this sequence obviously does not belong to M .

We construct now a sequence $\{p_n\}_{n<\omega}$ such that p_0 is an arbitrary element of P and p_{n+1} is related to p_n as p' was related to p in the lemma (the sequence D of the lemma is replaced by D^n). Moreover we require that $lh(p_{n+1})$ be greater than $\max(\rho_n, lh(p_n))$ which is of course possible.

If G is the class $\{p \in P: (En)\, p \geqslant p_n\}$ then $G[x_\beta]$ is generic in $P[x_\beta]$ over M for each β . Hence $M[G[x_\beta]] \models \underline{M} + WO_L$ and $M[G[x_\beta]]$ is a C-extension of M for each $\beta < \alpha$.

Put $M_\xi = M[\{\xi\}]$ for $\xi < \nu$ and $M_S = M[G[S]]$ for each S in $\underline{S}$. It follows that M_ξ is a C-extension of M and M_S is a C--extension of M_ξ for each ξ in S . Moreover all these models are models for $\underline{M} + WO_L$.

We shall now show that there exists no model $M_T \models GB$ such that for each ξ in T (where T is an element of $\underline{T}$) M_T is a C-extension of M_ξ . For suppose that such a model exists. Hence $G[\{\tau\}] \in M_T$ for each τ in T . It follows that $\bigcup_n \check{p}_n[\{\tau\}] \in M_T$ because η belongs to this union if and only if $\langle\tau, \langle\eta,1\rangle\rangle$ belongs to $G[\{\tau\}]$. Hence we obtain $\bigcap\{\bigcup\{\check{p}_n[\{\tau\}]: n \in \omega\}: \tau \in T\} \in M_T$. In view of the way the sequence p_n was constructed we see that the last intersection consists of terms of an increasing sequence of type ω which is cofinal with On_M . Since no model of GB can contain such a set, the existence of M_T is impossible. Theorem 2 is thus proved.

In the case of models of GBC treated in [5] it is possible to repeat transfinitely many times the above construction and obtain very complicated trees of infinite heights consisting of C-extensions of a given model. For models of $\underline{M} + WO_L$ such constructions are not possible because the union of an increasing sequence of models of this theory is, in general, not a model of $\underline{M}$.

To conclude we remark still that it is possible to adapt the proof given in [4] to the system $\underline{M}$ + C where C is the schema

$$(x)(EY)\,\varphi \rightarrow (EY)(x)(EX)[(X = Y^{(x)}) \wedge \varphi]$$

and establish in this way the following theorem: each model of the system $\underline{M}$ + C is a reduct of a model of a theory differing from $\underline{M}$ by containing an additional binary predicate S and axioms stating that S is a well ordering of the universe.

References

[1] R. Chuaqui, Forcing for the impredicative theory of classes. The Journal of Symbolic Logic 37 (1972), pp. 1-18.

[2] U. Felgner, Comparison of the axioms of local and universal choice, Fundamenta Mathematicae 71 (1971), pp. 43-62.

[3] A. Mostowski, Constructible sets with applications. North-Holland Publ. Co. - PWN 1969.

[4] ________, Models of second order arithmetic with definable Skolem functions, Fundamenta Mathematicae 75 (1972), pp. 223-234 and Errata to this paper, Fundamenta Mathematicae 84 (1974).

[5] ________, A remark on models of the Gödel - Bernays axioms for set theory (to appear).

CONTRIBUTED PAPERS

A REMARK ON POWERS OF SINGULAR CARDINALS

by B. Balcar and W. Guzicki

Abstract: In the paper we prove the following theorem: if for all regular $\lambda > \varkappa$, $\lambda^{\aleph_0} = \lambda$ holds, then for all singular strong limit $\lambda > \varkappa$, $2^\lambda = \lambda^+$ holds.

§ 0. Throughout the paper we use standard notation. If x is a set and $\varkappa$ a cardinal number, then by $\mathcal{P}_\varkappa(x)$ we denote the family of subsets of x of power at most $\varkappa$. We shall make use of a theorem of Přikry in the following form:

THEOREM. Assume $\omega < \varkappa = cf(\nu) < \nu$ and $\forall \rho < \nu\ [\rho^\varkappa < \nu]$. Let $\langle \tau_\xi ; \xi \in \varkappa \rangle$ be an increasing continuous sequence of cardinal numbers cofinal in ν. Let $T \subseteq \mathcal{P}(\varkappa \times \nu)$ and for each $\xi \in \varkappa$ set $T \restriction \tau_\xi = \{a \cap (\xi \times \tau_\xi) : a \in T\}$. Then if the set $\{\xi \in \varkappa : |T \restriction \tau_\xi| \leq \tau_\xi^+\}$ is stationary in $\varkappa$, then $|T| \leq \nu^+$ (cf [4]).

§ 1. Let $B(\varkappa,\lambda)$ denote the following statement:

$$\forall \nu > \varkappa\ [\nu > cf(\nu) = \lambda \wedge \forall \tau < \nu\ [\tau^\lambda < \nu] \rightarrow \nu^\lambda = \nu^+] .$$

Now we shall prove the following

THEOREM. $B(\varkappa, \aleph_0) \rightarrow \forall \lambda\ B(\varkappa,\lambda)$.

Proof. By induction on λ .

Case 1. $\lambda = \aleph_0$. This is just the assumption.

Case 2. $\lambda > \aleph_0$.

We assume that for all $\mu < \lambda$ we have $B(\varkappa,\mu)$ .In order to prove $B(\varkappa,\lambda)$ let us take $\nu > \varkappa$ such that $\nu > cf(\nu) = \lambda$ and $\forall\tau < \nu\ [\tau^\lambda < \nu]$. What we have to prove is $\nu^\lambda = \nu^+$.

From the assumptions on ν it easily follows that there exists an increasing continuous sequence $\langle \tau_\eta ; \eta \in \lambda\rangle$ of cardinal numbers such that $\sup \{\tau_\eta : \eta \in \lambda\} = \nu$ and for all $\eta \in \lambda$, $\tau_\eta > \varkappa$ and $\tau_\eta^\lambda < \tau_{\eta+1}$ hold.

Now let $\eta < \lambda$ be a limit number. We shall prove that $\tau_\eta^\eta = \tau_\eta^+$.

Since $\eta < \lambda$, we have $\mu = cf(\eta) < \lambda$. Thus we can use $B(\varkappa,\mu)$.

Obviously $\tau_\eta > \varkappa$ and $cf(\tau_\eta) = \mu$. If $\tau < \tau_\eta$, then there exists $\xi < \eta$ such that $\tau < \tau_\xi$. Then $\tau^\mu \leqslant \tau_\xi^\mu \leqslant \tau_\xi^\lambda < \tau_{\xi+1} < \tau_\eta$. By $B(\varkappa,\mu)$ we have $\tau_\eta^\mu = \tau_\eta^+$. Therefore we have to prove that $\tau_\eta^\mu = \tau_\eta^\eta$.

Let $\langle\eta(\xi) ; \xi \in \mu\rangle$ be an increasing continuous sequence cofinal in η . Let us consider the sequence $\langle\tau_{\eta(\xi)} ; \xi \in \mu\rangle$. Then $\tau_{\eta(\xi)}^\lambda < \tau_{\eta(\xi)+1} \leqslant \tau_{\eta(\xi+1)}$; in particular $\tau_{\eta(\xi)}^\eta < \tau_{\eta(\xi+1)}$.

Now for each $\xi \in \mu$ we select one function $\varphi_\xi : \mathcal{P}_\eta(\tau_{\eta(\xi)}) \xrightarrow{1:1} \tau_{\eta(\xi+1)}$. For each $A \subseteq \tau_\eta$ of power η we can define a function $g_A : \mu \to \tau_\eta$ in the following way: $g_A(\xi) = \varphi_\xi(A \cap \tau_{\eta(\xi)})$. Obviously if $A_1 \neq A_2$ then $g_{A_1} \neq g_{A_2}$, so $|\mathcal{P}_\eta(\tau_\eta)| \leqslant \tau_\eta^\mu$, i.e. $\tau_\eta^\eta \leqslant \tau_\eta^\mu$. The other inequality is trivial, so we have proved that for limit $\eta < \lambda$, $\tau_\eta^\eta = \tau_\eta^+$ holds.

Now take $T \subseteq \mathcal{P}(\lambda \times \nu)$ to be the family of all functions from λ into ν . Then for each limit $\eta < \lambda$, $f \cap (\eta \times \tau_\eta)$ is a function from a subset of η into τ_η . Since there are at most $\tau_\eta^\eta = \tau_\eta^+$ such functions, $|T \upharpoonright \tau_\eta| \leqslant \tau_\eta^+$. The set $\{\eta \in \lambda : \eta$ is a limit ordinal$\}$ is stationary in λ so by Přikry's theorem we conclude that $|T| \leqslant \nu^+$, thus $\nu^\lambda = \nu^+$, qed.

§ 2. Let $C(\varkappa,\lambda)$ be the following statement:

$$\forall\nu > \varkappa\ [\nu = cf(\nu) \to \nu^\lambda = \nu] .$$

Then we have the following.

PROPOSITION. $C(\varkappa, \aleph_0) \to B(\varkappa, \aleph_0)$.

Proof. Let $\nu > \kappa$ be such that $\nu > cf(\nu) = \aleph_0$. Then $\nu^+ \leq \nu^{cf(\nu)} = \nu^{\aleph_0} \leq (\nu^+)^{\aleph_0} = \nu^+$, so $\nu^{\aleph_0} = \nu^+$, qed.

COROLLARY. $C(\kappa, \aleph_0) \to \forall\lambda\, B(\kappa,\lambda)$.

The reader may easily observe that the full power of $C(\kappa,\aleph_0)$ was not used above. Namely it is enough to assume that $\nu^{\aleph_0} = \nu$ holds for successors of cardinals of cofinality ω . But in fact the last statement implies $C(\kappa,\aleph_0)$. The proof goes by straightforward induction.

PROPOSITION. If $\forall\lambda\, B(\kappa,\lambda)$ holds, then for all singular strong limit cardinal numbers $\lambda > \kappa$ we have $2^\lambda = \lambda^+$.

Proof. For singular strong limit λ it holds that $2^\lambda = \lambda^{cf(\lambda)}$. Since λ is strong limit, it satisfies the assumptions of $B(\kappa, cf(\lambda))$, so we can conclude that $\lambda^{cf(\lambda)} = \lambda^+$; therefore $2^\lambda = \lambda^+$, qed.

COROLLARY. If $C(\kappa,\aleph_0)$ holds, then the continuum hypothesis holds for singular strong limit cardinals bigger than κ .

§ 3. Ketonen's combinatorial lemma shows that if κ is a strongly compact cardinal, then $C(\kappa,\aleph_0)$ holds, thus giving us another proof of Solovay's theorem which says that the continuum hypothesis holds for singular strong limit cardinals bigger than a strongly compact cardinal (cf. [2]).

Keith Devlin observed that if $0^\#$ does not exist then $C(2^{\omega_1},\aleph_0)$ holds. The proof follows easily from the covering lemma of Jensen's marginalia to Silver's theorem (cf [1]). Therefore if $0^\#$ does not exist, the continuum hypothesis holds for all singular strong limit cardinals.

The only counterexample to $C(\kappa,\aleph_0)$ which the authors heard of is due to Magidor and uses the assumption of existence of a supercompact cardinal (cf [3]).

References

[1] K. Devlin and R. Jensen - Marginalia to Silver's theorem, Proceedings of Kiel Conference, Springer Lecture Notes in Mathematics.

[2] F. Drake - Set Theory, North-Holland, 1974.

[3] M. Magidor , A note on singular cardinals problem, mimeographed notes.

[4] K. Přikry - Another proof of Silver's theorem, mimeographed notes.

UNCOUNTABLE STANDARD MODELS OF ZFC + $V \neq L$

by J. L. Bell

Dedicated to the memory of A. Mostowski

A well-known result of Cohen ([1], p.109) asserts that in ZF + $V = L$ one can prove that there are no uncountable standard models of ZFC + "There is a non-constructible real". It is natural to ask what the situation is for uncountable standard models of ZFC + "There is a non-constructible set". In this paper we shall prove the following

THEOREM. ZFC + "There exists a natural model R_α of ZFC" $\vdash$ "There exist standard models of ZFC + $V \neq L$ of all cardinalities $< \alpha$."

This theorem has the following consequences. Let ZFI = ZFC + "There exists an inaccessible cardinal".

COROLLARY 1. ZFI $\vdash$ "There is a standard model of ZFC + $V \neq L$ of any cardinality less than the first inaccessible cardinal".

Let KMC be Kelley-Morse set theory with choice. Since it is known [5] that in KMC one can prove the existence of arbitrarily large natural models of ZFC , it follows immediately from the theorem that

COROLLARY 2. KMC $\vdash$ "There is a standard model of ZFC + $V \neq L$ of any cardinality" .

The proof of the theorem uses the technique of Boolean-valued models of set theory as presented, e.g. in [2]. For the theory of Boolean algebras we refer the reader to [6].

As usual, we write ZF for Zermelo-Fraenkel set theory, ZFC for ZF + axiom of choice, $V = L$ for the axiom of constructibility and $V \neq L$ for its negation.

By a standard model of ZF we understand a model of the form $\mathfrak{M} = \langle M, \in/M \rangle$, where M is a transitive set and $\in/M = \{\langle x,y \rangle \in M^2 : x \in y\}$. If $\mathfrak{M}$ is a standard model of ZFC and B is a complete Boolean algebra in $\mathfrak{M}$, we write, as usual $\mathfrak{M}^{(B)}$ for the B-extension of $\mathfrak{M}$ and $\|\sigma\|$ for the B-value of any sentence σ of set theory (which may contain names for elements of $\mathfrak{M}^{(B)}$). Well-known is the fact that $\|\sigma\| = 1$ for any theorem σ of ZFC . We recall that there is a canonical map $x \mapsto \hat{x}$ of $\mathfrak{M}$ into $\mathfrak{M}^{(B)}$. We shall also need the following fact ([2], Lemma 50).

LEMMA 1. For each formula $\varphi(x)$ of set theory (which may contain names for elements of $\mathfrak{M}^{(B)}$) there is $t \in \mathfrak{M}^{(B)}$ such that:

$$\|\exists x \varphi(x)\| = \|\varphi(t)\| .$$

Let B be a complete Boolean algebra; a subset P of B is said to be dense if $0 \notin P$ and $\forall x \in B[x \neq 0 \Rightarrow \exists p \in P\ (p \leq x)]$. If $\varkappa$ is a cardinal, P is said to satisfy the $\varkappa$-descending chain condition ($\varkappa$-dcc) if for each $\alpha < \varkappa$ and each descending α-sequence $p_0 \geq p_1 \geq \ldots \geq p_\xi \geq \ldots$ $(\xi<\alpha)$ from P there is $p \in P$ such that $p \leq p_\xi$ for all $\xi < \alpha$.

LEMMA 2. Suppose that B contains a dense subset satisfying the $\varkappa$-dcc, and let $\{A_\xi : \xi < \varkappa\}$ be a family of subsets of B such that $\bigvee A_\xi = 1$ for each $\xi < \varkappa$. Then there is an ultrafilter U in B such that $U \cap A_\xi \neq \emptyset$ for all $\xi < \varkappa$.

Proof. Let J be a set sufficiently large so that each A_ξ can be enumerated as $\{a_{\xi j} : j \in J\}$. We show that there is $f \in J^\varkappa$ such that, for each $\alpha < \varkappa$,

(1) $$\bigwedge_{\xi<\alpha} a_{\xi f(\xi)} \neq 0 \ .$$

We define f by recursion as follows. Let $\alpha < \varkappa$ and suppose that for each $\xi < \alpha$ we have selected $p_\xi \in P$ and $f(\xi) \in J$ in such a way that

(2) $$p_\xi \leq a_{\xi f(\xi)} \quad \text{for all } \xi < \alpha$$

(3) $$\eta \leq \xi < \alpha \Rightarrow p_\eta \geq p_\xi \ .$$

We show how to obtain p_α and $f(\alpha)$. Since P satisfies the $\varkappa$-dcc, there is $p \in P$ such that $p \leq p_\xi$ for all $\xi < \alpha$. We have

$$0 \neq p = p \wedge 1 = p \wedge \bigvee_{j\in J} a_{\alpha j} = \bigvee_{j \in J} p \wedge a_{\alpha j} \ ,$$

so there must be $j \in J$ such that $p \wedge a_{\alpha j} \neq 0$, and hence, since P is dense, $q \in P$ such that $q \leq p \wedge a_{\alpha j}$. We take $f(\alpha)$ to be such a $j \in J$, and p_α to be such a $q \in P$. It is now clear that (2) and (3) hold with "$\xi < \alpha$" replaced by "$\xi \leq \alpha$" and so by recursion we obtain p_α and $f(\alpha)$ to satisfy (2) and (3) for all $\alpha < \varkappa$. If $\alpha < \varkappa$, $\langle p_\xi : \xi < \alpha \rangle$ is a descending α-sequence in P and so there is (by dcc) a $p \in P$ such that $p \leq p_\xi$ for all $\xi < \alpha$. But then, by (2), we immediately obtain (1).

To complete the proof we observe that, by (1), the set $\{a_{\alpha f(\alpha)} : \alpha < \varkappa\}$ has the finite intersection property and hence can be extended to an ultrafilter in B. This ultrafilter clearly meets the requirements of the Lemma. ■

An ultrafilter U in B is said to <u>preserve</u> the family of joins $\bigvee A_\alpha$ $(\alpha < \varkappa)$, where $\{A_\alpha : \alpha < \varkappa\}$ is family of subsets of B, provided that for each $\alpha < \varkappa$,

$$\bigvee A_\alpha \in U \Rightarrow U \cap A_\alpha \neq \emptyset \ .$$

Lemma 2 gives the following generalization, for complete Boolean algebras, of the well-known Rasiowa-Sikorski lemma:

COROLLARY. Suppose that B contains a dense subset satisfying the $\varkappa$-dcc. Then for each family $\{A_\alpha : \alpha < \varkappa\}$ of subsets of B there is an ultrafilter in B which preserves the family of joins $\bigvee A_\alpha \quad (\alpha < \varkappa)$.

Proof. Put $a_\alpha = \bigvee A_\alpha$ and apply Lemma 2 to the family $\{A_\alpha \cup \{a_\alpha^*\} : \alpha < \varkappa\}$, where a_α^* is the complement of a_α in B. ■

Remark. I am grateful to Professor Vopěnka and others at the conference for suggesting the present version of this Corollary, which is stronger than my original version.

Now let $\varkappa$ be a regular cardinal and let $X_\varkappa$ be the space $2^\varkappa$ endowed with the $\varkappa$-topology, i.e. the topology whose basic open sets are of the form

$$U(\alpha, f) = \{g \in X_\varkappa : g(\xi) = f(\xi) \quad \text{for} \quad \xi \leq \alpha\}$$

where $f \in X_\varkappa$ and $\alpha < \varkappa$. We denote by $B_\varkappa$ the complete Boolean algebra of regular open subsets of $X_\varkappa$. ($B_\varkappa$ is the algebra which, in the corresponding Boolean extension, adds a new member to $\mathfrak{p}\varkappa$ but leaves $\mathfrak{p}\alpha$ undisturbed for all $\alpha < \varkappa$.)

It is clear that the family of all sets $U(\alpha, f)$ is dense in $B_\varkappa$ and that this family satisfies the $\varkappa$-dcc (since $\varkappa$ is regular). Hence, by the Corollary to Lemma 2 we have

LEMMA 3. If $\varkappa$ is a regular cardinal, then for each family $\{A_\alpha : \alpha < \varkappa\}$ of subsets of $B_\varkappa$ there is an ultrafilter in $B_\varkappa$ which preserves the family of joins $\bigvee A_\alpha \ (\alpha < \varkappa)$.

We now turn to

Proof of the Theorem. Let R_α be a natural model of ZFC. By [4], α is a limit cardinal, and so by the downward Löwenheim-Skolem theorem it will be enough to show that there is a standard model of ZFC + $V \neq L$ for each regular cardinal $< \alpha$. So let $\mathfrak{M} = \langle R_\alpha, \in / R_\alpha \rangle$ and let $\varkappa$ be a regular cardinal $< \alpha$. Put $B = B_\varkappa$. Then B is a complete Boolean algebra in $\mathfrak{M}$ and so we can form the B-extension $\mathfrak{M}^{(B)}$ of $\mathfrak{M}$.

Using Lemma 1, for each formula $\varphi(v_0,\ldots,v_n)$ of the language of set theory (<u>without</u> parameters from $\mathfrak{M}^{(B)}$) we let

$$f_\varphi : (\mathfrak{M}^{(B)})^n \to \mathfrak{M}^{(B)}$$

be a Skolem function for $\varphi(v_0,\ldots,v_n)$ in $\mathfrak{M}^{(B)}$, i.e. such that, for all $x_1,\ldots,x_n \in \mathfrak{M}^{(B)}$

$$(1) \qquad \|\exists v_0 \varphi(v_0,x_1,\ldots,x_n)\| = \|\varphi(f_\varphi(x_1,\ldots,x_n),\ x_1,\ldots,x_n)\| \ .$$

Let $\mathcal{A} \subseteq \mathfrak{M}^{(B)}$ be the closure of the set $\{\check{\xi}: \xi < \varkappa\}$ under the f_φ. Then $\mathcal{A}$ has cardinality $\varkappa$ and, using (1) we have

(2) for any formula $\varphi(v_0,\ldots,v_n)$ and any $a_1,\ldots,a_n \in \mathcal{A}$, there is $a_0 \in \mathcal{A}$ such that $\|\exists v_0 \varphi(v_0,a_1,\ldots,a_n)\| = \|\varphi(a_0,a_1,\ldots,a_n)\|$.

Let Ord(x) be the formula "x is an ordinal". It is well-known that, for any $x \in \mathfrak{M}^{(B)}$, we have $\|\mathrm{Ord}(x)\| = \bigvee_{\xi<\alpha} \|x = \check{\xi}\|$. Using Lemma 3, let U be an ultrafilter in B which preserves the joins

$$(3) \qquad \|\mathrm{Ord}(a)\| = \bigvee_{\xi<\alpha} \|a = \check{\xi}\| \qquad (a \in \mathcal{A}) \ .$$

Let $\mathcal{A}/U$ be the quotient of $\mathfrak{M}^{(B)}$ by U , i.e.

$$\mathcal{A}/U = \langle\{a^U: a \in \mathcal{A}\},\ \in_U\rangle$$

where a^U is the equivalence class of $a \in \mathcal{A}$ under the relation $\sim_U$ defined by $a \sim_U a' \iff \|a = a'\| \in U$ and $\in_U$ is defined by $a^U \in_U a'^U \iff \|a \in a'\| \in U$. Using (2), it is easy to show by induction on complexity of formulas that for any formula $\varphi(v_0,\ldots,v_n)$ of set theory and any $a_0,\ldots,a_n \in \mathcal{A}$,

$$\mathcal{A}/U \models \varphi\,[a_0^U,\ldots,a_n^U] \iff \|\varphi(a_0,\ldots,a_n)\| \in U \ .$$

It follows that $\mathcal{A}/U$ is a model of ZFC . Also, the $\check{\xi}^U$ for $\xi < \varkappa$ are all distinct, so $\mathcal{A}/U$ has cardinality $\varkappa$. Since B is atomless, we have $\|V \neq L\| = 1$, so $\mathcal{A}/U$ is also a model of $V \neq L$. Finally, since U preserves the joins (3), it quickly follows that the map $\xi \mapsto \hat{\xi}^U$ is order-preserving from (true) ordinals onto the ordinals of $\mathcal{A}/U$, so that the ordinals of $\mathcal{A}/U$ are well-ordered. The usual rank argument now implies that $\in_U$ is a well-founded relation, so that $\mathcal{A}/U$ is isomorphic to a standard model which meets the requirements of the theorem. This completes the proof. ∎

CONCLUDING REMARKS

1. Since $B_\varkappa$ is known to preserve cardinals, it is not hard to see that for a definable cardinal $\varkappa$ (e.g. $\aleph_0, \aleph_1, \ldots, \aleph_\omega$, etc.) the proof of the theorem yields a standard model $\mathcal{N}$ of cardinality $\varkappa^+$ such that

$$\mathcal{N} \models \text{ZFC} + \mathfrak{P}\varkappa \subseteq L + \mathfrak{P}\varkappa^+ \not\subseteq L \ .$$

Notice that in any theory consistent with ZF + V = L one <u>cannot</u> prove the existence of a standard model $\mathcal{N}$ of cardinality $\varkappa^+$ such that $\mathcal{N} \models \text{ZFC} + \mathfrak{P}\varkappa \not\subseteq L$, because in ZF + V = L one can prove that, for any such model, $\mathcal{N} \models \mathfrak{P}\varkappa \subseteq L$.

2. Both P. Vopěnka and J. Paris have pointed out that the assumption in the theorem that there exists a natural model of ZFC can be substantially weakened (thereby yielding, of course, a weaker conclusion). In fact one can prove the following

(*) ZFC + "There exists an uncountable standard model of ZFC"

$\vdash$ "There exists an uncountable standard model of ZFC + $V \neq L$".

The proof of (*) can be based on the following Lemma (which I have recently noticed resembles a result implicit in [3]):

LEMMA. Let $\varkappa$ be a regular uncountable cardinal and let $\mathfrak{M}$ be a standard model of ZFC such that (i) $|\mathfrak{M}| = \varkappa$, (ii) $\varkappa \in \mathfrak{M}$ and (iii) $\{x \subseteq \varkappa : |x| < \varkappa\} \subseteq \mathfrak{M}$. Then there is a standard model $\mathcal{N}$ of ZFC such that $\mathfrak{M} \subseteq \mathcal{N}$ and $\mathcal{N} \models \mathfrak{P}\varkappa \not\subseteq L$.

<u>Proof</u>. (Sketch). Let $B = B_\varkappa^{(\mathfrak{M})}$, i.e. the Boolean algebra B constructed in $\mathfrak{M}$. Since every subset of $\varkappa$ of cardinality $< \varkappa$ is in $\mathfrak{M}$, it quickly follows that B has a dense subset satisfying the $\varkappa$-dcc (consider the set of $U(\alpha, f)$ constructed in $\mathfrak{M}$). Hence, by the Corollary to Lemma 2 and the fact that $|\mathfrak{M}| = \varkappa$, there is an

$\mathfrak{M}$-generic ultrafilter U in B. Then $\mathcal{N} = \mathfrak{M}[U]$ meets the requirements of the lemma.

Now we can prove (*) á la Vopěnka and Paris. Suppose that there is an uncountable standard model $\mathfrak{M}$ of ZFC. If $\mathfrak{M} \models V \neq L$ then we are done, so assume $\mathfrak{M} \models V = L$. There are now two cases to consider.

<u>Case (a)</u> : $\omega_1 \in \mathfrak{M}$. We work in L until further notice, with the proviso that ω_1 is always the <u>true</u> ω_1, <u>not</u> $\omega_1^{(L)}$. By the Löwenheim-Skolem theorem we may assume $|\mathfrak{M}| = \omega_1$. It is now easy to see that (inside L), conditions (i) through (iii) of the above Lemma are satisfied by $\mathfrak{M}$ (with $\varkappa = \omega_1$). Therefore, applying the Lemma inside L, there is a standard model $\mathcal{N}$ of ZFC + $V \neq L$ such that $\mathfrak{M} \subseteq \mathcal{N}$, so that $\omega_1 \in \mathcal{N}$. But the property of being a standard model of ZFC + $V \neq L$ is L-absolute, so, emerging form L into the real world, $\mathcal{N}$ is truly a standard model of ZFC + $V \neq L$. Since $\omega_1 \in \mathcal{N}$, we have $|\mathcal{N}| \geq \omega_1$ and (*) follows.

<u>Case (b)</u>: $\omega_1 \notin \mathfrak{M}$. By the downward Löwenheim-Skolem theorem we may assume $|\mathfrak{M}| = \omega_1$. It is clear that every member of $\mathfrak{M}$ is countable, since if x were an uncountable member of $\mathfrak{M}$ it could (by AC in $\mathfrak{M}$) be put into one-one correspondence with an ordinal of $\mathfrak{M}$ which would have to be uncountable, contradicting the assumption that $\omega_1 \notin \mathfrak{M}$. It follows that there are only countably many subsets of ω in $\mathfrak{M}$, and so by the usual forcing argument we can find a generic extension $\mathcal{N}$ of $\mathfrak{M}$ which is a standard model of ZFC + $V \neq L$.

Thus in either case we have the conclusion of (*), completing the proof.

Notice that an argument similar to that used in case (a) also proves the following:

ZFC + "There exists an (uncountable) model of ZFC containing a regular uncountable cardinal $\varkappa$ " $\vdash$ "There exists a standard model of ZFC + $V \neq L$ of cardinality $\varkappa$ ".

<u>Acknowledgments</u>. I am grateful to several participants at the conference, in particular P.Vopěnka and W.Guzicki, for their stimulating observations, I would also like to thank Jeff Paris for his valuable comments on an earlier draft of this paper, Kenneth Kunen

for his timely assistance in proving (*), and George Wilmers for providing general aid.

References

[1] Cohen, P. J., Set Theory and the Continuum Hypothesis, Benjamin, New York, 1966.

[2] Jech, T. J., Lectures in Set Theory, Lecture Notes in Mathematics 217, Springer, Berlin, 1971.

[3] Levy, A., On the logical complexity of several axioms of set theory, AMC Proc. on Axiomatic Set Theory, Vol. XIII, Part I, 1971.

[4] Marek, W., On the metamathematics of impredicative set theory, Dissertationes Mathematicae XCVIII, Warszawa, 1973.

[5] Mostowski, A., Constructible Sets with Applications, North-Holland, Amsterdam , 1969.

[6] Sikorski, R., Boolean Algebras, 2nd ed., Springer, Berlin, 1964.

The London School of
Economics and Political
Science.

CHANGING COFINALITY OF $\aleph_2$

by Lev Bukovský

In the autumn of 1966, P. Vopěnka posed the question whether one can change the cofinality of a regular cardinal without collapsing smaller cardinals. During the Spring of 1966, I constructed a set of forcing conditions which changes the cofinality of $\aleph_2$ without collapsing $\aleph_1$ *). The result was presented at the Logic Colloquium' 69 in Manchester and preprints of this paper were distributed [1] . However, a few months later, a gap in the proof of an important lemma of [1] was found (B. Balcar was the first who called my attention to this fact). In 1970, K. Namba published the paper [6]. Namba's main theorem is identical with that of [1] (see also [2]). Namba was interested in the independence of $(\aleph_0,\alpha)$-distributive laws in complete Boolean algebras and his theorem is formulated in Boolean terminology. After the important paper of R. Jensen [5], appeared the theorem of [1] and [6] has become interesting since it shows that the Jensen's result is the best possible. That's why I decided to publish the paper [1] **). This paper should be considered as an improved version of [1]. Anyway, the presentation is influenced by [6], but I still hope it is worth publishing.

Namba's construction is different from mine. His proof is rather combinatorial. Namba uses ramification. My proof uses rather topological methods. I follow cardinality - see the notion of a $\varkappa$-discrete set. At first sight, the principal notions of Namba and me (poor and $\varkappa$-scanty) are different. However, our main results say that a set is poor ($\varkappa$-scanty) if and only if it contains a perfect subset. The notion of a $\varkappa$-scanty set is trivially $\varkappa$-additive and the corresponding distributive law may be obtained almost immediatelly.

The paper is organized as follows. The main body of the paper is contained in paragraph 1. In § 2, I prove the main theorem. In § 3, I propose some generalizations and related results. With the kind permission of Bohuslav Balcar, I present in § 3 some of his results. The corresponding theorems are indicated by his name.

We shall use standard set-theoretical notations and terminology (see e.g. [4], [8]). An ordinal is the set of smaller ordinals, a cardinal is an initial ordinal . (X) is the set of all subsets of X , ${}^{X}Y$ is the set of all functions from X into Y . The letters n,m,k denote natural numbers. The greek letters ξ,ζ,η denote ordinals, $\alpha,\beta,\varkappa,\lambda$ denote cardinals. If $T,\leq$ is a tree, then T_ξ is the ξ-th level. The operations in Boolean algebras are denoted by $\wedge, \vee, -$. If a,b denote the sequences $a_0,\dots,a_n$ and $b_0,\dots,b_m$ respectively, then $a \smile b$ denotes the sequence $a_0,\dots,a_n, b_0,\dots,b_m$.

§ 1. THE MAIN TOPOLOGICAL THEOREM.

In this paragraph, $\varkappa$ denotes a regular cardinal such that $\alpha^{\aleph_0} < \varkappa$ for each $\alpha < \varkappa$.

Let X,O be a topological space (O is the set of open subsets). We shall assume that there exists a system $\mathcal{B}_n$, $n \in \omega_0$ such that

(1) $\bigcup_n \mathcal{B}_n$ is an open basis of the topology O ,

(2) for each n , $\mathcal{B}_n$ consists of pairwise disjoint clopen sets and $\bigcup_n \mathcal{B}_n = X$,

(3) every $\mathcal{B}_{n+1}$ is a refinement of $\mathcal{B}_n$, i.e. each $A \in \mathcal{B}_{n+1}$ is a subset of some $B \in \mathcal{B}_n$,

(4) if $A_n \in \mathcal{B}_n$, $A_n \supseteq A_{n+1}$ for every $n \in \omega_0$, then $\bigcap_n A_n$ is a one-point set.

For a certain technical reason, we assume $\emptyset \in \mathcal{B}_n$ for every n . A typical example of such a topological space can be constructed as follows: $X = {}^{\omega_0}\varkappa$, O is the product topology (discrete on $\varkappa$) or equivalently, the topology induced by the Baire metric: $\rho(f,g) = \frac{1}{n+1}$ where $f,g \in X$, $f(n) \neq g(n)$ and $f(k) = g(k)$ for $k < n$. If $f \in X$, we denote $G(f,n) = \{g \in X \; ; \; \rho(f,g) < 1/n+1\}$. Let

$$\mathcal{B}_n = \{G(f,n) \; ; \; f \in X\} \cup \{\emptyset\} .$$

One can easily check that the conditions (1)-(4) are satisfied.

In the next sdction, we shall generalize some classical properties of perfect sets.

Let $\mathcal{A} \subseteq \mathcal{P}(X)$, $\mathcal{U} \subseteq \mathcal{P}(X)$. We say that $\mathcal{A}$ can be $\mathcal{U}$-<u>separated</u> if for each $A \in \mathcal{A}$, there is a $U_A \in \mathcal{U}$ such that $A \subseteq U_A$ and $U_{A_1} \cap U_{A_2} = \emptyset$ for $A_1 \neq A_2$.

"<u>Simultaneously separated</u>" means "O-separated". For example, a topological space is Hausdorff (normal), if every two different points (disjoint closed sets) can be simultaneously separated.

Now, <u>let</u> $A \subseteq X$, $\mathcal{A} \subseteq \mathcal{P}(A)$. The set $\mathcal{A}$ is a <u>λ-ramification</u> of A if $\overline{\overline{\mathcal{A}}} = \lambda$ and $\mathcal{A}$ can be simultaneously separated. If $\mathcal{A}$ can be $\mathcal{B}_n$-separated, then $\mathcal{A}$ is called <u>λ-n-ramification</u>. "Closed (open, etc.) λ-ramification" is a λ-ramification consisting of closed (open etc.) sets.

Since elements of $\mathcal{B}_n$ are open and closed, one can easily see that

(5) if $\mathcal{A}$ is a closed λ-n-ramification and $\mathcal{A}' \subseteq \mathcal{A}$, then $\bigcup \mathcal{A}'$ is a closed set.

If there exists a (closed) λ-ramification of a set A and λ is not cofinal with ω_0 , then one can construct a (closed) λ-n-ramification of A for some $n \in \omega_0$.

A non-empty closed set $A \subseteq X$ is called $\varkappa$-<u>perfect</u>, if for every open set U , $U \cap A \neq \emptyset$, there exists a closed $\varkappa$-ramification of $U \cap A$. In view of the preceding remark, in this definition "$\varkappa$-ramification" can be replaced by "$\varkappa$-n-ramification for some $n \in \omega_0$" . Moreover, evidently

(6) a non-empty closed set A is $\varkappa$-perfect if and only if for every $U \in \mathcal{B}_n$, $U \cap A \neq \emptyset$, there exists an integer $m > n$ such that the cardinality of the set $\{V \in \mathcal{B}_m \; ; \; V \cap U \cap A \neq \emptyset\}$ is at least $\varkappa$.

For constructing $\varkappa$-perfect sets we shall use a classical method formalized by the $\varkappa$-sieve. Let $T_n = {}^n\varkappa$, $(T_0 = {}^0\varkappa = \{\emptyset\})$, $T = \bigcup_n T_n$. T ordered by inclusion is a tree, T_n is the n-th level. A function ν from T with values closed subsets of X is called a $\varkappa$-sieve if

(7) $\nu(\emptyset) \neq \emptyset$,

(8) $x \subseteq y \to \nu(x) \supseteq \nu(y)$,

(9) if $x \in T_n$, $\nu(x) \neq \emptyset$, then $\{\nu(y) \,;\, y \in T_{n+1}, y \supseteq x\}$ is a $\varkappa$-m-ramification of $\nu(x)$ for some m .

By induction we can prove that in (9), $m \geq n$. Moreover, using (9), we can choose $U_y \in \mathcal{B}_m$ such that $\nu(y) \subseteq U_y$ and $U_{y_1} \cap U_{y_2} = \emptyset$ for $y_1, y_2 \in T_{n+1}$, $y_1 \neq y_2$. By (2), (3) and (9), for every $x, y \in T$ we have $U_x \subseteq U_y$ or $U_y \subseteq U_x$ or $U_x \cap U_y = \emptyset$.

Let $\{v_n; n \in \omega_0\}$ be a branch of T , $v_n \in T_n$. Assume that $\nu(v_n) \neq \emptyset$ for every n . By (4), $\bigcap_n U_{v_n}$ is a one-point set, say $\{x\}$. If we choose a point $x_n \in \nu(v_n)$, then $x_n \in U_{v_n}$ and therefore, $\lim_{n\to\infty} x_n = x$. Since every $\nu(v_n)$ is closed, $x_m \in \nu(v_n)$ for $m \geq n$, we obtain $x \in \nu(v_n)$. Thus, we have proved that

(10) if $\{v_n; n \in \omega_0\}$ is a branch of T and $\nu(v_n) \neq \emptyset$ for every $n \in \omega_0$, then $\bigcap_n \nu(v_n)$ is a one-point set.

<u>Lemma</u> 1. If ν is a $\varkappa$-sieve, then the set

$$L(\nu) = \bigcap_n \bigcup_{v \in T_n} \nu(v)$$

is $\varkappa$-perfect.

<u>Proof</u>. Let us define $C_n = \bigcup_{v \in T_n} \nu(v)$. Evidently $C_0 = \nu(\emptyset)$ is closed. Assume that C_n is closed. By (9), $C_{n+1} \subseteq C_n$. Let $x \notin C_{n+1}$. If $x \notin C_n$, then $X - C_n$ is an open neighborhood of x disjoint with C_{n+1} . If $x \in C_n$, then $x \in \nu(v)$ for some $v \in T_n$. By (5) and (9), the set

$$V = \bigcup \{\nu(\mu) \,;\, \mu \in T_{n+1} , \mu \supseteq v\}$$

is closed. Since $x \notin C_{n+1}$, $U_{\nu(v)} - V$ is an open neighborhood of x disjoint with C_{n+1}. Hence, C_{n+1} is also closed. Since $L(\nu) = \bigcap_n C_n$, $L(\nu)$ is closed.

We define a branch $\{v_n ; n \in \omega_0\}$ of T as follows: $v_0 = \emptyset$; if v_n is already defined in such a way that $\nu(v_n) \neq \emptyset$, then by (9), there exists $v_{n+1} \in T_{n+1}$, $v_{n+1} \supseteq v_n$ and $\nu(v_{n+1}) \neq \emptyset$. By (10), the intersection $\bigcap_n \nu(v_n)$ is non-empty. Thus, $L(\nu) \neq \emptyset$. Similarly, if $\nu(v) \neq \emptyset$, then $\nu(v) \cap L(\nu) \neq \emptyset$ (begin $v_0 = v$!).

Now, let $U \in \mathcal{B}_n$, $U \cap L(\nu) \neq \emptyset$. Thus, $U \cap \nu(v) \neq \emptyset$ for some $v \in T_n$. Since $\nu(v) \subseteq U_v \in \mathcal{B}_m$, $m \geq n$, we have $\nu(v) \subseteq U_v \subseteq U$. Hence by (9), the set

$$\{L(\nu) \cap \nu(\mu) \; ; \; \mu \in T_{n+1}, \; \mu \supseteq v\}$$

is a $\varkappa$-ramification of $U \cap L(\nu)$. q.e.d.

We shall prove the "$\varkappa$-additivy" of the property "does not contain a $\varkappa$-perfect subset". We introduce a new property of sets - "$\varkappa$-scanty". From the definition will follow that this property is $\varkappa$-additive. Thus, it suffices to prove that a set does not contain a $\varkappa$-perfect subset if and only if is $\varkappa$-scanty.

A set $A \subseteq X$ is called $\varkappa$-discrete, if there exists a $\varkappa$-ramification $\mathfrak{A}$ of A such that $A = \bigcup \mathfrak{A}$ and every $V \in \mathfrak{A}$ is of cardinality smaller than $\varkappa$. If $\mathfrak{A}$ is a $\varkappa$-n-ramification, A is called $\varkappa$-n-discrete. A set A is $\varkappa$-scanty, if A is a union of less than $\varkappa$ $\varkappa$-discrete sets.

Let us remark that a set A is $\varkappa$-n-discrete if and only if for every $U \in \mathcal{B}_n$, $\overline{\overline{A \cap U}} < \varkappa$. Evidently, a union of less than $\varkappa$ $\varkappa$-scanty sets is a $\varkappa$-scanty set. Also, a subset of a $\varkappa$-scanty set is $\varkappa$-scanty.

<u>Canonization sublemma</u>: If A is $\varkappa$-discrete, then $A = \bigcup_n A^{(n)}$, where $A^{(n)}$ is $\varkappa$-n-discrete and $A^{(n)} \subseteq A^{(n+1)}$ for every n.

It suffices to set

$$A^{(n)} = \bigcup\{V \cap A \; ; \; V \in \mathcal{B}_n \wedge \overline{\overline{V \cap A}} < \varkappa\} .$$

Lemma 2. (canonization). If A is $\varkappa$-scanty, then $A = \bigcup_n A^{(n)}$, where $A^{(n)}$ is $\varkappa$-n-discrete and $A^{(n)} \subseteq A^{(n+1)}$ for every n.

Proof. Let $A = \bigcup_{\xi<\alpha} A_\xi$, $\alpha < \varkappa$, A_ξ being $\varkappa$-discrete. By the sublemma, $A_\xi = \bigcup_n A_\xi^{(n)}$, $A_\xi^{(n)} \subseteq A_\xi^{(n+1)}$, $A_\xi^{(n)}$ is $\varkappa$-n-discrete. The set $A^{(n)} = \bigcup_{\xi<\alpha} A_\xi^{(n)}$ is also $\varkappa$-n-discrete, $A^{(n)} \subseteq A^{(n+1)}$ and $A = \bigcup_n A^{(n)}$. q.e.d.

Lemma 3. A $\varkappa$-perfect set is not $\varkappa$-scanty.

Proof by contradiction. Assume that the $\varkappa$-perfect set A is $\varkappa$-scanty. Then $A = \bigcup_n A^{(n)}$, where $A^{(n)}$ is $\varkappa$-n-discrete and $A^{(n)} \subseteq A^{(n+1)}$.

Let $U \in \mathcal{B}_0$ be such that $U \cap A \neq \emptyset$. Since A is $\varkappa$-perfect, there exists a $\varkappa$-n_0-ramification of $U \cap A$ for some n_0. Since $A^{(0)}$ is $\varkappa$-0-discrete, we have $\overline{\overline{U \cap A^{(0)}}} < \varkappa$. Thus, many elements of the ramification are disjoint with $A^{(0)}$; in particular, there exists a $U_0 \subseteq U$, $U_0 \in \mathcal{B}_{n_0}$ such that $U_0 \cap A \neq \emptyset$, $U_0 \cap A^{(0)} = \emptyset$. Now, since A is $\varkappa$-perfect, there exists a $\varkappa$-n_1-ramification of $U_0 \cap A$, $n_1 > n_0$, Since $\overline{\overline{U_0 \cap A^{(n_0)}}} < \varkappa$ ($A^{(n_0)}$ is $\varkappa$-n_0-discrete), there exists a $U_1 \in \mathcal{B}_{n_1}$ such that $U_1 \cap A \neq \emptyset$, $U_1 \subseteq U_0$, $U_1 \cap A^{(n_0)} = \emptyset$. Continuing this process we obtain a sequence $U_k \in \mathcal{B}_{n_k}$ such that $U_k \cap A \neq \emptyset$, $U_{k+1} \subseteq U_k$, $U_{k+1} \cap A^{(n_k)} = \emptyset$. By (4), there is a point x such that $x \in \bigcap_k U_k$. Since A is closed, we have $x \in A$. This is a contradiction, because $x \notin \bigcup_k A^{(n_k)} = \bigcup_n A^{(n)} = A$. q.e.d.

Lemma 4. If $\mathfrak{A}$ is a $\mathcal{B}_n$-separated set of $\varkappa$-scanty sets, then $\bigcup \mathfrak{A}$ is $\varkappa$-scanty.

Proof. By the canonization lemma, if suffices to show that the union of a $\mathcal{B}_n$-separated set $\mathfrak{A}$ of $\varkappa$-m-discrete sets, $m \geqslant n$, is a $\varkappa$-m-discrete set.

Denote $A = \bigcup \mathfrak{A}$. Let $U \in \mathcal{B}_m$. Then there exists a set $V \in \mathcal{B}_n$ such that $V \supseteq U$. Since $\mathfrak{A}$ is $\mathcal{B}_n$-separated, at most one element of $\mathfrak{A}$ is a subset of V, i.e. $A \cap V$ is $\varkappa$-m-discrete. Hence $U \cap A = U \cap (A \cap V)$ has cardinality less than $\varkappa$. q.e.d.

Lemma 5. If A is closed non-$\varkappa$-scanty set, then there exists a closed non-$\varkappa$-scanty $\varkappa$-n-ramification of A for some natural number n.

Proof. Assume that such a ramification does not exist, i.e. the set

$$\mathcal{C}_n = \{V \in \mathcal{B}_n \, ; \, A \cap V \text{ is not } \varkappa\text{-scanty}\}$$

has cardinality less than $\varkappa$ for every $n \in \omega_0$. By lemma 4, the set

$$C_n = \bigcup \{A \cap V \, ; \, V \in \mathcal{B}_n - \mathcal{C}_n\}$$

is $\varkappa$-scanty. Since

$$A \subseteq \bigcup \mathcal{C}_n \cup C_n ,$$

we obtain

$$A \subseteq \bigcap_n \bigcup \mathcal{C}_n \cup \bigcup_n C_n .$$

Since $\overline{\overline{\mathcal{C}}}_n < \varkappa$, $\varkappa$ is regular, the set $\bigcap\limits_n \bigcup \mathcal{C}_n$ is $\varkappa$-scanty (in fact of cardinality less than $\varkappa$). Thus A is also a $\varkappa$-scanty set - a contradiction. q.e.d.

Lemma 6. If A is closed non-$\varkappa$-scanty set, then there exists a $\varkappa$-perfect subset of A.

Proof. Using lemma 5, by induction, we can easily construct a $\varkappa$-sieve ν such that the values of ν are closed non-$\varkappa$-scanty subsets of A. Then $L(\nu)$ is a $\varkappa$-perfect subset of A. q.e.d.

From lemmas 3. and 6. we obtain directly

Theorem 1. Let A be a $\varkappa$-perfect set, $A = \bigcup\limits_{\xi<\alpha} A_\xi$, $\alpha < \varkappa$, A_ξ being closed. Then there exists a $\xi < \alpha$ such that A_ξ contains a $\varkappa$-perfect subset.

§ 2. THE INDEPENDENCE RESULT.

In this paragraph we shall prove the main result of this paper, namely

Theorem 2. If $\mathfrak{M}$ is a countable model of ZFC , $2^{\aleph_0} = \aleph_1$ holds true in $\mathfrak{M}$, then there exists an extension $\mathfrak{N}$ of $\mathfrak{M}$ such that $\aleph_2^{\mathfrak{M}}$ is cofinal with $\aleph_0$ inside $\mathfrak{N}$ and $\aleph_1^{\mathfrak{M}}$ is a cardinal of $\mathfrak{N}$.

We construct a complete Boolean algebra. We assume that $\varkappa$ is a regular cardinal and $\lambda^{\aleph_0} < \varkappa$ for every $\lambda < \varkappa$. Let $Pf(\varkappa)$ denote the set of all $\varkappa$-perfect subsets of ${}^{\omega_0}\varkappa$ with the corresponding topology. $Pf(\varkappa)$, $\subseteq$ is a separatively partial ordered set (see [4]). Thus, there exists (up to isomorphism) a unique complete Boolean algebra $B(\varkappa)$ such that $Pf(\varkappa)$ is a dense subset of $B(\varkappa)$.

Lemma 7. $B(\varkappa)$ is $(\aleph_0,\lambda)$-distributive for every $\lambda < \varkappa$.

Proof. Let $a_{n,\xi} \in B(\varkappa)$, $n \in \omega_0$, $\xi \in \lambda$, $\lambda < \varkappa$, $\bigwedge_n \bigvee_\xi a_{n,\xi} = 1$. To prove $(\aleph_0,\lambda)$-distributivity it suffices, for each $A \in Pf(\varkappa)$, to find a function $\psi \in {}^{\omega_0}\lambda$ and $B \in Pf(\varkappa)$ such that $B \subseteq A$ and $B \leq a_{n,\psi(n)}$ for each $n \in \omega_0$.

We start by constructing a $\varkappa$-sieve ν . Since $\bigvee_\xi a_{0,\xi} = 1$, there exista a ξ such that $a_{0,\xi} \wedge A \neq 0$, i.e. there is a $\varkappa$-perfect set $\nu(\emptyset) \subseteq A$, $\nu(\emptyset) \leq a_{0,\xi}$. Since $\nu(\emptyset)$ is $\varkappa$-perfect, there exists a $\varkappa$-n_0-ramification $\mathfrak{A}$ of $\nu(\emptyset)$. Let $\mathfrak{A} = \{A_\eta \ ; \ \eta < \varkappa\}$. Without loss of generality, we can assume that every A_η is $\varkappa$-perfect. Since $\bigvee_\xi a_{1,\xi} = 1$, for every $\eta < \varkappa$, there are an ordinal $\xi < \lambda$ and a $\varkappa$-perfect set $\nu(0,\eta)$ such that $\nu(0,\eta) \subseteq A_\eta$, $\nu(0,\eta) \leq a_{1,\xi}$.
If $\nu(x)$ is already defined for $x \in T_k$, we shall continue as follows since $\nu(x)$ is $\varkappa$-perfect, there exists a $\varkappa$-n_k-ramification $\mathcal{C}$ of $\nu(x)$. Let $\mathcal{C} = \{C_\eta \ ; \ \eta < \varkappa\}$, C_η being $\varkappa$-perfect. Since $\bigvee_\xi a_{k+1,\xi} = 1$, for every $\eta < \varkappa$, there is an ordinal $\xi < \lambda$ and a $\varkappa$-perfect set $\nu(x \smile \{\eta\})$ such that $\nu(x \smile \{\eta\} \subseteq C_\eta$, $\nu(x \smile \{\eta\}) \leq$ $\leq a_{k+1,\xi}$.

Evidently, ν is a $\varkappa$-sieve and by the lemma 1. $L(\nu)$ is a $\varkappa$-perfect set, $L(\nu) \subseteq A$.

For $n \in \omega_0$, $\xi \in \lambda$, we set

$$B_{n,\xi} = \bigcup \{\nu(x) ; x \in T_n , \nu(x) \leq a_{n,\xi}\} .$$

Evidently $L(\nu) = \bigcap_n \bigcup_{\xi<\lambda} B_{n,\xi}$. By (5), every $B_{n,\xi}$ is closed. Thus, for every function $\psi \in {}^{\omega_0}\lambda$, the set $\bigcap_n B_{n,\psi(n)}$ is closed. Since

$$L(\nu) = \bigcup_{\psi \in {}^{\omega_0}\lambda} \bigcap_n B_{n,\psi(n)} ,$$

$\lambda^{\aleph_0} < \varkappa$, by theorem 1. there exists a function $\psi \in {}^{\omega_0}\lambda$ and a $\varkappa$-perfect set B such that $B \subseteq \bigcap_n B_{n,\psi(n)}$. Now, it is easy to see that $B \leq \bigwedge_n a_{n,\psi(n)}$. q.e.d.

The proof of theorem 2. is now almost trivial. If we define $A_{n,\xi} = \{f \in {}^{\omega_0}\varkappa ; f(n) = \xi\}$ $n \in \omega_0$, $\xi \in \varkappa$, then $\bigvee_\xi A_{n,\xi} = 1$ for every n and $\bigwedge_n A_{n,\psi(n)} = 0$ for every $\psi \in {}^{\omega_0}\varkappa$. Thus, $B(\varkappa)$ is not $(\aleph_0,\varkappa)$-distributive.

Now, we set $N = M[G]$, where G is M-generic over $B(\aleph_2)$ (constructed in M). Using well-known relations between the distributivity of $B(\aleph_2)$ and the properties of N (see e.g. [8]), the theorem follows directly from lemma 7.

Since $\overline{\overline{Pf(\varkappa)}} \leq 2^{\varkappa^{\aleph_0}}$, the algebra $B(\varkappa)$ satisfies the $(2^{\varkappa^{\aleph_0}})^+$-chain condition. Therefore, in the extension N of M, all cardinals greater than $2^{\aleph_2^{\aleph_0}}$ are also preserved, As we shall see later, we cannot say this about $2^{\aleph_2}$.

A regular complete subalgebra B_1 of a complete Boolean algebra B_2 is called <u>locally proper subalgebra</u> if for every $\mu \in B_2$, $\mu \neq 0$, the sets $\{v \in B_2, v \leq \mu\}$, $\{v \in B_1, v \leq \mu\}$ are different. Using the standard argument due to G. Sacks (see [4], p.109), one can easily prove

<u>Theorem</u> 3. Every locally proper subalgebra of $B(\varkappa)$ is $(\omega_0,\varkappa)$-distributive (assuming $\lambda^{\aleph_0} < \varkappa$ for $\lambda < \varkappa$). Thus, under the

assumption of theorem 2, if $M \subseteq N' \subseteq N$, N' is a model of ZFC, $N' \models cf(\aleph_2^M) = \aleph_0$, then $N' = N$.

§ 3. GENERALIZATIONS

A closed subset of ${}^{\omega}\omega_\varkappa$ can be considered as the set of all branches of a subtree of the tree $T = \bigcup_n {}^n\varkappa$. If $S \subseteq T$, we denote by $br(S)$ the set of all branches of S. The set $br(S)$ is $\varkappa$-perfect if the tree is sufficiently ramified, i.e. if for every $x \in S$, there exist $\varkappa$ incomparable elements of S, all greater than x. K. Namba considers trees which are still more ramified, more precisely, if there is a natural number n such that for every $x \in S_m$, $m \geqslant n$, already the set $\{y \in S_{m+1} ; y > x\}$ is of cardinality $\varkappa$.

The most natural generalization is to replace the words "is of cardinality $\varkappa$" by the words "does not belong to the ideal J". More precisely, let J be uniform ideal of subsets of $\varkappa$, i.e., $\varkappa \notin J$, if $A,B \in J$, $C \subseteq A$, then $A \cup B$, $C \in J$, if $\overline{\overline{A}} < \varkappa$, then $A \in J$. A non-empty subtree S of T is called an J-tree if for every $x \in S_n$, there is a natural number $m > n$ such that the set

$$\{y(m-1) \in \varkappa ; y \in S_m \wedge y > x\}$$

does not belong to the ideal J. A (closed) set $A \subseteq {}^{\omega}\omega_\varkappa$ is called J-perfect if $A = br(S)$ for some J-tree S. If J is the ideal of all sets of cardinality smaller than $\varkappa$, then "J-perfect" coincides with "$\varkappa$-perfect".

The notion of a $\varkappa$-scanty set does not lend itself to generalization. Namba's poor set is more convenient for this purpose (because the generalization is rather combinatorial - not topological). A set $A \subseteq {}^{\omega}\omega_\varkappa$ is called J-n-poor if there exists a tree $S \subseteq T$ such that $A \subseteq br(S)$ and for every $m > n$, the set

$$\{y(m-1) \in \varkappa ; y \in S_m \wedge y > x\}$$

belongs to the ideal J. A set A is called J-poor if $A = \bigcup_n A^{(n)}$, $A^{(n)}$ is J-n-poor. A set is J-rich if it is not J-poor. One can easily prove the following lemmas:

Lemma 3'. Every J-perfect set is J-rich.

Lemma 5'. If $A = br(S)$, A is J-rich, then there exists n such that the set

$\{y \in S_n :$ the set of branches of S going through y is J-rich$\}$ does not belong to the ideal J .

Lemma 6'. If A is closed J-rich, then A contains a J-perfect subset.

As a corollary we obtain

Theorem 4. Let J be λ^+-additive, i.e. if $X_\xi \in J$, $\xi < \lambda$, then $\bigcup_\xi X_\xi \in J$. If $A = \bigcup_{\xi<\lambda} A_\xi$, A_ξ is closed, A is J-perfect then there exists a $\xi < \lambda$ such that A_ξ contains a J-perfect subset.

Proof. It suffices to prove that the notion of a J-n-poor set is λ^+-additive. Let $B_\xi = br(S_\xi)$, $\xi < \lambda$, B_ξ being J-n-poor. Then $\bigcup_\xi B_\xi$ is a subset of $br(\bigcup_\xi S_\xi)$ and by the additivity of J , one can easily see that $\bigcup_\xi B_\xi$ is J-n-poor. q.e.d.

We denote by $B(J)$ the complete Boolean algebra obtained from the separatively ordered set $Pf(J)$ of J-perfect subsets of ${}^{\omega}\omega_\varkappa$.

Corollary. If J is $(\lambda^{\aleph_0})^+$-additive, then $B(J)$ is $(\aleph_0,\lambda)$-distributive.

Let $A = br(S)$ be a J-perfect set. A is called regular if there exists a set of natural numbers X such that for each n , for each $x \in S_n$, n belongs to X if and only if the set $\{\xi \in \varkappa ; x \cup \{\xi\} \in S\}$ does not belong to J . Thus, X is the set of levels of simultaneous ramification of S .

By corresponding modifications of the preceding reasonings, one can prove

Theorem 5. (B.Balcar) Let J be $(2^{\aleph_0})^+$-additive. Then every J-perfect set contains a regular J-perfect subset.

Corollary. (B. Balcar) Let J be $(2^{\aleph_0})^+$-additive. Let F denote the ideal of finite subsets of ω_0. The normal completion of $\mathcal{P}(\omega_0)/F$ is isomorphic to a subalgebra of $B(J)$.

Proof. If $X \subseteq \omega_0$, we denote by $[x]$ the corresponding element of $\mathcal{P}(\omega_0)/F$. We denote by $\nu([x])$ the Boolean union of those regular J-perfect sets, for which the set of levels of simultaneous ramification is equal to X up to a finite set. ν induces the desired isomorphism. q.e.d.

Our results on cardinals in a generic extension $M[G]$, G being M-generic over $B(J)$, are poor. Under some assumption, $\aleph_1$ is preserved. B.Balcar has mode the following simple observation: If M_2 is an extension of M_1, the Kurepa hypothesis (see [3]) $KH(\aleph_2)$ holds true in M_1, and $\aleph_2^{M_1}$ is cofinal with $\aleph_0$ in M_2, then the cardinality of $\aleph_3^{M_1}$ is at most $\aleph_1^{\aleph_0}$ in M_2. In fact, by $KH(\aleph_2)$ (see [3]), there exists a set $\{h_\xi ; \xi \in \aleph_3^{M_1}\}$ of mappings from $\aleph_2^{M_1}$ into $\aleph_1^{M_1}$ such that for $\xi_1 \neq \xi_2$, there is η such that $(\forall \zeta > \eta)\ h_{\xi_1}(\zeta) \neq h_{\xi_2}(\zeta)$. Let φ be the mapping from $\aleph_0$ into $\aleph_2^{M_1}$, $\varphi \in M_2$ such that $\sup\{\varphi(n) ; n \in \omega_0\} = \aleph_2^{M_2}$. The one-to-one function

$$\psi(\xi) = \varphi \circ h_\xi$$

maps $\aleph_3^{M_1}$ into ${}^{\omega_0}\omega_1$.

As a corollary we have

Theorem 6. (B.Balcar) Let J be $\aleph_2$-additive, $2^{\aleph_0} = \aleph_1$, and $KH(\aleph_2)$ hold true, all relative to a model M. Let G be an M-generic ultrafilter over $B(J)$. Then in the model $M[G]$, The cardinality of $\aleph_3^M$ is at most $\aleph_1$.

The corresponding generalization of this theorem is left to the reader.

The paper [7] by K. Namba contains important related results.

FOOTNOTES TO PAGE 1.

*) This work has been done during my stay at the University of Leeds, England.

**) Jensen's result was discussed during the Colloquium and that is one of my reasons for publishing the paper here.

References

[1] L. Bukovský , A changing of cofinality of $\aleph_2$, mimeographed

[2] L. Bukovský , Review of [6] , Zentralblatt für Mathematik, 263, (1974), 02035

[3] K. Devlin , Aspects of Constructibility, Springer 1973.

[4] T. Jech , Lectures in Set Theory, Springer 1971.

[5] R. Jensen, K. Devlin , Marginalia to a theorem of Silver, to appear in the Proc. of Kiel Conference, Springer.

[6] K. Namba , Independence proof of $(\omega_0, \omega_\alpha)$-distributive law in Complete Boolean algebras, Comment. Math. Univ. St. Pauli , 19 (1970), pp. 1-12.

[7] K. Namba , $(\omega_1,2)$-distributive law and perfect sets in generalized Baire space, Comment. Math. Univ. St. Pauli., 20 (1971), pp. 107-126,

[8] P. Vopěnka, P. Hájek , The Theory of Semisets, Nort Holland and Academia 1972.

Katedra Matematiky Univerzity P.J. Šafárika
04154 Kosice, Komenského 14, Czechoslovakia

INDUCTIVE DEFINITIONS : POSITIVE AND MONOTONE

by Douglas Cenzer
University of Florida.

Introduction. The concept of inductive definability is central to mathematical logic. The set of well-formed formulas of the predicate calculus, the subset of those formulas provable from a given set of axioms as well as the subset of those formulas true in a given model are all inductively defined sets. (Similarly, the sets of Gödel numbers of formulas belonging to the above sets can be given by inductive definitions.) Other examples of inductively defined sets are the largest perfect subset of a given set of reals, and the class of Borel subsets of the continuum $({}^{\omega}\omega)$.

The particular inductive definition central to this paper is that of recursion in an object of finite type, as introduced by Kleene in [5]. This definition will be outlined in § 2 along with the similar definition of ordinal recursion in an object of finite type.

In the abstract, an inductive operator Γ over a set A is a map from $P(A^k)$ to $P(A^k)$ (for some finite k) such that for all X , $X \subseteq \Gamma(X)$. Γ determines a transfinite sequence $\{\Gamma^\sigma : \sigma$ an ordinal$\}$, where $\Gamma^\sigma = \bigcup\{\Gamma(\Gamma^\tau) : \tau < \sigma\}$. Thus $\Gamma^0 = \emptyset$, $\Gamma^{\sigma+1} = \Gamma(\Gamma^\sigma)$ and for limit ordinals λ , $\Gamma^\lambda = \bigcup\{\Gamma^\sigma : \sigma < \lambda\}$. The closure ordinal (or length) $|\Gamma|$ of Γ is the least ordinal σ such that $\Gamma^{\sigma+1} = \Gamma^\sigma$; clearly $|\Gamma|$ always has cardinality less than or equal to Card(A) . The closure $Cl(\Gamma)$ of Γ is $\Gamma^{|\Gamma|}$, the set inductively defined by Γ .

For a class Ξ of operators, the closure ordinal $|\Xi|$ is defined to be the supremum of the $|\Gamma|$ for Γ in Ξ and the closure algebra $Cl(\Xi)$ is the class of subsets of A^k (for any finite k)

which are reducible to $Cl(\Gamma)$ for some Γ in Ξ. ($X \subseteq A^k$ is reducible to $Y \subseteq A^{k+1}$ if and only if there are $a_1,\ldots,a_k = \underset{\sim}{a}$ such that, for all $x_1,\ldots,x_k = \underset{\sim}{x}$, $\underset{\sim}{x} \in X \leftrightarrow (\underset{\sim}{x},\underset{\sim}{a}) \in Y$.)

Typically, Ξ will be the class of operators definable in a certain manner. Given a particular language $\mathcal{L}$ appropriate to the set A, subsets of A^k may be defined inductively in the language $\mathcal{L}(R)$, where R is a k-place relation symbol intended to represent the closure of Γ. A formula $\phi(\underset{\sim}{x},R)$ is interpreted to mean that $\underset{\sim}{x} \in \Gamma_\phi(R)$. For example, the formula ϕ, defined by $\phi(x,R) \leftrightarrow (\forall y)\ (y < x \rightarrow Rx)$, when interpreted in A = the natural numbers ω with the usual "<", defines an operator Γ_ϕ with $(\Gamma_\phi)^n = \{0,1,\ldots,n-1\}$ for all n and therefore with $|\Gamma_\phi| = Cl(\Gamma_\phi) = \omega$.

The positive formulas $\phi(R)$ of a given language $\mathcal{L}(R)$ compose the smallest class closed under conjunction, disjunction and quantification and containing (1) all formulas in which R does not occur and (2) all formulas $R(\underset{\sim}{t})$, where $\underset{\sim}{t}$ is a sequence of terms. Γ is a positive inductive operator if and only if there is a positive formula ϕ of some language $\mathcal{L}(R)$ such that $\Gamma = \Gamma_\phi$. Elementary positive inductive definitions over abstract structures were studied in depth by Moschovakis in [6].

Closely related is the concept of a <u>monotone</u> inductive operator. While the notion of a positive operator is syntactic, the notion of monotone is purely combinatorial: Γ is monotone if and only if, for all X and Y, $X \subseteq Y$ implies $\Gamma(X) \subseteq \Gamma(Y)$. It is easily seen that all positive operators are monotone.

The problem of reversing this inclusion under certain conditions is an interesting one. As will be outlined in § 2, the applicability of the recursion theorem makes positive operators rather manageable in most cases. Thus it would be helpful to know that certain monotone operators are in fact positive.

For the classic cases of Σ^0_1, Π^0_1, Σ^1_1 and Π^1_1 inductive definitions over the natural numbers, monotone and positive agree. Inductive definability over the natural numbers is studied in § 3.

Inductive definitions over the continuum are examined using the techniques of [1]. In this setting, monotone agrees with positive for Σ^1_1 and Π^1_1 operators.

For a time it seemed possible that in the most general setting all monotone operators would be effectively positive. Recently, however, Harrington and Kechris [3] have strongly disproved this conjecture. They showed, in fact, that under some quite reasonable conditions on the class Ξ, monotone inductive definability from Ξ will agree with general inductive definability from Ξ . These conditions allow for a large number of cases in which general (non-monotone) inductive definability is known to be more powerful that positive inductive definability. An example is the class of Σ^1_3 operators over the continuum.

The problem of characterizing the cases in which monotone and positive inductive definability agree is considered at the end.

2. Positive inductive definability and the Recursion Theorem.

In this section we will consider positive inductive definitions of finite type which are (ordinal) semirecursive relative to a functional of finite type.

Recall Kleene's inductive definition Ω of recursion relative to a fixed type two functional F and a fixed type three functional U . The functions recursive in F,U are assigned natural number indices "e" , with $Cl(\Omega)$ being the set of sequences $(e,\underset{\sim}{m},\underset{\sim}{\alpha},n)$ such that $\{e\}(\underset{\sim}{m},\underset{\sim}{\alpha}) \simeq n$.

For all natural numbers $a,b,i<k$, $j<l$, t, $b_0,\dots,b_{t-1}=\underset{\sim}{b},n,p$ and $m_0,\dots,m_{k-1}=\underset{\sim}{m}$, and all reals $\alpha_0,\dots,\alpha_{l-1}=\underset{\sim}{\alpha}$:

$\{<0,k,l,n>\}(\underset{\sim}{m},\underset{\sim}{\alpha}) \simeq n$ [Constant] ;

$\{<1,k,l,i>\}(\underset{\sim}{m},\underset{\sim}{\alpha}) \simeq m_i$ [Projection] ;

$\{<2,k,l,i>\}(\underset{\sim}{m},\underset{\sim}{\alpha}) \simeq m_i+1$ [Successor] ;

$\{<3,k,l,a,\underset{\sim}{b}>\}(\underset{\sim}{m},\underset{\sim}{\alpha}) \simeq \{a\}(\{b_0\}(\underset{\sim}{m},\underset{\sim}{\alpha}),\dots,\{b_{t-1}\}(\underset{\sim}{m},\underset{\sim}{\alpha}),\underset{\sim}{\alpha})$
[Composition] ;

$\{c\}(0,\underset{\sim}{m},\underset{\sim}{\alpha}) \simeq \{a\}(\underset{\sim}{m},\underset{\sim}{\alpha})$ and

$\{c\}(p+1,\underset{\sim}{m},\underset{\sim}{\alpha}) \simeq \{b\}(p,\underset{\sim}{m},\{c\}(p,\underset{\sim}{m},\underset{\sim}{\alpha}),\underset{\sim}{\alpha})$,

where $c=<4,k+1,l,a,\underset{\sim}{b}>$ [Primitive Recursion] ;

$\{<5,k+1,l>\}(b,\underset{\sim}{m},\underset{\sim}{\alpha}) \simeq \{b\}(\underset{\sim}{m},\underset{\sim}{\alpha})$ [Evaluation] ;

$\{<6,k,l,i,j>\}(\underset{\sim}{m},\underset{\sim}{\alpha}) \simeq \alpha_j(m_i)$ [Application] ;

$\{<7,k,l,b>\}(\underset{\sim}{m},\underset{\sim}{\alpha}) \simeq F(\lambda p\{b\}(p,\underset{\sim}{m},\underset{\sim}{\alpha}))$ [Type 2 Application] ;

$\{<8,k,l,b>\}(\underset{\sim}{m},\underset{\sim}{\alpha}) \simeq U(\lambda\beta\{b\}(\underset{\sim}{m},\beta,\underset{\sim}{\alpha}))$ [Type 3 Application] .

($\langle\underset{\sim}{m}\rangle$ is a code for m , such as $2^m 3^m \dots p_k^{m_{k-1}}$.)

This definition can be extended to higher types by adding further clauses similar to (6,7) and (8). All function arguments (the $\underset{\sim}{\alpha},\beta,F$ and U) are considered to be partial. Thus if f has domain $\{3\}$ and $f(3)=7$, then $\{<0,0,1,0>\}(f) \simeq 0$ and $\{<6,1,1,0,0>\}(3,f) \simeq 7$.

Since, however, inductively defined sets of total objects are of primary interest, total functions will be of great importance. As it stands, the inductive operator Ω is Σ_1^0 - because clauses (3) and (4) require the existence of intermediate outputs such as $\{b_0\}(\underset{\sim}{m},\underset{\sim}{\alpha})$. Suppose now that F is restricted to total arguments; then clause (7) requires that, for all p, $\{b\}(p,\underset{\sim}{m},\underset{\sim}{\alpha})\downarrow$ (converges). This makes Ω Π_1^0. Similarly, if U is restricted to total arguments, then Ω will be Π_1^1.

As canonical objects of this nature, consider, for each n, the type $n+2$ functional C_{n+2}, defined by

$$C_{n+2}(F) \simeq 0 \iff F \text{ is total on objects of type } n .$$

These may be viewed as weaker versions of the functionals ${}^{n+2}E^{\#}$ defined by

$${}^{n+2}E^{\#}(F) \simeq \begin{cases} 1, & \text{if } (\forall x)\, F(x) > 0 \text{ and} \\ 0, & \text{if } (\exists x)\, F(x) = 0 . \end{cases}$$

For $n=0$, the two are equivalent.

Notice that if the type three functional U is removed from consideration, the real arguments in the definition of Ω are reduced to the status of parameters - that is, they can be deleted without any adverse consequences. (See [2] for a discussion of parametrized inductive definitions.)

On the other hand, in order to introduce a type four functional, it is necessary to insert type two arguments and new clauses corresponding to (6) and (8) lifted one type.

Now suppose that Γ is a positive operator which inductively defines a set of type n and is semirecursive relative to some fixed functional (hereinafter ignored). This means that for some index c, for all x and B:

$$(*) \qquad x \in \Gamma(B) \iff \{c\}(x,B)\downarrow .$$

(B is identified here with its positive characteristic function, having domain B and range $\{0\}$.)

If Γ is applied to the typical semirecursive set B with positive characteristic function $\{b\}$, then

(*) $$x \in \Gamma(B) \longleftrightarrow \{c\}(x, \lambda y\{b\}(y))\downarrow .$$

The goal is to show that $Cl(\Gamma)$ is itself semirecursive (relative to C_n for $n > 1$). The following two lemmas are essential.

LEMMA 2.1. (The Recursion Theorem). For any a, there is an index $\bar{e}$ such that for all $\underset{\sim}{x}$:

$$\{\bar{e}\}(\underset{\sim}{x}) \simeq \{a\}(\bar{e}, \underset{\sim}{x}) .$$

Furthermore, whenever $\langle \bar{e}, \underset{\sim}{x}, n\rangle \in \Omega^{\sigma+1}$, $\langle a, \bar{e}, \underset{\sim}{x}, n\rangle \in \Omega^{\sigma}$. □

(See Hinman [4] for a proof.)

Let $\{b\}^{\sigma}$ stand for $\{\langle b, \underset{\sim}{x}, n\rangle\}$ in Ω^{σ}. The first application of the Recursion Theorem is the following Full Substitution Theorem. The reader is referred to [10] for a similar proof.

LEMMA 2.2. For each n, there is a primitive recursive function θ_n such that for any indices b, c and variables $\underset{\sim}{x}$ of any type, if y varies over type n objects, then:

$$\{\theta_n(b,c)\}(\underset{\sim}{x}) \simeq \{c\}(\underset{\sim}{x}, \lambda y\{b\}(\underset{\sim}{x}, y))$$

Furthermore, whenever $\langle \theta_n(b,c), x, n\rangle \in \Omega^{\sigma+1}$, $\{c\}(x, \{b\}^{\sigma}) \simeq n$. □

In applying this lemma to the study of inductive definability, we must be careful (for $n > 0$) to keep the variable y total. In this case, Lemma 2.2 holds only relative to the functional C_{n+1}. As an illustration, let $n = 1$ and consider the computation of $\{c\}(\lambda\alpha\{b\}(\alpha))$, where $\{b\}$ is to represent a set of reals and therefore can only be applied to total arguments. In particular, let $c = \langle 7,0,0,d\rangle$ for some index d; then $\{c\}(\{b\}) \simeq \{b\}(\lambda p\{d\}(p))$, but only if $\{d\}$ is total. The functional C_2 can be used to test for this requirement.

The goal of demonstrating that the closure of Γ is semirecursive is now in view.

Recalling the equations (*) above, let $a = \theta_n(b,c)$. It is clear that for all x

(*) $\{a\}(b,x) \simeq \{c\}(x, \lambda y\{b\}(y))$.

The domain B of $\{b\}$ will be a fixed point for Γ if and only if, for all x, $\{b\}(x) \simeq \{a\}(b,x)$. The existence of such an index b follows immediately from the Recursion Theorem. It remains to be shown that the index indicated yields the minimal fixed point or closure of the operator Γ.

Let C now be the actual closure of Γ and let B equal the domain of the partial recursive function $\{\bar{e}\}$ given by the recursion theorem to satisfy $\{\bar{e}\}(x) \simeq \{a\}(\bar{e},x)$ for all total type-n objects x. The proof that $B = C$ is in two parts.

(i) $C \subseteq B$: A simple induction argument shows that for all ordinals σ, $\Gamma^{\sigma} \subseteq B$. For $\sigma = 0$ or a limit ordinal, this is obvious. Now suppose that $\Gamma^{\sigma} \subseteq B$ and that $x \in \Gamma^{\sigma+1}$. Then $\{c\}(x,\Gamma^{\sigma})\downarrow$ and therefore $\{c\}(x, \lambda y\{\bar{e}\}(y))\downarrow$, since Γ^{σ} is included in the domain of $\{\bar{e}\}$ and Γ is positive. However, $\bar{e}$ was chosen to satisfy $\{\bar{e}\}(x) \simeq \{c\}(x, \lambda y\{\bar{e}\}(y))$, so $\{\bar{e}\}(x)\downarrow$ and $x \in B$. It follows that $\Gamma^{\sigma+1} \subseteq B$.

(ii) $B \subseteq C$: Another induction argument shows that for all σ, the domain of $\{\bar{e}\}^{\sigma}$ is included in C. Again, this is trivial for $\sigma = 0$ or a limit ordinal. Now suppose that $\{\bar{e}\}^{\sigma} \subseteq C$ and that $\{\bar{e}\}^{\sigma+1}(x) \simeq 0$. This means that $\langle \bar{e},x,0\rangle$ is in $\Omega^{\sigma+1}$. Recalling the equations (*) and applying the "furthermore" clauses of Lemmas 2.1-2, we see that $\langle a,\bar{e},x,0\rangle$ and $\langle \theta_n(\bar{e},c),x,0\rangle$ are in Ω^{σ} and therefore $\{c\}(x, \{\bar{e}\}^{\sigma}) \simeq 0$. By our induction hypothesis, it follows that $\{c\}(x,C) \simeq 0$, so that $x \in \Gamma(C) = C$. Thus, as desired, $\{\bar{e}\}^{\sigma+1} \subseteq C$. In fact, we have shown $\{\bar{e}\}^{\sigma} \subseteq \Gamma^{\sigma}$ for all σ.

This completes the proof of the following Strong Transitivity Theorem.

THEOREM 2.3. If a positive operator Γ inductively defines a set of type n and is semirecursive relative to a fixed functional F, then the closure of Γ is semirecursive in F and the functional C_n. Furthermore, $|\Gamma| \leq |\Omega[F]|$.

Call a type n functional <u>standard</u> if its domain is precisely the set of total type (n-1) arguments. C_n is the canonical standard type n object and is recursive in any other standard type $\geq n$ functional.

This is strictly weaker than the usual concept of <u>normality</u>. A type n functional is normal if ${}^{n}E$ is recursive in it, where ${}^{n}E$

is just the restriction of ${}^{n}E^{\#}$ to total arguments. Of course ${}^{n}E$ is itself standard, as are all normal objects, whereas C_n is only normal for $n = 2$. The study of higher type functionals is often restricted to those which are normal, in order that certain basic principles (such as Selection) will obtain. See Moschovakis [7] and Sacks [8] for more information.

Nevertheless, as demonstrated above, the concept of standardness is suitable to the study of positive inductive definability.

COROLLARY 2.4. If a positive operator Γ inductively defines a set of type n and is semirecursive relative to a fixed standard functional F of type $\geq n$, then the closure of Γ is also semirecursive in F. □

Suppose that the functional F is deleted from the statement of Theorem 2.3. For $n = 1$, C_1 may also be deleted. For $n = 2$, semirecursive in C_2 is the same as semirecursive in ${}^{2}E$ or Π^1_1. For $n = 3$, note that, putting C_3 in the place of U in the definition of Ω above, recursion in C_3 can be given by a Π^1_1 positive inductive definition over the continuum, which must have Π^1_1 closure (see 2.8 below). Therefore all type ≤ 3 objects semirecursive in C_3 are Π^1_1. Conversely, all Π^n_1 relations of any type are obviously semirecursive in C_{n+2}.

COROLLARY 2.5. Let Γ be positive and semirecursive (SR) :

(a) If Γ is type 2, then $Cl(\Gamma)$ is SR (Σ^0_1) $[Cl(\Gamma) \subseteq \omega]$.

(b) If Γ is type 3, then $Cl(\Gamma)$ is Π^1_1.

(c) If Γ is type 4, then $Cl(\Gamma)$ is Π^1_1.

(d) If Γ is type $n+3$, then $Cl(\Gamma)$ is SR in C_{n+2}. □

Recall the usual conventions. A functional which is (partial) recursive in $[{}^{2}E, {}^{3}E, {}^{3}E^{\#}]$ is called hyper-arithmetic (HA), hyperanalitic (HAn), hyperprojective (HP)]. The sets of type ≤ 2 semirecursive in ${}^{2}E$ turn out to be exactly the Π^1_1 sets; all type 3 sets SR in ${}^{2}E$ are also Π^1_1. Substituting the functionals ${}^{n}E^{(\#)}$ for F in Corollary 2.4, we obtain the following.

COROLLARY 2.6. Let Γ be a positive inductive operator:

(a) If Γ is type 2 and Π^1_1, then $Cl(\Gamma)$ is Π^1_1.

(b) If Γ is type 3 and SHAn, then $Cl(\Gamma)$ is SHAn.

(c) If Γ is type 3 and SHP, then $Cl(\Gamma)$ is SHP.

(d) If Γ is type $n+3$ and SR in ${}^{n+2}E(\#)$, then $Cl(\Gamma)$ is SR in ${}^{n+2}E(\#)$. □

Corollary 2.4 applies equally well to ordinal recursion. Since ${}^{2}E$ is always σ-recursive and since ∞-SR agrees with Σ_2^1 for type ≤ 2 objects, we have the following.

COROLLARY 2.7. Let Γ be a positive inductive operator:

(a) If Γ is type 2 and Σ_2^1, then $Cl(\Gamma)$ is Σ_2^1.

(b) If Γ is type 2 or 3 and σ-SR, then $Cl(\Gamma)$ is σ-SR (σ is not required to be admissible here). □

Let Γ be a Π_1^1 positive inductive operator over the continuum; it is easily seen that if B is Π_1^1, then $\Gamma(B)$ will also be Π_1^1. Thus Γ behaves as an ω_1-SR operator, where ω_1 is the first admissible. Similarly, for Σ_1^1 Γ, if B is Σ_2^1, then $\Gamma(B)$ will also be Σ_2^1, so that Γ behaves as an ∞-SR operator. The techniques of Theorem 2.3 can be adapted to prove the following.

COROLLARY 2.8. Let Γ be a positive inductive operator over the continuum:

(a) If Γ is Π_1^1, then $Cl(\Gamma)$ is also Π_1^1.

(b) If Γ is Σ_1^1, then $Cl(\Gamma)$ is Σ_2^1. □

Many of the above results are of course well-known, having been proven by other means. 2.5a is the basic result of Kleene [5]: 2.5b and 2.6a are classic theorems of Spector [9]: 2.6c,e and 2.8a are just a small part of the work of Moschovakis on inductive definability [6,7].

3. Monotone inductive definability.

In this section, the basic properties of monotone inductive definitions are developed. Several cases are considered in which all monotone operators are positive, so that the results of the previous section can be applied.

Recall that a set B is a fixed point for the operator Γ if $\Gamma(B) = B$. The most basic fact about monotone inductive operators is the existence of a unique minimal fixed point — the intersection of all fixed points.

LEMMA 3.1. If Γ is a monotone inductive operator, then

$x \in Cl(\Gamma)$ IFF $(\forall B)[\Gamma(B) \subseteq B \rightarrow x \in B]$.

<u>Proof</u>. Let $C = Cl(\Gamma)$ and let $D = \bigcap\{B: \Gamma(B) = B\}$. (Note that $\Gamma(B) \subseteq B$ if and only if $\Gamma(B) = B$.) Then whenever $\Gamma(B) = B$, $D \subseteq B$; thus $D \subseteq C$ is immediate. On the other hand, suppose that $\Gamma(B) \subseteq B$. Since $\Gamma^{0} = \emptyset \subseteq B$ and $\Gamma^{\sigma} \subseteq B$ implies $\Gamma^{\sigma+1} = \Gamma(\Gamma^{\sigma}) \subseteq \Gamma(B) = B$, it follows by induction that for all ordinals σ , $\Gamma^{\sigma} \subseteq B$, so that $C = \Gamma^{\infty} \subseteq B$. Thus $C \subseteq D$. □

This lemma yields immediately that the closure of a Π^1_1 monotone operator over the natural numbers is itself Π^1_1 . More generally, we have the following.

<u>Proposition</u> 3.2. Let Γ be a monotone inductive operator defining a set of type n :

(a) If Γ is Π^n_m $(m > 0)$, then $Cl(\Gamma)$ is also Π^n_m .

(b) If Γ is Σ^k_m (Π^k_m) for $k > n$, then $Cl(\Gamma)$ is also Σ^k_m (Π^k_m). □

The other basic principle of monotone inductive definability is the one which demonstrates that all monotone operators Γ can be viewed as positive. Unfortunately, this often requires increasing the complexity of the definition of Γ . The challenge is to refine the lemma to avoid this higher complexity.

LEMMA 3.3. Let Γ be a monotone inductive operator. Then for all x and B ,

$x \in \Gamma(B)$ IFF (i) $(\exists Y \subseteq B)\ x \in \Gamma(Y)$

IFF (ii) $(\forall Y \supseteq B)\ x \in \Gamma(Y)$. □

It follows immediately that if Γ is Σ^1_1 or Π^1_1 over the natural numbers and monotone, then Γ is positive. More generally, we have the following.

<u>Proposition</u> 3.4. Let Γ be a monotone inductive operator defining a set of type n . If Γ is Σ^k_m (Π^k_m) for $k \geq n$ and $m > 0$,

then Γ is Σ^k_m (Π^k_m) positive. □

The interesting cases arise when $k < n$. The first examples are Σ^0_1 and Π^0_1 monotone operators over the natural numbers.

Let f be any (partial) function mapping natural numbers to natural numbers. Recursion relative to f is given by the Σ^0_1 (in f) positive inductive operator $\Omega[f]$ over the natural numbers. Since $|\Omega[f]| = \omega$, any particular computation is only finite in length and therefore uses only finitely much of the information in f. This implies the following lemma.

LEMMA 3.5. For any type one f and any index a:

$\{a\}(f)\downarrow$ IFF $(\exists$ finite $g \subseteq f)$ $\{a\}(g)\downarrow$. □

Let K_B equal the characteristic function of a set B of natural numbers. Putting K_B in place of f in Lemma 3.5, we obtain the following improvement.

LEMMA 3.6. For any index a and any B:

$\{a\}(K_B)\downarrow$ IFF (i) $(\exists$ finite $Y \subseteq B)$ $\{a\}(K_Y)\downarrow$

IFF (ii) $(\exists$ cofinite $Y \supseteq B)$ $\{a\}(K_Y)\downarrow$.

Proof. Let g be the function given by Lemma 3.5. In (i) let $Y = \{m: g(m) = 0\}$; in (ii), let $Y = \{m: \neg(g(m) = 1)\}$. □

If Γ is a Σ^0_1 inductive operator over ω, then it is in fact semirecursive. Thus $m \in \Gamma(B) \leftrightarrow \{a\}(m,K_B)\downarrow$ for some fixed index a. Similarly, for Π^0_1 Γ, $m \in \Gamma(B) \leftrightarrow \{a\}(m,K_B)\uparrow$. Applying lemma 3.6, we can obtain the following.

Proposition 3.7. Let Γ be a monotone inductive operator over the natural numbers.

(a) If Γ is Σ^0_1, then

$m \in \Gamma(B)$ IFF $(\exists$ finite $X \subseteq B)$ $m \in \Gamma(X)$.

(b) If Γ is Π^0_1, then

$m \in \Gamma(B)$ IFF $(\forall$ cofinite $X \supseteq B)$ $m \in \Gamma(X)$.

In either case, Γ is positive and still Σ^0_1 (Π^0_1).

Proof. Parts (a) and (b) are immediate from the lemma. Now (a) can be rewritten in the form:

$$m \in \Gamma(B) \quad \text{IFF} \quad (\exists n)(\exists s)[(\forall i < n)(s)_i \in B \wedge m \in \Gamma(\{(s)_i : i < n\})] .$$

This is clearly Σ_1^0 and positive; part (b) is similar. □

Let F be any (partial) function from the continuum to ω, such as the characteristic function K_B of a set B of reals. Recursion in F is given by the operator $\Omega[F]$ over natural numbers, which must be countable in length. It follows that all possible computations together use only countably much of the information in F .

LEMMA 3.8. For any index a and any type 2 functional F :
$\{a\}(F)\downarrow$ IFF $(\exists$ countable $G \subseteq F)$ $\{a\}(G)\downarrow$. □

Any Σ_1^1 operator Γ over the continuum can be written:

$$\alpha \in \Gamma(X) \iff (\exists \gamma)(\forall p) \ \{a\}(p,\gamma,\alpha,K_X)\downarrow .$$

Similarly, a Π_1^1 operator Γ can be written:

$$\alpha \in \Gamma(X) \iff (\forall \gamma)(\exists p) \ \{a\}(p,\gamma,\alpha,K_X)\uparrow .$$

The following result can be obtained using lemma 3.8 with only a little difficulty - see [1] for details.

THEOREM 3.9. Let Γ be a monotone inductive operator over the continuum.

(a) If Γ is Σ_1^1, then

$$\alpha \in \Gamma(B) \quad \text{IFF} \quad (\exists \text{ countable } X \subseteq B) \quad \alpha \in \Gamma(X) .$$

(b) If Γ is Π_1^1, then

$$\alpha \in \Gamma(B) \quad \text{IFF} \quad (\forall \text{co-countable } X \supseteq B) \quad \alpha \in \Gamma(X) .$$

In either case, Γ is positive and still Σ_1^1 (Π_1^1) . □

Results (3,7) and (3.9) are refinements of the basic lemma 3.3. For Π_1^1 operators Γ over the continuum, we can also refine lemma 3.1.

THEOREM 3.10. Let Γ be a Π^1_1 monotone inductive operator over the continuum:

$$\alpha \in \mathrm{Cl}(\Gamma) \quad \text{IFF} \quad (\forall \text{co-countable } B)[\Gamma(B) \subseteq B \to \alpha \in B].$$

Proof. The direction $(\to)$ is immediate from Lemma 3.1. For the converse, let C be $\mathrm{Cl}(\Gamma)$, so that $C = \Gamma(C)$, and suppose that $\alpha \notin C$. We want a co-countable fixed point B such that $\alpha \notin B$. Since $\alpha \notin \Gamma(C)$, there exists by (3.9) a co-countable $B_0 \supseteq C$ such that $\alpha \notin \Gamma(B_0)$. Let $\{\gamma_i : i<\omega\}$ enumerate $\Gamma(B_0) - B_0$; each $\gamma_i \notin \Gamma(C)$, so as above there exist co-countable $D_i \supseteq C$ such that $\gamma_i \notin \Gamma(D_i)$. Set $B_1 = B_0 \cap \bigcap\{D_i : i<\omega\}$; then $B_1 \supseteq C$, B_1 is co-countable and for each i, $\gamma_i \notin \Gamma(B_1)$, that is, $B_1 \cap [\Gamma(B_0) = B_0] = \emptyset$. Since $\Gamma(B_1) \subseteq \Gamma(B_0)$ by monotonicity, it follows that $\Gamma(B_1) \subseteq B_0$.

Continuing in this manner we obtain a descending sequence $\{B_n : n<\omega\}$ of co-countable sets such that, for all n, $\alpha \notin B_n$ and $C \subseteq \Gamma(B_{n+1}) \subseteq B_n$. Finally, let $B = \bigcap\{B_n : n<\omega\}$. Then $\alpha \notin B$, B is co-countable, and $\Gamma(B) \subseteq \bigcap\{\Gamma(B_{n+1}) : n<\omega\} \subseteq \bigcap\{B_n : n<\omega\} = B$. □

More generally, let F be a type $(n+2)$ functional. Then recursion in F is defined by the operator $\Omega[F]$ over type n objects x with type $(n+1)$ arguments α as parameters. The key clause is the definition of $F(\lambda y\,\{b\}(y,\underset{\sim}{x},\underset{\sim}{\alpha}))$; this means that at a particular step, a computation may have one new branch for each type $(n+1)$ object. Since in a convergent computation each branch is finite in height, it follows that the entire computation has only as many steps as there are type $(n+1)$ objects. Thus, as shown above for $n=1$, the type $(n+2)$ functional F can be coded by a type $(n+1)$ functional.

Following the same outline as in (3.8-9-10), the general result is obtained.

THEOREM 3.11. Let Γ be a monotone inductive operator defining a set of type $(n+1)$ for $n>1$. If Γ is Π^n_1 (Σ^n_1) then Γ is actually Π^n_1 (Σ^n_1) positive. Furthermore:

(a) If Γ is Π^n_1, then $\mathrm{Cl}(\Gamma)$ is also Π^n_1.

(b) If Γ is Σ^n_1, then $\mathrm{Cl}(\Gamma)$ is also Σ^n_1.

Proof. The only result of a new character is (b). Recall from Corollary 2.8. that the closure of a Σ^1_1 monotone operator over the continuum is only Σ^1_2 (and in fact need not be Σ^1_1). This, however, is just because the notion of well-foundedness is Π^1_1 . To obtain result (b), we first note that any F being in the closure of Γ depends only on a "small" set of objects already being in - small meaning codable by a type n object. Then F is in the closure of Γ IFF there is a type n object which codes a chain of small sets, each included in Γ of its predecessor and F in one of them.

For $k > 1$, the closures of the Σ^n_k positive inductive operators over type n object are just the sets semi-hyper-elementary over those objects, or semirecursive in ${}^{n+2}E^{\#}$. The Σ^n_2 (and possibly also the Π^n_2) monotone operators have the same set of closures. This cannot be extended any further, as shown by the following recent result of Harrington and Kechris [3].

<u>Proposition</u> 3.12. For inductive definitions over objects of type $n > 0$, for $k > 1$, $Cl(\Sigma^n_k) = Cl(\Sigma^n_k\text{-monotone})$.

Since there will always be Π^n_2 (non-monotone) operators with are not semihyperelementary, it is clear that there will always be Σ^n_3 monotone inductive operators which are not positive.

References

[1] D. Cenzer, Monotone inductive definitions over the continuum, J.Symbolic Logic, to appear.

[2] ________, Parametrized inductive definitions and recursive inductive operators over the continuum, Fund.Math., to appear.

[3] L.A. Harrington and A.S. Kechris, On Monotone vs. Non-Monotone Induction, mimeographed notes.

[4] P. Hinman, Recursion-Theoretic Hierarchies, Springer, to appear.

[5] S.C. Kleene, Recursive functionals and quantifiers of finite types, I, Trans. Amer. Math. Soc. 91, (1959), 1-52.

[6] Y.N. Moschovakis, Elementary Induction on Abstract Structures, North-Holland, Amsterdam (1974).

[7] ____________, Structural characterizations of classes of relations, Generalized Recursion Theory (J. Fenstad and P. Hinman, Ed.), North-Holland, Amsterdam, (1974), 53-80.

[8] G. Sacks, The 1-section of a type n object, Gen.Rec.Theory, 81-93.

[9] C. Spector, Inductively defined sets of natural numbers, Infinitistic Methods, Pergamon, Oxford, 97-102 (1961).

[10] D. Cenzer, Ordinal recursion and inductive definitions, in Gen. Rec, Theory 221-264.

AN ALTERNATIVE TO MARTIN'S AXIOM

by Keith J. Devlin [1]

§ 0. Introduction

We work in ZFC, and use the usual notations and conventions. In particular, ordinals are von Neumann ordinals, cardinals are initial ordinals, the cardinality of the set X is denoted by $|X|$, and $\alpha, \beta, \gamma, \ldots$ are reserved for ordinal numbers. We assume the reader is familiar with forcing (as described in Jech (1971), for example), the statement of, and elementary consequences of, Martin's axiom (as described in Martin - Solovay (1970), Solovay - Tennenbaum (1971), or Jech (1971)), the statement of the Souslin Hypothesis and its formulation in terms of trees (as in Devlin - Johnsbråten (1974)), and the elementary theory of $\aleph_1$ - trees (also in Devlin-Johnsbråten (1974)).

In the mid 1960's, work on the Souslin problem by Solovay, Tennenbaum, Martin and Rowbottom (and others?) led Martin to formulate a general principle which came to be known as Martin's axiom. This can be formulated as an assertion about the continuum, and indeed in a very real sense that is just what it is. However, for our purposes, we shall restrict ourselves to the original formulation, and, starting from that point, use the original motivation for Martin's axiom in order to formulate an alternative principle.

---

[1] The results in this paper have only a marginal connection with our talk at the meeting, the details of which will be published elsewhere. However, we obtained the main theorem presented here, theorem B, during the course of the meeting, so these proceedings seem to be a suitable resting place for our paper.

Recall that the Souslin Hypothesis (SH) says (in one of its formulations) that there is no ω_1-tree having no uncountable antichains (such a tree being called a Souslin tree if it were to exist). [By ω_1-tree, we mean what some authors call a normal ω_1-tree.] In order to establish the consistency of SH with ZFC by forcing, one observes that if T were a Souslin tree, then, inverted and regarded as a forcing poset, T is a c.c.c. poset, and $\Vdash_T$ " $\check{T}$ has an uncountable antichain", and the idea now is to carry out an iterated forcing argument, adding uncountable antichains to all Souslin trees in sight (by forcing with the Souslin trees themselves, as above). We assume that the details of this are well known to the reader. In any case they can be found in both Solovay-Tennenbaun (1971) and Jech (1971). Perhaps the most crucial part of the entire argument is the verification that ω_1 is not collapsed during the iteration. This is done by showing that the c.c.c. holds at each stage. Consideration of the required result (i.e. the consistency of ZFC + SH) and the method of proof just outlined leads to the formulation of Martin's axiom (in its simplest form).

Martin's Axiom (MA) says that: if P is a poset of cardinality $\aleph_1$ which satisfies c.c.c., and if $\mathcal{F}$ is a family of $\aleph_1$ dense initial sections of P, there is a set G which is $\mathcal{F}$-generic on P.

In order to establish the implication MA $\rightarrow$ SH, we simply suppose that $\langle T, \leqslant_T\rangle$ were a Souslin tree and consider the poset $P = \langle T, \geqslant_T\rangle$ and the family $\mathcal{F} = \{\bigcup_{\alpha\leqslant\beta<\omega_1} T_\alpha \mid \alpha < \omega_1\}$ of dense initial sections of P, since in this case, an $\mathcal{F}$-generic subset of P would be a cofinal branch of T, leading at once to an uncountable antichain of T.

In order to establish the consistency of MA with ZFC, we simply carry out iterated forcing argument, adding generic subsets to all c.c.c. posets of cardinality $\aleph_1$. Again, the proof hinges upon the preservation of the c.c.c.
A feature of the original Solovay-Tennenbaum iteration to obtain SH was that new real were introduced at limit stages, and consequently CH failed in the final model. In fact, with MA we can prove outright that MA $\rightarrow 2^{\aleph_0} > \aleph_1$. [If $\langle f_\alpha \mid \alpha < \omega_1\rangle$ enumerated 2^ω, consider the c.c.c. poset P of all finite maps from ω_1 into 2 and the family $\mathcal{F} = \{Y_\alpha \mid \alpha < \omega_1\}$ of dense initial sections $Y_\alpha = \{f \in P \mid \alpha \in \mathrm{dom}(f),\ f \neq f_\alpha \upharpoonright \mathrm{dom}(f)\}$.]

Let us consider now the implication MA $\rightarrow$ SH, the original motivation for the principle MA, If T is a Souslin tree, then, as we remarked earlier, T is (modulo an inversion of the partial ordering) a c.c.c. poset. Hence, forcing with T preserves cardinals (and cofinalities). But T preserves cardinals for a different reason, as well. Recall that a poset P is σ-<u>dense</u> if the intersection of any countable family of dense initial sections of P is again a dense initial section of P, and that if P is a σ-dense poset, then $\Vdash_P$ "$\mathcal{P}(\check{\omega}) = \check{\mathcal{P}(\omega)}$", so not only does P not collapse ω_1, it in fact introduces no new real numbers. Clearly then, if P is a σ-dense poset of cardinality $\aleph_1$, then P preserves cardinals (and cofinalities) in forcing extensions. But it is easily seen that a Souslin tree T is (when inverted) a σ-dense poset. Hence for this reason also, Souslin trees do not collapse cardinals. Indeed, more is shown. When we iterate just to obtain SH, thereby introducing new reals as we mentioned earlier, the new reals only appear at limit stages of the iteration, and hence only "incidentally". (indeed, as is described in <u>Devlin-Johnsbråten</u> (1974), <u>Jensen</u> has shown that one can iterate to obtain SH + CH without introducing any new reals, though the argument is extremely complicated). The same motivation which led to MA thus leads to an alternative principle:

<u>Devlin's Axiom</u>[1] (DA) says that : if P is a σ-dense poset of cardinality $\aleph_1$, and if $\mathcal{F}$ is a family of $\aleph_1$ dense initial sections of P, there is a set G which is $\mathcal{F}$-generic on P.

Clearly, DA $\rightarrow$ SH, by virtually the same argument as for MA. But since DA only deals with posets which cannot introduce new reals, one might hope that DA is consistent with CH, thereby obtaining the consistency of SH + CH. Unfortunately, one can prove outright that DA $\rightarrow 2^{\aleph_0} > \aleph_1$, and this we shall do in § 1. What about the consistency of DA with ZFC ? Well, since DA implies the failure of CH, one cannot hope to carry out an analogue of the MA proof, preserving σ-density at each stage in the argument. And since the posets involved do not necessarily satisfy c.c.c. (but see our concluding remarks concerning this point), we cannot show that ω_1 is preserved because of the c.c.c. The proof of the consistency of DA with ZFC would thus appear to present us with some awe-some difficulties. Nevertheless DA is consistent, and by virtue of a perhaps unexpected reason. In § 2 we show that MA $\rightarrow$ DA.

---

1. Since it would appear that DA is destined to be the world's most useless axiom of set theory, we feel that this title does not indicate any undue amount of immodesty.

§ 1. $DA \rightarrow 2^{\aleph_0} > \aleph_1$.

Since DA only deals with σ-dense posets, which do not affect $\mathcal{P}(\omega)$, we cannot expect to prove $DA \rightarrow 2^{\aleph_0} > \aleph_1$ by means of a direct application of DA , as was the case with MA . The idea is to proceed by assuming DA + CH and deriving a contradiction. Using CH , we can easily construct σ-dense posets. Then, by means of an argument of the Jensen-Johnsbråten type, we obtain the required contradiction. In fact, our argument is not the original one of Jensen and Johnsbråten (which appeared in their (1974) paper, as well as in Devlin-Johnsbråten (1974) under a slightly different guise), but rather an ingenious modification of it due to, Gregory (1973). Gregory was actually concerned with a problem concerning boolean algebras, but his argument carries over virtually unchanged to the present situation (a fact which was pointed out to us by Jech).

We assume DA + CH from now on. The aim of this section is to derive a contradiction from this assumption, thereby establishing $DA \rightarrow 2^{\aleph_0} > \aleph_1$.

Let $\mathcal{F}$ be the set of all functions F such that:

(i) $\mathrm{dom}(F) \subseteq \omega_1^{\underset{\smile}{\omega_1}}$ (= $\bigcup_{\alpha<\omega_1} \omega_1^{\alpha}$) ;

(ii) $\mathrm{ran}(F) \subseteq \omega_1^{\underset{\smile}{\omega_1}}$;

(iii) if F(f) is defined, then $\mathrm{dom}(F(f)) = \mathrm{dom}(f) + 1$;

(iv) if F(f) is defined, so is $F(f \upharpoonright \alpha)$ for all $\alpha < \mathrm{dom}(f)$;

(v) if F(f) is defined, then $F(f \upharpoonright \alpha) = F(f) \upharpoonright (\alpha + 1)$ for all $\alpha \leq \mathrm{dom}(f)$.

By CH , let $\langle E_\alpha \mid \alpha < \omega_1 \rangle$ enumerate all pairs $\langle f,t \rangle$ such that $f \in \omega_1^{\underset{\smile}{\omega_1}}$ and $t \in [\omega_1^{\underset{\smile}{\omega_1}}]^{\leq\omega}$, with each pair appearing cofinally often in this enumeration.

By recursion on $\alpha < \omega_1$, we define a sequence $\langle F_\alpha \mid \alpha < \omega_1 \rangle$ such that:

(i) $F_\alpha \in \mathcal{F}$

(ii) $|\mathrm{dom}(F_\alpha)| \leq \omega$;

(iii) $\alpha < \beta \rightarrow F_\alpha \subseteq F_\beta$.

Set $F_0 = \emptyset$. For $\lim(\alpha)$, set $F_\alpha = \bigcup_{\beta<\alpha} F_\beta$. Finally, suppose that F_α is defined. We define $F_{\alpha+1}$ by cases, as follows.

<u>Case</u> 1. $E_\alpha = \langle f, \emptyset \rangle$, where $\mathrm{dom}(f) = 2$.

Since $\mathrm{dom}(F_\alpha)$ is countable, there is $h \in \omega_1^{\omega_1}$, $\mathrm{dom}(h) = 1$, such that $F_\alpha(h)$ is not defined. Set $F_{\alpha+1} = F_\alpha \cup \{\langle f,h \rangle\}$. (Thus $F_{\alpha+1}(h) = f$.)

<u>Case</u> 2. $E_\alpha = \langle f, \{g\}\rangle$, where $\mathrm{dom}(f) = \gamma + 1$, $\mathrm{dom}(g) = \delta < \gamma$, $F(g) = f \upharpoonright (\delta + 1)$.

Pick $h \in \omega_1^\gamma$ such that $g = h \upharpoonright \delta$ and $F_\alpha(h \upharpoonright \delta + 1)$ is not defined, and set $F_{\alpha+1} = F_\alpha \cup \{\langle F \upharpoonright \beta + 1 , h \upharpoonright \beta\rangle | \delta < \beta \leq \gamma\}$.

<u>Case</u> 3. $E_\alpha = \langle f,t\rangle$, where $\mathrm{dom}(f) = \gamma + 1$, $t \subseteq \mathrm{dom}(F_\alpha)$, $k \in t \rightarrow F_\alpha(k) \subseteq f$, and there are uncountably many $h \in \omega_1^\gamma$ such that h is a union of elements of t .

Pick $h \in \omega_1^\gamma$ a union of elements of t with $F_\alpha(h)$ not defined, and set $F_{\alpha+1} = F_\alpha \cup \{\langle f,h\rangle\}$.

<u>Case</u> 4. Otherwise.

Set $F_{\alpha+1} = F_\alpha$.

That completes the definition of $\langle F_\alpha \mid \alpha < \omega_1\rangle$ .Set $F = \bigcup_{\alpha<\omega_1} F_\alpha$. Clearly, $F \in \mathcal{F}$.

For $b \in \omega_1^{\omega_1}$, set $T_b = \{f \mid F(f) \subseteq b\}$. Clearly, T_b is a tree (under function extension). Moreover, $f \in T_b \wedge 0 < \alpha < \mathrm{dom}(f) \rightarrow$ $\rightarrow f \upharpoonright \alpha \in T_b$.

Fix some $b \in \omega_1^{\omega_1}$ for the time being, and set $T = T_b$.

<u>Lemma</u> 1. $T \neq \emptyset$.

<u>Proof</u>. Pick α with $E_\alpha = \langle b \upharpoonright 2, \emptyset \rangle$. Then Case 1 held when $F_{\alpha+1}$ was defined. So, for some h , $F_{\alpha+1}(h) = b \upharpoonright 2$. Thus $F(h) \subseteq b$. Thus $h \in T$. □

<u>Lemma</u> 2. Suppose $g \in T_\delta$ and $\delta < \gamma < \omega_1$. Then g has uncountable many extensions in T_γ .

<u>Proof</u>. Since $g \in T_\delta$, $F(g) = b \upharpoonright (\delta + 1)$. Pick β with $F_\beta(g) = F(g)$. For uncountably many $\alpha > \beta$, $E_\alpha = \langle b \upharpoonright (\gamma + 1), \{g\} \rangle$. For each such α , Case 2 held when $F_{\alpha+1}$ was defined, so there is $h \in \omega_1^\gamma$ with $h \supseteq g$, $F_\alpha(h)$ undefined, and $F_{\alpha+1}(h) = b \upharpoonright \gamma + 1$. Then $h \in T_\gamma$. And since $F_\alpha(h)$ is undefined, h cannot be associated to any smaller α in this manner, so distinct α's give distinct h's . □

<u>Lemma</u> 3. Let $t \subseteq T$ be countable, and suppose there are uncountable many $h \in \omega_1^\gamma$ such that h is the union of elements of t . Then there is $h \in T_\gamma$ which is the union of elements of t .

<u>Proof</u>. For each $k \in t$, $F(k) \subseteq b$, so we can find $\beta(k)$ with $F_{\beta(k)}(k)$ defined. Set $\beta = \sup_{k \in t} \beta(k)$. Thus $F_\beta(k)$ is defined for all $k \in t$. For some $\alpha > \beta$, $E_\alpha = \langle b \upharpoonright (\gamma + 1), t \rangle$. Thus Case 3 held when $F_{\alpha+1}$ was defined, so there is $h \in \omega_1^\gamma$ such that h is a union of elements of t and $F_{\alpha+1}(h) = b \upharpoonright (\gamma + 1)$. Thus $F(h) \subseteq b$, giving $h \in T_\gamma$. □

<u>Lemma</u> 4. T is σ-dense (under the inverse ordering).

<u>Proof</u>. Let D_m , $m < \omega$, be dense initial sections of T (under $\supseteq$). Let $f \in T$ be given. Define $h(s) \in T$ for $s \in 2^m$ by induction on m . Set $h(\emptyset) = f$. Suppose now $h(s) \supseteq f$ are defined for all $s \in 2^m$. Let $\beta(m) = \sup \{\mathrm{dom}(h(s)) \mid s \in 2^m\}$. Let $s \in 2^m$ be given. Set $u = s \cup \{\langle 0,m \rangle\}$, $v = s \cup \{\langle 1,m \rangle\}$. Pick $h(u)$, $h(v) \in T$, $h(u), h(v) \supseteq h(s)$, with $h(u), h(v) \in D_m$, $\mathrm{dom}(h(u)) = \mathrm{dom}(h(v)) > \beta(m)$. Set $t = \{h(s) \mid s \in 2^{\omega}\}$. Let $\gamma = \sup_{m<\omega} \beta(m)$. By Lemma 3, there is $h \in T$ such that h is a union of elements of t . Clearly, $h \supseteq f$ and $h \in \bigcap_{m<\omega} D_m$. The Lemma follows at once. □

Thus, for every $b \in \omega_1^{\omega_1}$, T_b is a σ-dense tree of cardinality $\aleph_1$. So, applying DA , every such T_b has an uncountable branch.

So we may make the following definitions. Let $b \in \omega_1^{\omega_1}$. Let $T_b^0 = \omega_1^{\underset{\smile}{\omega_1}}$ and set $b_0 = b$. By induction, for $n < \omega$, let $T_b^{n+1} = \{f \in \omega_1^{\underset{\smile}{\omega_1}} \mid F(f) \subseteq b_n\}$ and let b_{n+1} be an uncountable branch of T_b^{n+1}. (Thus $T^{n+1} = T_{b_n}$ for each n).

<u>Lemma</u> 5. Let $b, b' \in \omega_1^{\omega_1}$. Suppose that $(\forall n \in \omega)[b_n \restriction 1 = b'_n \restriction 1]$. Then $b = b'$.

<u>Proof</u>. We prove by induction on β that for all $\beta < \omega_1$, $(\forall n \in \omega)[b_n \restriction \beta = b'_n \restriction \beta]$. This implies that $(\forall n \in \omega)[b_n = b'_n]$, so in particular, $b = b_0 = b'_0 = b'$.

For $\beta = 1$ there is nothing to prove. And limit stages in the induction are immediate of course. So assume now the result for β. Let $n < \omega$. Set $f = b_{n+1} \restriction \beta$. By induction hypothesis, $f = b'_{n+1} \restriction \beta$. Thus $F(f) \subseteq b_n$ and $F(f) \subseteq b'_n$. But $\mathrm{dom}(F(f)) = \beta + 1$. Hence $b_n \restriction \beta + 1 = F(f) = b'_n \restriction \beta + 1$. □

By lemma 5 now, the mapping $b \to \langle b_n \restriction 1 \mid n < \omega \rangle$ is a one-one mapping from $\omega_1^{\omega_1}$ into $(\omega_1^{\ 1})^{\omega}$. Hence $\aleph_1^{\aleph_1} \leq \aleph_1^{\aleph_0}$. Hence $2^{\aleph_0} > \aleph_1$, an we have obtained our desired contradiction. We have thus proved:

<u>Theorem</u> A. $\mathrm{DA} \to 2^{\aleph_0} > \aleph_1$. □

§ 2. MA → DA .

If P is a poset, we let $BA(P)$ denote the complete boolean algebra of all regular open subsets of P (with P endowed with the order topology), isomorphed so that P is a subposet of $BA(P)$, dense in $BA(P)$. It is a standard fact that if P is σ-dense, then $BA(P)$ is (ω, ∞)-distributive (see <u>Devlin-Johnsbråten</u> (1974), p.68)

We assume MA from now on, and prove DA. Let P be a given σ-dense poset, $|P| = \aleph_1$, and let $\mathcal{F} = \{D_\alpha \mid \alpha < \omega_1\}$ be a collection of dense initial sections of P. We show that P carries an $\mathcal{F}$-generic subset.

Let $B = BA(P)$. Let $\langle p_\alpha \mid \alpha < \omega_1 \rangle$ enumerate P. We define a set $T \subseteq B$ such that:

(i) T is a tree under $\geqslant_B$;

(ii) For all $\alpha < \omega_1$, $T_{\alpha+1} \subseteq D_\alpha$;

(iii) $(\forall \alpha < \beta < \omega_1)(\forall x \in T_\alpha)(\exists y_1, y_2 \in T_\beta)(y_1 \neq y_2 \wedge x <_T y_1, y_2)$.

(iv) T is dense in B ;

(v) For all $\alpha < \omega_1$, $\bigvee T_\alpha = \mathbb{1}$ (in B).

(vi) For all $\alpha < \omega_1$, $(\forall x, y \in T_\alpha)(x \neq y \to x \wedge y = \mathbb{O}$ (in B)).

The definition of T is by induction on the levels of T. Let $T_0 = \{\mathbb{1}_B\}$. Suppose now that T_α is defined. Let $x \in T_\alpha$. Pick $p, q \in P$, $p, q \leq_B x$, p incompatible with q. (As usual in such situations, we assume that all our posets are such that every element has an incompatible pair of extensions – otherwise forcing is not possible, of course). If $x \wedge p_\alpha > \mathbb{O}$ (in B), we pick p here with $p \leq_B x \wedge p_\alpha$. Since P is dense in B, this causes no problems. Now pick $p', q' \in D_\alpha$, $p' \leq_P p$, $q' \leq_P q$. Extend $\{p', q'\}$ to a maximal pairwise incompatible subset of $\{r \in D_\alpha \mid r \leq_B x\}$, say $A(x)$. Let $A(x)$ be the set of all successors of x in $T_{\alpha+1}$. Notice that $\bigvee A(x) = x$ (in B). Finally, suppose $\lim(\alpha)$ and T_β, $\beta < \alpha$ are all defined. Let $T_\alpha = \{\bigwedge b$ (in B) $\mid b$ is an α-branch of $T \upharpoonright \alpha$ and $\bigwedge b > \mathbb{O}$ (in B)$\}$. To show that T is as required is quite straightforward, except perhaps for one point. We must show that if $\lim(\alpha)$ and $x \in T \upharpoonright \alpha$, then x has an extension on T_α, and moreover $x = \bigvee\{y \in T_\alpha \mid y <_B x\}$ (in B). Well, let $X = \{b \mid b$ is an α-branch of $T \upharpoonright \alpha$ through $x\}$. Clearly, for each $b \in X$, $x = x \wedge (\bigvee b)$. Thus $x = \bigwedge\{x \wedge (\bigvee b) \mid b \in X\} =$ $= x \wedge \bigwedge\{\bigvee b \mid b \in X\}$. So, as B is (ω, ∞)-distributive, $x = x \wedge \bigvee\{\bigwedge b \mid b \in X\}$. In other words, $x = x \wedge \bigvee\{\bigwedge b \mid b \in X \;\&\; \bigwedge b > \mathbb{O}\}$ $= x \wedge \bigvee\{y \in T_\alpha \mid y <_B x\}$. Hence $x = \bigvee\{y \in T_\alpha \mid y <_B x\}$, as required.

Lemma 1. If T has an uncountable branch, then P has an $\mathcal{F}$-generic subset.

Proof. Let b be an uncountable branch of T. Set $H = \{p \in P \mid (\exists \alpha < \omega_1)(p \in b \cap T_{\alpha+1})\}$. Clearly, H is a pairwise compatible subset of P such that $(\forall \alpha < \omega_1)(H \cap D_\alpha \neq \emptyset)$. Let G be the final section of P determined by H. Then G is an $\mathcal{F}$-generic subset of P. □

By virtue of lemma 1, it suffices to show that T has an uncountable branch. We assume otherwise, and derive a contradiction.

Lemma 2. (i) T has height ω_1.
(ii) For all $\alpha < \omega_1$, $|T_\alpha| \leq \omega_1$.

Proof. (i) Immediate from the assumption that T has no uncountable branch.
(ii) Since $x,y \in T_\alpha$ & $x \neq y \rightarrow x \wedge y = \emptyset$ (in B) and P is a dense subset of B of cardinality $\aleph_1$. □

The following lemma, due to Kunen, Baumgartner, and Reinhardt-Malizt, is proved in Devlin-Johnsbråten (1974), pp.65-67.

Lemma 3. Let A be a tree of height ω_1, having countable levels, and having no uncountable branches, and let $\mathcal{P}(A)$ be the poset of all finite order-preserving mappings from A into Q (the rationals). Then $\mathcal{P}(A)$ satisfies the c.c.c. □

For our present purposes, we require an extension of this result.

Lemma 4. Let $\mathcal{P}$ be the poset of all finite order-preserving mappings from T into Q. Then $|\mathcal{P}| = \aleph_1$ and $\mathcal{P}$ satisfies the c.c.c.

Proof. Clearly $|\mathcal{P}| = \aleph_1$. In order to show that $\mathcal{P}$ satisfies c.c.c., it suffices to show that some (particular) dense subposet of $\mathcal{P}$ satisties c.c.c. Say a finite subset t of T is neat iff, regarding t as a tree under the induced ordering, elements of the same height in t are of the same height in T and every element of t has extensions on all higher levels of t and predecessors on all lower levels of t. Let $\tilde{\mathcal{P}} = \{p \in \mathcal{P} \mid \text{dom}(p) \text{ is neat}\}$. Clearly, $\tilde{\mathcal{P}}$ is dense in $\mathcal{P}$. We show that $\tilde{\mathcal{P}}$ satisfies c.c.c. Suppose not. Let X be an uncountable pairwise incompatible subset of $\tilde{\mathcal{P}}$. We may assume that for all $p,q \in X$, dom(p) and dom(q) are isomorphic (as trees). For $p \in \tilde{\mathcal{P}}$, let $\sigma(p)$ be the lowest level of dom(p). We may assume that X is chosen so that $p,q \in X \rightarrow \sigma(p) \neq \sigma(q)$ and the common value of $|\sigma(p)|$ for $p \in X$ is as small as possible. For $p \in \tilde{\mathcal{P}}$, set $t(p) = \{x \in T \mid (\exists y \in \sigma(p))(x \leq_T y)\}$.

Claim. For all $\alpha < \omega_1$, $\{t(p) \cap T_\alpha \mid p \in X\}$ is countable.

Proof of claim. Suppose that $\{t(p) \cap T_\alpha \mid p \in X\}$ is uncountable for some $\alpha < \omega_1$. Let $Y \subseteq X$ be uncountable with

$p,q \in Y \to t(p) \cap T_\alpha \neq t(q) \cap T_\alpha$. For $p \in Y$, set $\delta(p) = t(p) \cap T_\alpha$. By a well-known and standard argument, we can find a fixed finite set $\delta \subseteq T_\alpha$ and an uncountable set $Z \subseteq Y$ such that $p,q \in Z \to |\delta(p)| = |\delta(q)| \ \& \ \delta(p) \cap \delta(q) = \delta$. For $p \in Z$ now, let $u(p) = \{x \in \mathrm{dom}(p) \mid (\exists y \in \delta)(y \leq_T x)\}$, $v(p) = \mathrm{dom}(p) - u(p)$. Notice that as the $\delta(p)$ are all distinct, $\sigma(p) \cap v(p) \neq \emptyset$ for all $p \in Z$.

And clearly, $u(p)$, $v(p)$ are neat subtrees of T , so $p \upharpoonright u(p)$, $p \upharpoonright v(p) \in \tilde{\mathfrak{P}}$. But look, if $p \upharpoonright u(p)$ and $q \upharpoonright u(q)$ are compatible, $p,q \in Z$, then p and q are compatible (this is easily seen). Hence $\{p \upharpoonright u(p) \mid p \in Z\}$ is pairwise incompatible. But $|\sigma(p \upharpoonright u(p))| = |\sigma(p)| - |\sigma(p) \cap v(p)| < |\sigma(p)|$ for all $p \in Z$. This leads at once to a contradiction with one of our assumptions. This proves the claim.

By the claim, it follows at once that for each $\alpha < \omega_1$, there are at most countably many $p \in X$ with $\sigma(p) \subseteq T_\alpha$, and that $\{\alpha < \omega_1 \mid (\exists p \in X)(\sigma(p) \subseteq T_\alpha)\}$ is cofinal in α . Thus, using the claim again, we see that $A = \{x \in T \mid (\exists p \in X)(\exists y \in \mathrm{dom}(p))(x \leq_T y)\}$, regarded as a tree under the induced ordering, has height ω_1 and countable levels only. But clearly, X is an uncountable pairwise incompatible subset of $\mathfrak{P}(A)$, contrary to lemma 3. This proves lemma 4. □

Let $\mathcal{F} = \{U_x \mid x \in T\}$, where $U_x = \{p \in \mathfrak{P} \mid x \in \mathrm{dom}(p)\}$, a collection of $\aleph_1$ dense initial sections of $\mathfrak{P}$. By MA , let G be an $\mathcal{F}$-generic subset of $\mathfrak{P}$. Let $f = \bigcup G$. Clearly, f is an order-preserving map from T into Q .

<u>Lemma</u> 5. $\|\check{\omega}_1 < \omega_1\|^B = 1$.

<u>Proof</u>. Since T is dense in B , it suffices to show that $\Vdash_T$ " $\check{\omega}_1 < \omega_1$" . Let G be V-generic on T . Clearly, G is a cofinal branch of T in V^B . Let $g : \omega_1 \to G$ be the canonical enumeration of G . Then $\check{f} \circ g = \check{\omega}_1 \to \check{Q}$ is one-one. Hence $\check{\omega}_1$ is countable. □

But B is (ω, ∞)-distributive, so $\|\check{\omega}_1 = \omega_1\|^B = 1$. This contradiction means that we have now proved:

<u>Theorem</u> B. MA $\to$ DA . □

§ 3. What Use is DA ?

At the moment, we know of no application of DA other than the Souslin hypothesis. Indeed, in the absence of CH , we do not know of an example of a σ-dense poset of cardinality $\aleph_1$ other than a Souslin tree, or modifications to a Souslin tree. (For example, the disjoint union of $\aleph_1$ Souslin trees is a non c.c.c. σ-dense poset of cardinality $\aleph_1$, but hardly a useful example.) We exclude, of course, σ-closed posets here, by virtue of:

<u>Theorem</u> C. If P is a σ-closed poset (i.e. if every decreasing countable chain from P has a lower bound in P), and if $\mathfrak{F}$ is a collection of $\aleph_1$ dense initial sections of P , then P has an $\mathfrak{F}$-generic subset.

<u>Proof</u>. The standard construction of generic sets generalises to this case with no problems. □

Thus, if DA is to have a non-trivial application, then it must involve σ-dense posets which are not σ-closed.

The requirement that the poset involved in DA should have cardinality $\aleph_1$ is not really necessary. It suffices that the poset be σ-dense and satisfy the $\aleph_2$-c.c. (i.e. the usual cardinal preserving condition at the "$\aleph_1$-level"). More precisely, let DA* say that: if P is a σ-dense poset satisfying $\aleph_2$-c.c. , and if $\mathfrak{F}$ is a collection of $\aleph_1$ dense initial sections of P , then P has an $\mathfrak{F}$-generic subset. Then:

<u>Theorem</u> D. MA $\rightarrow$ DA* $\rightarrow$ DA .

<u>Proof</u>. Clearly, DA* $\rightarrow$ DA . To prove MA $\rightarrow$ DA* ,we argue much as before. There were only two places where we used the fact that $|P| = \aleph_1$. The first place was to ensure that the tree T constructed in $B = BA(P)$ was dense in B . This fact will no longer be true in the pressent case, but we shall be able to get by without it, as will become clear in a moment. The second place was to ensure that T has only levels of cardinality at most $\aleph_1$. Since, in this case, B will certainly satisfy the $\aleph_2$-c.c. (since P is dense in B) , this will still be true in the present case. Thus, we can carry through the previous proof as far as lemma 2.5, the only difference

being that T is not dense in B. But look, for each $\alpha < \omega_1$, $X_\alpha = \{b \in B | (\exists x \in T_\alpha)(b \leq_B x)\}$ is a dense initial section of B (since T_α is always a maximal antichain of B). Thus, a V-generic subset of B generates a cofinal branch of T, and we can complete the proof as in the case of DA now. □

DA* does, of course, apparently broaden our range of possible candidate posets. For instance, a Souslin $\aleph_2$-tree is now admissible. But, of course, we still only allow $\aleph_1$ many dense sets, so perhaps we really gain nothing new here!

Or maybe DA → MA ? We see no reason why this should be the case, but nor do we see how to show that the implication is not valid. Indeed, are even DA* and DA equivalent? Certainly, we can express DA* in boolean terms, as a sort of generalised Rasiowa-Sikorski lemma, thus:

Theorem E. DA* $\longleftrightarrow$ For each (ω, ∞)-distributive $\aleph_2$-c.c. complete boolean algebra B, and for each family $\mathcal{F}$ of at most $\aleph_1$ suprema from B, there is an ultrafilter U on B which preserves all of the suprema in $\mathcal{F}$ (i.e. if $b = \bigvee_{i \in I} b_i$ is in $\mathcal{F}$, then $b \in U$ iff $b_i \in U$ for some $i \in I$).

Proof. Analogous to the case of MA and c.c.c. complete boolean algebras, described in Martin-Solovay (1970). □

References.

Devlin-Johnsbråten (1974). The Souslin Problem. Springer: Lecture Notes in Mathematics 405.

Gregory (1973). A Countably Distributive Complete Boolean Algebra Not Uncountably Representable, Proceedings of the Am.Math.Soc. 42 (1974), pp.42-46.

Jech (1971). Lectures in Set Theory, Springer: Lecture Notes in Mathematics 217.

Jensen-Johnsbråten (1974). A New Construction of a Non-Constructible Δ^1_3 set in Integers. Fund.Math. 81, pp.279-290.

Martin-Solovay (1970). Internal Cohen Extensions. Annals of Mathematical Logic 2, pp.143-178.

Solovay-Tennenbaum (1971). Iterated Cohen Extensions and Souslin's Problem. Annals of Mathematics 94, pp.201-245.

ELEMENTARY EXTENSIONS OF β-MODELS OF A_2 WITH FIXED HEIGHT

by M. Dubiel
University of Warsaw

In this paper the letters M , N with or without subscripts are restricted to denote standard models of $ZFC^- + V = HC$. Let $\preccurlyeq$, $\prec$ denote elementary inclusion and strict elementary inclusion. By $S(M)$ we denote the collection of all $N \succcurlyeq M$ such that N has the same height as M . We are interested in the partial ordering $\mathfrak{S}(M) = \langle S(M), \preccurlyeq \rangle$. We shall prove two results. The first says that for any countable M_0 there exists countable $M \supsetneq M_0$ of the same height such that $\mathfrak{S}(M)$ contains an element minimal $\succ M$. The second says that given M_0 there exists $M \supsetneq M_0$ of the same height such that there is an embedding of $O = \langle P(^{\omega}2) , \subsetneq \rangle$ in $\mathfrak{S}(N)$ for any countable N in $S(M)$. This result is best possible in the sense that, since models of $ZFC^- + V = HC$ are equivalent to models of second order arithmetic A_2 (see [1], [8]) for every M there is an embedding of $\mathfrak{S}(M)$ in O .

The following is our starting point:
Recall that a notion of forcing $P \in M$ is called homogeneous if for all $p, q \in P$ there exists an automorphism $f: P \to P$, $f \in M$, such that $f(p)$ is compatible with q .

Theorem 1. (Guzicki [4], Theorem 13 and Lemma 16). Let M_0 be countable and P be a homogeneous notion of forcing in M_0 then there exists countable $M \supset M_0$ with the same height such that $N[G] \succ N$ for every N in $\mathfrak{S}(M)$ and G P-generic over N .

Guzicki used this theorem to deduce that any countable model M_0 has an uncountable extension of the same height. Using Martin's Axiom he found an extension of M_0 of the same height of power $2^{\aleph_0}$.

<u>Corollary</u>. Let M_0 be countable. There exists $M \supseteq M_0$ such that $\mathfrak{S}(M)$ contains an element minimal $> M$.

Proof. Let P be any suitable notion of forcing in M_0 and M_1 be the model given by Theorem 1. Consider $M_1[G]$ where G is any set P-generic over M_1. Since $M_1[G]$ is transitive model of $ZFC^- + V = HC$ then every elementary submodel of $M_1[G]$ is transitive (see [5], Corollary 1.3). Let M be any elementary submodel of $M_1[G]$ such that $M_1 \prec M$ and M is maximal with respect to the property $G \notin M$ - exists by Zorn's Lemma. It is clear that $M_1[G]$ is a maximal elementary extension of M.

<u>Theorem 2</u>. Let M_0 be countable. There exists $M \supset M_0$ such that if $N \in \mathfrak{S}(M)$ is countable then there exists $F\colon P({}^{\omega}2) \to \mathfrak{S}(N)$ such that for all $X, Y \subseteq {}^{\omega}2$, $X \subseteq Y \iff F(X) \preccurlyeq F(Y)$. Further for all $X \subseteq {}^{\omega}2$, $|F(X)| = |X| + \aleph_0$.

Proof. Let P be Cohen's notion of forcing, i.e. P consists of all finite functions from ω into 2 ordered by inverse inclusion. In the following argument what is important is that P be homogeneous and absolute, and $P \times P$ be isomorphic to P. For any $f_0,\dots,f_{n-1} \in P$ let $f_0 \times \dots \times f_{n-1}(x) = f_i(y)$, where $y = [x/n]$ and $i = x - ny$. For $G_0,\dots,G_{n-1} \subseteq P$ we define $G_0 \times \dots \times G_{n-1} = \{f_0 \times \dots \times f_{n-1} : f_0 \in G_0,\dots,f_{n-1} \in G_{n-1}\}$. Let M be the model obtained from M_0 by Guzicki's theorem and $N \in \mathfrak{S}(M)$ be countable.

<u>Claim</u>. There exists a mapping $G\colon {}^{\omega}2 \to {}^{\omega}2$ such that for any distinct $f_0,\dots,f_{n-1}, g_0,\dots,g_{m-1} \in {}^{\omega}2$ $G(f_0) \times \dots \times G(f_{n-1})$ is P-generic over $N[G(g_0) \times \dots \times G(g_{m-1})]$.

We shall first explain why the claim suffices to prove the theorem. Recall that for any $H \subseteq P$, H- P-generic over N, $N[H]$ is the least model of ZFC^- such that $N \subset N[H]$ and $H \in N[H]$. Thus if $f_0,\dots,f_{n-1}, g_0,\dots,g_{m-1}$ are distinct members of ${}^{\omega}2$ we have

$$N[G(f_0) \times \ldots \times G(f_{n-1}) \times G(g_0) \times \ldots \times G(g_{m-1})] =$$
$$= (N[G(f_0) \times \ldots \times G(f_{n-1})])[G(g_0) \times \ldots \times G(g_{m-1})]$$

and

$$N[G(f_0) \times \ldots \times G(f_{n-1})] = N[G(f_{\pi(0)}) \times \ldots \times G(f_{\pi(n-1)})]$$

where π is any permutation of n,

For $X \subseteq {}^{\omega}2$ define

$$F(X) = \bigcup \{N[G(f_0) \times \ldots \times G(f_{n-1})] : n < \omega,\ f_0, \ldots, f_{n-1} \text{ distinct members of } X\}$$

where, when $n = 0$, by $N[G(f_0) \times \ldots \times G(f_{n-1})]$ we mean N.

Let $X, Y \subseteq {}^{\omega}2$ be finite where $X = \{f_0, \ldots, f_{n-1}\}$, $Y = \{f_0, \ldots, f_{n-1}, g_0, \ldots, g_{m-1}\}$, then

$$F(X) = N[G(f_0) \times \ldots \times G(f_{n-1})]$$

and

$$F(Y) = F(X)[G(g_0) \times \ldots \times G(g_{m-1})]$$

where $G(g_0) \times \ldots \times G(g_{m-1})$ is P-generic over $F(X)$ from the claim. From the application of Theorem 1 $F(X) \preccurlyeq F(Y)$. Next observe that if X is finite and Y infinite then $F(X) \preccurlyeq F(Y)$, because the union of a set of models directed upwards by elementary inclusion is an elementary extension of each member of the set. Further if X, Y are both infinite, $a_0, \ldots, a_{n-1} \in F(X)$ and $\phi(x_0, \ldots, x_{n-1})$ is a formula of set theory then there exists some finite $X' \subseteq X$ such that $a_0, \ldots, a_{n-1} \in F(X')$ whence

$$F(X) \models \phi(a_1, \ldots, a_{n-1}) \Leftrightarrow F(X') \models \phi(a_0, \ldots, a_{n-1})$$
$$\Leftrightarrow F(Y) \models \phi(a_0, \ldots, a_{n-1})$$

Thus $X \subseteq Y$ always implies $F(X) \preccurlyeq F(Y)$. If $g \in X - Y$ then $G(g) \in F(X)$. However for all distinct $f_0, \ldots, f_{n-1}$ in Y, $G(g)$ is generic over $N[G(f_0), \ldots, G(f_{n-1})]$ whence $G(g) \notin N[G(f_0) \times \ldots$

$\ldots \times G(f_{n-1})]$ and $G(g) \in F(X) - F(Y)$. Thus $X \not\subseteq Y$ implies $F(X) \not\subseteq F(Y)$. Since $|F(X)| = |X| + \aleph_0$ is obvious this completes the proof of the theorem.

For the proof of the claim we must recall some facts about forcing. Let N and P be fixed as above : then there is a canonical enumeration $\langle a_i(H): i < \omega\rangle$ of $N(H)$ which is uniform for $H \subseteq P$, H P-generic over N. Further, for any formula $\psi(x_0,\ldots,x_{n-1})$, $N[H] \models \psi(a_0(H),\ldots,a_{n-1}(H))$ implies that there exists $p \in H$ such that for every $H' \subseteq P$, H' P-generic over N and $p \in H'$, $N[H'] \models \psi(a_0(H'),\ldots,a_{n-1}(H'))$. It is this phenomenon of compactness which makes the claim true. To be more specific we construct a mapping $J : {}^{\omega>}2 \to P$ by induction on the length of the argument as follows:

$J(\langle\ \rangle) = 1_P$, the completely undefined function. Suppose that J has been defined on all sequences of length k. We now find $J(\eta)$ simultaneously for all $\eta \in {}^{(k+1)}2$ in a finite number of steps. Initially for $\eta = \sigma^\frown\langle i\rangle$ we let $J(\eta,0) = J(\sigma)^\frown\langle i\rangle$ where $p^\frown\langle i\rangle \in P$ is obtained from $p \in P$ by :
$\mathrm{Dom}\,(p^\frown\langle i\rangle) = \mathrm{Dom}\,p \cup \{\mu x\, (x \notin \mathrm{Dom}\,p)\}$, $p^\frown\langle i\rangle \supseteq p$ and $(p^\frown\langle i\rangle)(\mu x\,(x \notin \mathrm{Dom}\,p)) = i$.. Now suppose that after j steps of the process we have obtained $J(\eta, j)$ for all $\eta \in {}^{(k+1)}2$. There is to be one step for each triple $\langle i, \langle\eta_0,\ldots,\eta_{n-1}\rangle,\langle\sigma_0,\ldots,\sigma_{m-1}\rangle\rangle$ where $i < k$ and $\eta_0,\ldots,\eta_{n-1}$, $\sigma_0,\ldots,\sigma_{m-1}$ are distinct members of ${}^{(k+1)}2$. Suppose that the $(j+1)$-st step corresponds to the triple displayed. If possible we extend the functions $J(\eta,j)$, $\eta \in {}^{(k+1)}2$, to new finite functions $J(\eta, j+1)$ in P with the following property: there exists $p, q \in P$ such that

$$J(\eta_0, j+1) \times \ldots \times J(\eta_{n-1}, j+1) \leq p$$

$$J(\sigma_0, j+1) \times \ldots \times J(\sigma_{m-1}, j+1) \leq q$$

and for every $H \subseteq P$, H P-generic over N and $p \in H$, we have $N[H] \models q \in a_i[H]$. If there are no such extensions we let $J(\eta,j+1) = J(\eta,j)$ after the final step. Now we can define G by

$$G(f) = \{p: \exists n\, \exists \eta\ [\eta \in {}^n2\ \&\ \eta \subseteq f\ \&\ J(\eta) \leq p]\}$$

It is easy to check that G has the required property.

Theorem 2 implies that if an ordering $\langle S, \leq\rangle$ is embeddable in $\langle P({}^{\omega}2), \subseteq\rangle$ then it is embeddable in $\mathfrak{J}(N)$ for every $N \in \mathfrak{J}(M)$. In particular the order types η, λ, $\eta\cdot\omega_1$ and $\lambda\cdot\omega_1$ are embeddable in these structures and many more besides. Here η, λ denote the respective order types of the rationals and the real numbers.

The present paper was inspired by the investigations of A.Mostowski[6],[7]. He considered various natural partial orderings of the ω-models of A_2 including $\preccurlyeq$. Let $\mathcal{G}$ denote the family of ω-models of A_2 which are not β-models but which are elementarily equivalent to the "principial model" consisting of all integers and sets of integers. From the results of [7] it easily follows that the order type $\lambda\cdot\omega_1$ is embeddable in $\langle\mathcal{G}, \leq\rangle$. Notice that, if $\mathcal{G}'\subseteq\mathcal{G}$ is linearly ordered by $\preccurlyeq$, then all the members of $\mathcal{G}'$ have the same height, i.e. the same ordinals are representable in every member of $\mathcal{G}'$. (see [3]). Thus our results closely parallel those of Mostowski except that where he was concerned with members of $\mathcal{G}$ of fixed height we are concerned with β-models of fixed height.

References

[1] K.R. Apt and W. Marek, Second order arithmetics and related topics, Annals of Mathematical Logic 6(1974), pp.177-229.

[2] P.J. Cohen, Set theory and the continuum hypothesis, Benjamin, New York 1966.

[3] M. Dubiel, Nonstandard well orderings in ω-models of A_2, Bull. Acad. Polon. Sci. 21(1973), pp.

[4] W. Guzicki, Uncountable β-models with countable height, Fund. Math. LXXXIII (1974), pp.143-152.

[5] W. Marek and M. Srebrny, On transitive models for fragments of set theory, Bull. Acad. Polon. Sci. 21(1973), pp. 389-392.

[6] A. Mostowski, Partial orderings of the family of ω-models, preprint of the University of Warsaw, 1971.

[7] A. Mostowski, A transfinite sequence of ω-models, Journal of Symbolic Logic 37(1972), pp.96-102.

[8] P. Zbierski, Models for higher order arithmetics, Bull. Acad. Polon. Sci. 19(1971), pp.557-562.

DAS PROBLEM VON SOUSLIN

FÜR GEORDNETE ALGEBRAISCHE STRUKTUREN

Ulrich Felgner

Mathematisches Institut der Universität Tübingen

Im Jahre 1920 stellte Souslin [21] das folgende Problem:

"Falls $\leqq$ eine lineare Ordnung auf der Menge M ist, derart daß $\langle M,\leqq\rangle$ die folgenden drei Bedingungen erfüllt:

(S I): $\langle M,\leqq\rangle$ *besitzt kein erstes und kein letztes Element;*

(S II): $\langle M,\leqq\rangle$ *ist in Bezug auf die Intervall-Topologie, die durch die lineare Ordnung* $\leqq$ *induziert wird, ein zusammenhängender topologischer Raum;*

(S III): *Jede Menge von paarweise disjunkten offenen Intervallen ist höchstens abzählbar;*

ist dann $\langle M,\leqq\rangle$ ordnungs - isomorph mit der Menge $\mathbb{R}$ aller reellen Zahlen ? "

Die Bedingungen (S I) , (S II) und (S III) werden wir im Folgenden die Bedingungen von Souslin nennen. Wir wissen nicht, ob Souslin eine positive oder negative Antwort auf seine Frage vermutet hat. Aber vielleicht dürfen wir doch Souslin unterstellen, daß er an eine positive Antwort auf seine Frage geglaubt hat. Wir werden daher, wie es heute üblich geworden ist, die folgende Annahme die Souslin'sche Hypothese nennen:

(SH) *"Jede linear-geordnete Menge* $\langle M,\leqq\rangle$ *, welche die drei Bedingungen*

(S I), (S II) *und* (S III) *von Souslin erfüllt, ist topologisch die reelle Zahlengerade* $\mathbb{R}$ ".

Für eine lange Zeit gelang es weder Topologen noch Mengentheoretikern, die Souslin'sche Hypothese (SH) entweder zu beweisen oder zu widerlegen. Im Jahre 1954 zeigte A.S.Jesenin-Vol'pin [13], daß (SH) im System ZFU nicht beweisbar ist. Dabei ist ZFU das Axiomen-System der Mengenlehre von Zermelo-Fraenkel, welches die Existenz von unendlich vielen Urelementen zuläßt. P.Hájek und P.Vopěnka [7] bewiesen, daß (SH) auch in ZF nicht beweisbar ist. Th.Jech [12] und S.Tennenbaum [22] verallgemeinerten dieses Ergebnis und zeigten, daß (SH) auch nicht auf der Basis ZF + AC beweisbar ist. Dabei ist AC das Auswahl-Axiom. Es gilt sogar: (SH) ist weder im System ZF + AC + $2^{\aleph_0} = \aleph_1$ noch im System ZF + AC + $2^{\aleph_0} > \aleph_1$ beweisbar. R.B.Jensen bewies 1969, daß die Negation von (SH) aus dem Gödelschen Konstruktibilitäts-Axiom V = L folgt. Daß aber auch (SH) selbst mit den Axiomen der Mengenlehre ZF+AC konsistent ist, wurde erst im Jahre 1968 von R.Solovay und S.Tennenbaum in einer gemeinsamen Arbeit nachgewiesen. R.Solovay und S.Tennenbaum [20] konstruierten ein Modell $\mathfrak{M}$ der Mengenlehre ZF + AC, sodaß (SH) in $\mathfrak{M}$ gilt. Allerdings ist in $\mathfrak{M}$ die Kontinuums-Hypothese verletzt. Die Konstruktion eines Modelles $\mathfrak{N}$ von ZF + AC + $2^{\aleph_0} = \aleph_1$ + (SH) gelang erst R.B.Jensen im Jahre 1970.

Es sind auch einige Ergebnisse bekannt, denenzufolge die Annahme der Existenz starrer und auch homogener Souslin-Kontinua konsistent ist. Insbesondere folgt aus V = L die Existenz eines Souslin-Kontinuums $\langle M, \leqq \rangle$ derart, daß die Automorphismengruppe von $\langle M, \leqq \rangle$ 2-fach transitiv ist (das heißt: alle abgeschlossenen Intervalle von $\langle M, \leqq \rangle$ sind paarweise ordnungs-isomorph). Dabei wird $\langle M, \leqq \rangle$ Souslin-Kontinuum genannt, falls $\langle M, \leqq \rangle$ die drei Souslin'schen Bedingungen (S I), (S II), (S III) erfüllt, aber nicht mit $\langle \mathbb{R}, \leqq \rangle$ ordnungs-isomorph ist. Für eine Darstellung dieser und weiterer Resultate verweisen wir auf das Buch

[4] von K.Devlin und H.Johnsbråten (siehe auch M.E.Rudin [18]).

Sei $\leqq$ eine lineare Ordnung auf der Menge M und seien A und B Teilmengen von M. Dann wird das geordnete Paar $\langle A, B\rangle$ ein *Dedekindscher Schnitt* in $\langle M,\leqq\rangle$ genannt, falls $A \cap B = \emptyset$, $A \cup B = M$, $A \neq \emptyset$, $B \neq \emptyset$ und $a < b$ für alle $a \in A$ und alle $b \in B$ gilt.

Der Dedekindsche Schnitt $\langle A, B\rangle$ wird "Sprung" genannt, falls A ein größtes Element und B ein kleinstes Element hat. Der Dedekindsche Schnitt $\langle A, B\rangle$ wird "Lücke" genannt, falls A kein größtes und B kein kleinstes Element hat. Schließlich nennen wir den Dedekindschen Schnitt $\langle A, B\rangle$ "stetig", falls entweder A ein größtes und B kein kleinstes Element hat, oder A kein größtes aber B ein kleinstes Element hat. Eine linear geordnete Menge $\langle M, \leqq\rangle$ heißt *stetig* (oder: *stetig geordnet*), falls jeder Dedekindsche Schnitt in $\langle M, \leqq\rangle$ stetig ist.

Mit dieser Terminologie können wir die zweite Souslin'sche Bedingung (S II) wie folgt äquivalent umformulieren:

(S II): $\langle M, \leqq\rangle$ *ist eine dicht geordnete und stetig-geordnete Menge.*

Die Menge $\mathbb{R}$ aller reellen Zahlen ist nicht nur linear geordnet, sie bildet auch in Bezug auf die Addition eine geordnete abelsche Gruppe und in Bezug auf die Operationen $+$ und $\cdot$ einen reell-abgeschlossenen Körper. Es ist daher sehr natürlich, das Problem von Souslin wie folgt auszudehnen:

> "Sei $\langle G, \cdot, \leqq\rangle$ eine linear geordnete Gruppe, so daß die lineare Ordnung $\leqq$ auf der Menge G die drei Bedingungen (S I), (S II) und (S III) von Souslin erfüllt. Sind dann $\langle G, \cdot, \leqq\rangle$ und $\langle \mathbb{R}, +, \leqq\rangle$ als geordnete Gruppen isomorph?"

Die positive Antwort auf diese Frage folgt leicht aus Ergebnissen von O.Hölder [10] und F.Loonstra [16] (vergl. L.Fuchs [6] p.74 und p.76). Wir wollen analoge Fragen für geordnete Körper, geordnete Ringe, rechtsgeordnete Fast-Körper und reguläre Halbgruppen stellen. Wir wollen die folgenden Sätze beweisen (dabei ist $\mathbb{R}^+ = \{x \in \mathbb{R} \; ; \; x > 0\}$) :

<u>SATZ 1 :</u> *Es sei* $\langle F,+,\cdot,\leqq\rangle$ *ein rechts-geordneter Fastkörper derart, daß die lineare Ordnung* $\leqq$ *die zweite Bedingung* (S II) *von Souslin erfüllt, Dann ist* $\langle F,+,\cdot\rangle$ *ein kommutativer Körper und* $\langle F,+,\cdot,\leqq\rangle$ *und* $\langle \mathbb{R},+,\cdot,\leqq\rangle$ *sind isomorph als geordnete Körper.*

Insbesondere sind daher auch geordnete Schiefkörper, welche (S II) erfüllen, mit dem Körper der reellen Zahlen isomorph. Auch ein geordneter Ring R, der (S II) erfüllt, ist ein kommutativer Körper und daher mit dem Körper der reellen Zahlen isomorph, sofern er kein Zero-Ring ist. Bei geordneten Halbgruppen ist auch die Bedingung (S I) bedeutungsvoll. Wir werden beweisen, daß *jede natürlich-geordnete Halbgruppe, die archimedisch geordnet ist und sowohl* (S I) *als auch* (S II) *erfüllt, mit* $\langle \mathbb{R}^+,+,\leqq\rangle$ *isomorph ist.* Demgegenüber gilt jedoch:

<u>SATZ 2:</u> (In ZF + V = L) *Es existiert eine positiv-geordnete archimedische kommutative Halbgruppe* $\langle H,\cdot,\leqq\rangle$ *derart, daß* $\langle S,\leqq\rangle$ *die drei Bedingungen* (S I), (S II), (S III) *erfüllt, aber nicht mit* $\langle \mathbb{R}^+,\leqq\rangle$ *ordnungs-isomorph ist.*

Die für das Souslin'sche Problem typische Bedingung (SIII) betrachten wir in § 5 , und beweisen unter anderem den folgenden Satz:

<u>SATZ 3:</u> *Eine dicht-geordnete abelsche Gruppe, welche* (S III) *erfüllt, besitzt eine abzählbare dichte Teilmenge.*

In unserer mengentheoretischen Notation folgen wir Drake [5]. Insbesondere ist $\omega = \{0,1,\ldots\}$ die kleinste unendliche Ordinalzahl und $\mathbb{N} = \{1,2,3,\ldots\}$ die Menge aller von Null verschiedenen natürlichen Zahlen. $\langle x,y\rangle$ bezeichnet das geordnete Paar von x und y . In unserer Notation für Strukturen folgen wir Chang-Keisler [2] p.20 und p.266.

§ 1. Souslins Problem für geordnete Gruppen, Ringe und Körper

Sei $\langle G,\cdot\rangle$ eine (multiplikativ geschriebene) Gruppe und $\leq$ eine lineare Ordnung auf der Menge G. Dann wird $\langle G,\cdot,\leq\rangle$ eine *rechts-geordnete Gruppe* genannt, falls für alle $a,b,c \in G$ gilt: $a\leq b \Rightarrow a\cdot c\leq b\cdot c$. Wir nennen $\langle G,\cdot,\leq\rangle$ *links-geordnet*, falls für alle $a,b,c \in G$ gilt: $a\leq b \Longrightarrow c\cdot a\leq c\cdot b$. Wir nennen $\langle G,\cdot,\leq\rangle$ eine *geordnete Gruppe* (oder auch: *linear-geordnete Gruppe*), falls $\langle G,\cdot,\leq\rangle$ sowohl rechts- als auch links-geordnet ist.

Im Folgenden möge e stets das Eins-Element von $\langle G,\cdot\rangle$ bezeichnen. Die rechts-geordnete Gruppe $\langle G,\cdot,\leq\rangle$ wird *archimedisch* genannt, falls $\forall a\in G\, \forall b\in G\, \exists n\in \mathbb{N} \left[e<a \wedge e<b \Longrightarrow a<b^n \right]$ gilt.

LEMMA 1: *Sei $\langle G,\cdot,\leq\rangle$ eine rechts-geordnete Gruppe derart, daß $\langle G,\leq\rangle$ Souslins Bedingung* (S II) *erfüllt. Dann sind $\langle G,\cdot,\leq\rangle$ und $\langle \mathbb{R},+,\leq\rangle$ isomorph als geordnete Gruppen. Insbesondere ist $\langle G,\cdot,\leq\rangle$ dann abelsch und archimedisch.*

Beweis. Wir beweisen zunächst die Archimedizität von $\langle G,\cdot,\leq\rangle$. Wähle ein beliebiges Element $a\in G$ mit $e\leq a$, und definiere

$$B_a = \{ x\in G \,;\, \exists n\in\mathbb{N} :\ x\leq a^n \}$$

$$C_a = \{ x\in G \,;\, \forall n\in\mathbb{N} :\ a^n < x \}.$$

Angenommen $C_a \neq \emptyset$. Dann ist $\langle B_a,C_a\rangle$ ein Dedekindscher Schnitt und nach (S II) existiert ein $s\in G$ mit $b\leq s\leq c$ für alle $b\in B_a$ und alle $c\in C_a$. Wir behaupten, daß $a=e=s$ gilt. Fall 1: es gelte $s\cdot a = s$. Dann folgt $a=e$ sofort. Fall 2: es gelte $s\cdot a<s$. Dann gilt $s\cdot a \in B_a$ und folglich $s\cdot a \leq a^n$ für ein $n\in\mathbb{N}$. Setze $m=n-1$. Aus $b\leq s$ für alle $b\in B_a$ folgt $a^m\leq s$ und daher $a^n = a^{m+1} = a^m\cdot a \leq s\cdot a$, da $\langle G,\cdot,\leq\rangle$ rechts-geordnet ist. Aus $s\cdot a\leq a^n$ folgt jetzt $a^n = s\cdot a$, also $s=a^m$. Aus $s=a^m\leq a^{m+k}\leq\ldots\leq s$ für alle $k\in\mathbb{N}$ folgt $a^m = a^{m+1}$, also $a=e$. Fall 3: es gelte $s<s\cdot a$. Aus der Rechts-Monotonie der Multiplikation folgt $s\cdot a^{-1} < (s\cdot a)\cdot a^{-1} = s$, also $s\cdot a^{-1}\in B_a$. Daher existiert ein $n\in\mathbb{N}$ so,daß $s\cdot a^{-1}\leqq a^n$. Also folgt $s\leqq a^{n+1}$. Aus $a^k\leqq s$ für alle $k\in\mathbb{N}$ folgt $s=a^{n+1}=a^{n+2}=s$, also $a=s$, $a=e$. Wir haben gezeigt: wenn $e<a$, dann $C_a=\emptyset$. Also ist $\langle G,\cdot,\leqq\rangle$ archimedisch. Nach einem Satz von P.Conrad [3] sind archimedische rechts-geordnete Gruppen auch links-geordnet. Nach einem Satz von Hölder [10] ist G daher eine abelsche Gruppe, $G\subseteqq\mathbb{R}$. Aber $\langle G,\cdot,\leqq\rangle$ ist eine in sich dichte, stetig geordnete Gruppe, Daher muß $\langle G,\cdot,\leqq\rangle$ mit der Gruppe $\langle\mathbb{R},+,\leqq\rangle$ ordnungsisomorph sein (siehe F.Loonstra [16],K.Iseki [11] und L.Fuchs [6] p.76), Q.E.D.

Nach Lemma 1 ist insbesondere auch jede linear-geordnete Gruppe $\langle G,\cdot,\leqq\rangle$, welche die Souslin'schen Bedingungen erfüllt,mit der additiven Gruppe aller reellen Zahlen ordnungsisomorph.Auf die Voraussetzung der Bedingungen (S I) und (S III) kann dabei verzichtet werden. (S I) folgt bereits aus der Torsionsfreiheit von $\langle G,\cdot\rangle$.Die Ergebnisse dieser Arbeit zeigen ferner, daß (S III) vom algebraischen Standpunkt aus irrelevant ist. Allerdings ist (S III) gerade die typische Bedingung des ursprünglichen Souslin'schen Problems.

LEMMA 2 : *Sei $\langle F,+,\cdot,\leqq\rangle$ ein linear geordneter Schiefkörper derart, daß $\langle F,\leqq\rangle$ die Souslinsche Bedingung* (S II) *erfüllt. Dann ist $\langle F,+,\cdot\rangle$ kommutativ und $\langle F,+,\cdot,\leqq\rangle$ und $\langle\mathbb{R},+,\cdot,\leqq\rangle$ sind als geordnete Körper isomorph.*

Beweis. Für $x\in F$ sei $|x|$ das Maximum von x und $-x$. Sei 1 das Einselement des Körpers $\langle F,+,\cdot\rangle$. Setze $2=1+1$, $3=2+1$ etc. Für $k\in F$ setze

$$A(k) = \{ x\in F \ ; \ \exists n\in\mathbb{N} : |k-x|<n \}$$

$$B(k) = \{ x\in F \ ; \ \forall y\in A(k): y<x \} .$$

Falls $B(k)\neq\emptyset$, dann existiert nach (S II) ein $s\in F$ mit $a\leqq s\leqq b$ für alls $a\in A(k)$ und alle $b\in B(k)$. Dann gilt $s-1<s$ und folglich entweder $s-1\in A(k)$ oder $s-1<a$ für alle $a\in A(k)$.In beiden Fällen ergibt sich

sofort ein Widerspruch. Also gilt $A(k) = F$ für alle $k \in F$ und F ist archimedisch. Nach einem Satz von D.Hilbert [9] p.105 (siehe auch L. Fuchs [6] p.185) ist F kommutativ. Die Isomorphie von $\langle F,+,\cdot,\leq\rangle$ und $\langle \mathbb{R},+,\cdot,\leqq\rangle$ als geordnete Körper folgt jetzt aus R.Baer [1],Satz 1.3, Q.E.D.

Eine Struktur $\langle A,+,\cdot\rangle$ mit zwei binären Operationen "+" und "·" wird *Ring* genannt, falls $\langle A,+\rangle$ eine abelsche Gruppe ist und beide Distributiv-Gesetze gelten. Die Assoziativität oder Kommutativität der Multiplikation wird nicht gefordert! Wir verlangen auch nicht die Existenz eines Eins-Elementes für die Multiplikation. Ein Ring heißt *Zero-Ring*, falls $x \cdot y = 0$ für alle x und y gilt.

$\langle A,+,\cdot,\leqq\rangle$ ist ein linear-geordneter Ring, falls $\langle A,+,\cdot\rangle$ ein Ring ist , $\leqq$ eine lineare Ordnung auf A, und die folgenden drei Monotonie Gesetze gültig sind: $\forall x,y,z \in A\ [\,x \leqq y \Longrightarrow x+z \leqq y+z\,]$, $\forall x,y,z \in A\,[(x \leqq y \wedge 0 \leqq z) \Rightarrow x\cdot z \leqq y\cdot z]$ und $\forall x,y,z \in A\,[(x \leqq y \wedge 0 \leqq z) \Rightarrow z\cdot x \leqq z\cdot y]$. Ferner nennen wir einen linear- geordneten Ring *archimedisch*, falls seine additive Gruppe archimedisch geordnet ist.

<u>LEMMA 3:</u> *Sei* $\langle A,+,\cdot,\leqq\rangle$ *ein linear-geordneter Ring derart, daß* $\langle A,\leqq\rangle$ *Souslins Bedingung* (S II) *erfüllt. Dann ist* $\langle A,+,\cdot\rangle$ *entweder ein Zero-Ring und* $\langle A,+,\leqq\rangle$ *und* $\langle \mathbb{R},+,\leqq\rangle$ *sind isomorph als geordnete Gruppen, oder* $\langle A,+,\cdot,\leqq\rangle$ *und* $\langle \mathbb{R},+,\cdot,\leqq\rangle$ *sind isomorph als geordnete Körper.*

Beweis. Zunächst folgt aus Lemma 1, daß $\langle A,+,\leqq\rangle$ und $\langle \mathbb{R},+,\leqq\rangle$ isomorphe geordnete Gruppen sind. Wir nehmen an, daß $\langle A,+,\cdot\rangle$ kein Zero-Ring ist. Nach einem Satz von G.Pickert und A.B.Hion (siehe L.Fuchs [6] p. 185) existiert ein Unterring $\langle B,+,\cdot\rangle$ des Ringes $\langle \mathbb{R},+,\cdot\rangle$,so daß $\langle A,+,\cdot,\leqq\rangle$ und $\langle B,+,\cdot,\leqq\rangle$ isomorph als geordnete Ringe sind. Wir behaupten, daß notwendig $B = \mathbb{R}$ gelten muß. Sei $a \in B$ mit $a > 0$ beliebig gewählt. Multiplikation mit a^{-1} ist ein Ordnungs-Automorphismus der additiven Gruppe $\langle \mathbb{R},+,\leqq\rangle$. Daher sind $\langle B,+,\leqq\rangle$ und $\langle C,+,\leqq\rangle$ isomorph als geordnete Gruppen, wobei $C = \{ a^{-1}\cdot b \ ; \ b \in B \}$. Offenbar ist C nicht abzählbar. Also existiert eine irrationale Zahl t mit $t \in C$. Nach einem Satz von Kronecker (siehe Hardy-Wright [8] Satz 439, p.423) ist die von 1 und t erzeugte Gruppe eine dichte Teilmenge von $\mathbb{R}$. Aus $1 \in C$ und $t \in C$ folgt: C liegt in $\mathbb{R}$ dicht. Also liegt auch B in $\mathbb{R}$ dicht. Aber $\langle B,\leqq\rangle$ erfüllt Souslins Bedingung (S II). Daher gilt $B = \mathbb{R}$. Daher sind also $\langle A,+,\cdot,\leqq\rangle$ und $\langle \mathbb{R},+,\cdot,\leqq\rangle$ isomorph als geordnete Ringe, Q.E.D.

§ 2. Souslins Problem für Rechts-geordnete Fastkörper

Das Studium von Gruppen $\langle G,+\rangle$, die eine Gruppe A von Automorphismen besitzen derart, daß die identische Abbildung das einzige Element von A ist, welche ein von Null verschiedenes Element von G festläßt, führte H.Zassenhaus im Jahre 1935 zur Definition des Fastkörper-Begriffes. Fastkörper spielen eine wichtige Rolle in der Theorie der Permutations-Gruppen und in der Geometrie. Der Begriff des Fastkörpers wird heute etwas allgemeiner definiert, als ihn Zassenhaus ursprünglich einführte.

DEFINITION. Eine Struktur $\langle F,+,\cdot\rangle$ wird *Fastkörper* (*nearfield*) genannt, falls sie die folgenden sechs Eigenschaften hat:

(*i*) Die Grundmenge F besitzt wenigstens zwei verschiedene Elemente,

(*ii*) "+" und "·" sind total definierte binäre Operationen,

(*iii*) $\langle F,+\rangle$ ist eine Gruppe mit dem neutralen Element 0,

(*iv*) $\langle F - \{0\},\cdot\rangle$ ist eine Gruppe mit dem neutralen Element 1,

(*v*) $\forall x \in F \ (0\cdot x = 0)$,

(*vi*) $\forall a,b,c \in F \ (a\cdot(b+c) = a\cdot b + a\cdot c)$.

Fastkörper erfüllen definitionsgemäß das Rechts-Distributivgesetz (*vi*). Ein Fastkörper, der auch das Links-Distributiv-Gesetz $(a+b)\cdot c =$

$= a\cdot c + b\cdot c$ erfüllt, wird Schiefkörper genannt. Die additive Gruppe $\langle F,+\rangle$ eines jeden Fastkörpers ist kommutativ (B.H. Neumann [17], *siehe auch* Karzel [14] und Zemmer [23]).

In einem Fastkörper $\langle F,+,\cdot\rangle$ ist definitionsgemäß für alle $x \in F$ und alle $y \in F$ das Produkt $x\cdot y$ erklärt. Axiomatisch wurde nur $0\cdot x = 0$ gefordert. Man folgert leicht, daß auch $y\cdot 0 = 0$ für alle $y \in F$ gelten muß, denn $y\cdot 0 = y\cdot(0+0) = y\cdot 0 + y\cdot 0$, also $0 = y\cdot 0$.

Es gilt: $x = x\cdot 1$ für alle $x \in F$, denn weil $\langle F-\{0\},\cdot\rangle$ eine Gruppe ist, ist $x = x\cdot 1$ für alle $x \neq 0$ richtig. Nach Axiom (v) gilt aber auch $0 = 0\cdot 1$. Es gilt ferner $x\cdot(-1) = -x$, denn $x - x = 0$ und:

$$x + x\cdot(-1) = x\cdot 1 + x\cdot(-1) = x\cdot(1-1) = x\cdot 0 = 0 .$$

Es gilt auch $(-1)\cdot x = -x$ (für einen Beweis siehe H. Karzel [14]). Es folgt schließlich: $(-1)^2 = 1$.

<u>DEFINITION:</u> $\langle F,+,\cdot,\leqq\rangle$ ist ein *rechts-geordneter* Fastkörper, falls $\leqq$ eine lineare Ordnung auf F und $\langle F,+,\cdot\rangle$ ein Fastkörper ist und die folgenden beiden Eigenschaften erfüllt sind:

(†) $\forall x,y,z \in F\,(x<y \Rightarrow x+z < y+z)$,

(††) $\forall x,y,z \in F\,(x<y \wedge 0<z \Rightarrow x\cdot z < y\cdot z)$.

Den Begriff des *links-geordneten* Fastkörpers definiert man analog: statt der Regel (††) fordert man dann jedoch die Gültigkeit der folgenden Regel:

(†††) $\forall x,y,z \in F\,(x<y \wedge 0<z \Rightarrow z\cdot x < z\cdot y)$.

Falls $\langle F,+,\cdot,\leqq\rangle$ ein rechts-geordneter Fastkörper ist, dann gilt $0<x \Leftrightarrow (-x)<0$. Es gilt ferner $-1<0<1$; denn wäre etwa $1<0$, dann wäre $0<-1$, also $0 = 0\cdot(-1) < (-1)\cdot(-1) = (-1)^2 = 1$ (nach (††)), ein Widerspruch zur Annahme von $1<0$.

Daraus folgt: $0<x \Leftrightarrow 0<x^{-1}$; denn wäre etwa $0<x$ und $x^{-1}<0$, dann wäre $0<-x^{-1}$ und nach (††) erhielten wir dann den folgenden Widerspruch:

$$0 = 0\cdot(-x^{-1}) < x\cdot(-x^{-1}) = x\cdot(-1)\cdot(x^{-1}) = -(x\cdot x^{-1}) = -1 .$$

Als Konsequenz aus diesen Rechengesetzen erhalten wir: $F^{+} = \{x \in F ; 0 < x\}$ ist eine Untergruppe von $\langle F - \{0\}, \cdot \rangle$. Offenbar ist F^{+} ein Normalteiler von $\langle F - \{0\}, \cdot \rangle$ vom Index 2 .

SATZ 1 : *Sei $\langle F, +, \cdot, \leqq \rangle$ ein rechts-geordneter Fastkörper derart, daß $\langle F, \leqq \rangle$ die Souslin'sche Bedingung* (S II) *erfüllt. Dann ist $\langle F, +, \cdot \rangle$ ein kommutativer Körper und $\langle F, +, \cdot, \leqq \rangle$ und $\langle \mathbb{R}, +, \cdot, \leqq \rangle$ sind isomorph als geordnete Körper.*

Beweis. Die Menge $F^{+} = \{ x \in F ; x < 0 \}$ ist eine rechts-geordnete Untergruppe und offenbar erfüllt $\langle F^{+}, \cdot, \leqq \rangle$ auch die Souslin'sche Bedingung (S II) (Man bemerke, daß $\langle F - \{0\}, \cdot, \leqq \rangle$ die Bedingung (S II) keineswegs erfüllt) . Nach Lemma 1 ist $\langle F^{+}, \cdot \rangle$ kommutativ. Aus $-x = x\cdot(-1)$ und $(-1)^2 = 1$ folgt, daß dann auch $\langle F - \{0\}, \cdot \rangle$ eine abelsche Gruppe ist. Dann erfüllt aber $\langle F, +, \cdot \rangle$ auch das Links-Distributiv-Gesetz und folglich ist $\langle F, +, \cdot, \leqq \rangle$ ein geordneter Körper. Die Behauptung des Satzes folgt jetzt sofort aus Lemma 2, Q.E.D.

Das folgende Korollar folgt zwar nicht durch ein Symmetrie-Argument aus Satz 1, denn $\langle F, +, \cdot, \leqq \rangle$ erfüllt nur das Rechts-Distributivgesetz und ist in Satz 1 rechts-geordnet. Aber auch für links-geordnete Fastkörper $\langle F, +, \cdot, \leqq \rangle$ ist $F^{+} = \{ x \in F ; x < 0 \}$ eine Untergruppe von $\langle F - \{0\}, \cdot \rangle$ vom Index 2, sodaß sich aus Lemma 1 und Lemma 2 das folgende Korollar auf analoge Weise ergibt:

KOROLLAR : *Sei $\langle F, +, \cdot, \leqq \rangle$ ein links-geordneter Fastkörper, so daß $\langle F, \leqq \rangle$ die Bedingung* (S II) *von Souslin erfüllt. Dann sind $\langle F, +, \cdot, \leqq \rangle$ und $\langle \mathbb{R}, +, \cdot, \leqq \rangle$ als geordnete Körper isomorph.*

§ 3. Souslins Problem für Halbgruppen

In den vorangegangenen Abschnitten haben wir gesehen,daß das Problem von Souslin für Gruppen, Körper und Fastkörper eine positive Lösung hat. Ausschlaggebend dafür war nicht so sehr, daß die algebraischen Operationen eine gewisse Homogenität der geordneten Menge $\langle G,\leqq\rangle$, bzw. $\langle F,\leqq\rangle$, bewirken, denn, wie bereits zu Anfang erwähnt, gibt es mit $\langle\mathbb{R},\leqq\rangle$ nicht isomorphe Souslin-Kontinua $\langle S,\leqq\rangle$ dergestalt, daß je zwei abgeschlossene Intervalle von $\langle S,\leqq\rangle$ ordnungs-isomorph sind.Ausschlaggebend waren vielmehr die Existenz von Inversen ($-x$, bzw. x^{-1}) und die Kürzungsregeln. Davon wurde im Beweis der Archimedizität wesentlicher Gebrauch gemacht. Da in Halbgruppen weder die Kürzungsregeln zu gelten brauchen, noch Inverse zu existieren brauchen, ist zu erwarten, daß das Souslinsche Problem für Halbgruppen eine negative Lösung haben kann. Dies ist in der Tat der Fall. Es interessiert uns hier aber, welche zusätzlichen interessanten Eigenschaften eine geordnete Halbgruppe $\langle S,\cdot,\leqq\rangle$ noch haben darf, so daß $\langle S,\leqq\rangle$ zwar (S I),(S II) und (S III) erfüllt, aber nicht mit $\langle\mathbb{R},\leqq\rangle$ ordnungs-isomorph ist.Um die Aufklärung solcher Sachverhalte geht es uns in den nun folgenden Abschnitten.

Eine Struktur $\langle H,\cdot\rangle$ wird *Halbgruppe* genannt, falls "$\cdot$" eine total definierte assoziative binäre Operation auf der nicht-leeren Menge H ist. Eine Halbgruppe $\langle H,\cdot\rangle$ heißt *regulär*, falls in ihr die beiden folgenden Kürzungs-Regeln gelten:

$$(*)\quad \forall a\in H\,\forall x\in H\,\forall y\in H\,(\,x\cdot a=y\cdot a\Longrightarrow x=y\,)\;,$$

$$(**)\quad \forall a\in H\,\forall x\in H\,\forall y\in H\,(\,a\cdot x=a\cdot y\Longrightarrow x=y\,)\;.$$

Eine Struktur $\langle H,\cdot,\leqq\rangle$ wird *geordnete Halbgruppe* genannt, falls $\langle H,\cdot\rangle$ eine Halbgruppe ist, $\leqq$ eine lineare Ordnung auf H, und aus $x\leqq y$ sowohl $x\cdot z\leqq y\cdot z$ als auch $z\cdot x\leqq z\cdot y$ folgt $(x,y,z\in H)$.

BEISPIEL : Es sei $\langle S, \leqq \rangle$ eine beliebige linear geordnete Menge. Wir können auf S eine Multiplikation $x \cdot y$ (für $x \in S$, $y \in S$) wie folgt einführen: $x \cdot y = \mathrm{Max}\{x,y\}$, wenn $\mathrm{Max}\{x,y\}$ das größte der beiden Elemente x und y ist. Dann ist offenbar $\langle S, \cdot, \leqq \rangle$ eine geordnete Halbgruppe. Diese ist nicht regulär, wenn S mehr als ein Element besitzt.

Falls $\mathfrak{M}$ ein Modell der Mengenlehre ist und $\langle S, \leqq \rangle$ in $\mathfrak{M}$ ein Souslin-Kontinuum ist, dann kann auf $\langle S, \leqq \rangle$ wie im Beispiel oben angegeben eine Multiplikation "$\cdot$" so definiert werden, daß $\langle S, \cdot, \leqq \rangle$ eine geordnete Halbgruppe ist, welche (S I) , (S II) und (S III) erfüllt.Alle in der Einleitung geschilderten mengentheoretischen Resultate können daher auch für souslinsch-geordnete Halbgruppen ausgesprochen werden.Das Problem von Souslin für Halbgruppen kann für uns also erst dann interessant werden, wenn wir von der geordneten Halbgruppe noch weitere Eigenschaften fordern.

DEFINITION (O.Hölder [10]). Eine geordnete Halbgruppe $\langle H, \cdot, \leqq \rangle$ ist *positiv geordnet*, falls für alle $x, y \in H$ gilt: $x \leqq x \cdot y$ und $y \leqq x \cdot y$. Die geordnete Halbgruppe $\langle H, \cdot, \leqq \rangle$ ist *natürlich geordnet*, falls sie positiv geordnet ist und das folgende Axiom erfüllt:

$$\forall x \in H \, \forall y \in H \left[x < y \Longrightarrow \exists a \in H \, \exists b \in H \left(x \cdot a = b \cdot x = y \right) \right] .$$

DEFINITION. Eine geordnete Halbgruppe $\langle H, \cdot, \leqq \rangle$ nennen wir *archimedisch* geordnet, falls sie positiv geordnet ist und die folgende Bedingung erfüllt: $\forall a \in H \left[C_a \neq \emptyset \Longrightarrow \forall x \in H \left(a \cdot x = x \cdot a = x \right) \right]$, wobei $C_a = \{ y \in H ; \forall n \in \mathbb{N} : a^n < y \}$.

SATZ 2: *In* ZF + V = L *gilt: Es gibt eine positiv-geordnete kommutative archimedische Halbgruppe* $\langle H, \cdot, \leqq \rangle$ *derart, daß* $\langle H, \leqq \rangle$ *nicht mit* $\langle \mathbb{R}, \leqq \rangle$ *ordnungs-isomorph ist, aber trotzdem die Bedingungen* (S I) , (S II) *und* (S III) *von Souslin erfüllt.*

Beweis. Jensen bewies, daß aus $V = L$ die Existenz eines normalen Souslin-Baumes $\langle T, \leqq \rangle$ folgt, der die Eigenschaft hat, daß jeder Punkt $x \in T$ genau $\aleph_0$ unmittelbare Nachfolger hat (siehe Devlin-Johnsbråten [4],p. 12-18). Für $x \in T$ sei $\hat{x} = \{ y \in T \; ; \; y < x \}$ und $ht(x)$ der Ordnungstyp von $\hat{x}$. Als Zweig in $\langle T, \leqq \rangle$ bezeichnen wir jede bezüglich $\leqq$ linear-geordnete Teilmenge b von T, welche die Eigenschaft hat, daß aus $y < x$ und $x \in b$ stets $y \in b$ folgt. Wenn b ein Zweig in $\langle T, \leqq \rangle$ ist, ξ der Ordnungstyp von b und $\alpha < \xi$, dann sei $b(\alpha)$ dasjenige Element $y \in b$ mit $ht(y) = \alpha$. Wenn $x \in T$, dann ist $\hat{x}$ ein Zweig in T und daher ist $\hat{x}(\alpha)$ dasjenige Element $y \in T$ welches die Eigenschaften $y \leq x$ und $ht(y) = \alpha$ hat, falls $\alpha \leq ht(x)$ (dabei ist $\hat{x}(\alpha)$ für $\alpha > ht(x)$ undefiniert).

Für $x \in T$ sei $Succ(x) = \{ y \in T \; ; \; x < y \wedge ht(y) = ht(x) + 1 \}$. Sei $\mathbb{Q}$ die Menge aller rationalen Zahlen und $\mathbb{Q}^+ = \{ x \in \mathbb{Q} ; \; 0 < x \}$. Da für jedes $x \in T$ die Menge $Succ(x)$ abzählbar unendlich ist, gibt es für jedes $x \in T$ eine ein-eindeutige Abbildung f_x von $\mathbb{Q}^+$ auf $Succ(x)$. Auf $Succ(x)$ können wir daher eine lineare Ordnung $\preccurlyeq$ vom Ordnungstyp η_0 definieren, indem wir für $y_1, y_2 \in Succ(x)$ setzen:

$$y_1 \prec y_2 \Longleftrightarrow f_x^{-1}(y_1) < f_x^{-1}(y_2) .$$

Sei B die Menge aller maximalen Zweige von $\langle T, \leqq \rangle$. Für $b_1 \in B$ und $b_2 \in B$, mit $b_1 \neq b_2$ definiere: $b_1 \lhd b_2 \Leftrightarrow b_1(\alpha) \prec b_2(\alpha)$, wo α der Ordnungstyp von $b_1 \cap b_2$ ist. Man bemerke dabei, daß stets $b_1 \cap b_2 \neq \emptyset$ gilt und daß der Ordnungstyp α von $b_1 \cap b_2$ nie eine Limeszahl ist. Dies folgt sofort aus der Normalität von $\langle T, \leqq \rangle$ (siehe [4] p.12, Eigenschaften (ii) und (v)). B ist in Bezug auf $\trianglelefteq$ eine dicht-geordnete Menge.

Sei $\mathbf{On}$ die Klasse aller Ordinalzahlen und sei

$$\Sigma = \{ \alpha \in \mathbf{On} \; ; \; \alpha < \omega_1 \wedge \exists \xi \in \mathbf{On} : \alpha = \xi + 1 \} .$$

Es sei $\lambda \neq 0$ eine abzählbare Limeszahl und ψ eine auf $\lambda \cap \Sigma$ definierte Funktion mit Werten in $\mathbb{Q}^+$. Dann bezeichnen wir mit $\tilde{\psi}$ die periodische Fortsetzung von ψ, die auf ganz Σ definiert ist. Das heißt, für $\delta \in \Sigma$ ist $\tilde{\psi}(\delta) = \psi(\alpha)$ wenn $\delta = \lambda \cdot \zeta + \alpha$ mit $\zeta \in \mathbf{On}$ und $\alpha < \lambda$ ($\tilde{\psi}$ ist eindeutig definiert; vergleiche dazu W.Sierpinski [19] p.298, Theorem 2).

Falls $\lambda \neq 0$ eine abzählbare Limeszahl ist und ψ eine auf $\lambda \cap \Sigma$ definierte Funktion mit Werten in $\mathbb{Q}^+$, dann sei ψ^* diejenige Funktion, die auf ganz Σ definiert ist, und für die gilt: $\psi^*(\delta) = \psi(\delta)$ falls $\delta \in \lambda \cap \Sigma$, und $\psi^*(\delta) = 0$ falls $\delta \in \Sigma$, $\lambda < \delta$.

Es sei jetzt $b \in B$ und $\lambda = \lambda_b$ der Ordnungstyp von b. Dann ist λ_b eine Limeszahl. Sei ψ_b diejenige auf $\lambda_b \cap \Sigma$ definierte Funktion mit Werten in $\mathbb{Q}^+$ für die gilt: wenn $\delta \in \lambda_b \cap \Sigma$, dann $\psi_b(\delta) = f_z^{-1}(b(\delta))$ wo $z = b(\gamma)$ für $\delta = \gamma + 1$.

Auf der linear geordneten Menge $\langle B, \trianglelefteq \rangle$ können wir jetzt eine Multiplikation "$\bullet$" wie folgt einführen: es seien $b \in B$ und $c \in B$, λ_b der Ordnungstyp von b, λ_c der Ordnungstyp von c, und $\psi_b \# \psi_c$ die auf $\Sigma \cap Max\{\lambda_b, \lambda_c\}$ wie folgt definierte Funktion:

$$(\psi_b \# \psi_c)(\delta) = \psi_b^*(\delta) + \psi_c^*(\delta) \qquad \textit{für} \quad \delta \in \Sigma \cap Max\{\lambda_b, \lambda_c\},$$

dann sei $b \bullet c$ dasjenige Element $d \in B$ für das $\psi_d \subseteq \widetilde{\psi_b \# \psi_c}$ gilt. Klar ist, daß höchstens ein solches d existieren kann ($\psi_d \subseteq \psi_b \# \psi_c$ besagt, daß ψ_d und $\psi_b \# \psi_c$ auf $\Sigma \cap \lambda_d$ übereinstimmen, wenn λ_d der Ordnungstyp von d ist). Es existiert aber auch mindestens ein derartiges d in B, denn für jede Funktion $\Phi : \omega \rightarrow \mathbb{Q}^+$ existiert mindestens ein Zweig $a \in B$ derart, daß $\psi_a(n) = \Phi(n)$ für alle $n \in \omega$ gilt.

$\langle B, \bullet, \trianglelefteq \rangle$ ist eine positiv-geordnete archimedische Halbgruppe. In $\langle B, \bullet, \trianglelefteq \rangle$ gelten die Kürzungsregeln nicht, denn der Souslin-Baum $\langle T, \leqq \rangle$ enthält bereits $2^{\aleph_0}$ Zweige vom Ordnungstyp ω, während $\{x \in T;\ ht(x) = \omega\}$ abzählbar ist.

Die gesuchte Halbgruppe $\langle H, \odot, \lesssim \rangle$ erhalten wir, wenn wir den Dedekind'schen Abschluß von $\langle B, \bullet, \trianglelefteq \rangle$ bilden, $H = \{\langle X, Y \rangle ;\ \langle X, Y \rangle$ *ist Dedekind'scher Schnitt in* $\langle B, \trianglelefteq \rangle$ *und* Y *hat kein kleinstes Element*$\}$. Dann ist $\langle X_1, Y_1 \rangle \odot \langle X_2, Y_2 \rangle = \langle X_3, Y_3 \rangle$ genau dann wenn $X_3 = \{a \in B;$ *es gibt* $x_1 \in X_1$ *und* $x_2 \in X_2$ *mit* $a \trianglelefteq x_1 \bullet x_2\}$. Eine lineare Ordnung kann auf H wie folgt erklärt werden: $\langle X_1, Y_1 \rangle \lesssim \langle X_2, Y_2 \rangle \Longleftrightarrow X_1 \subseteq X_2$. Es ist unmittelbar klar, daß H in Bezug auf $\odot$ und $\preceq$ eine positiv- geordnete archimedische Halbgruppe ist. $\langle H, \preceq \rangle$ erfüllt die Bedingungen (S I),(S II) und (S III) und ist nicht in $\langle \mathbb{R}, \leqq \rangle$ einbettbar (vergl. [4] p.14),QE D.

§ 4 . Souslins Problem für reguläre Halbgruppen

LEMMA 4 : *Sei $\mathcal{H} = \langle H, \cdot, \leqq \rangle$ eine natürlich geordnete, kommutative reguläre Halbgruppe. Falls $\langle H, \leqq \rangle$ die Bedingungen* (S I) *und* (S II) *von Souslin erfüllt, dann sind $\langle H, \cdot, \leqq \rangle$ und $\langle \mathbb{R}^+, +, \leqq \rangle$ isomorph als geordnete Halbgruppen.*

Beweis. Zunächst betten wir die Halbgruppe $\mathcal{H}$ wie folgt in eine Gruppe $G = Q(H)$ ein. Für $a,b \in H$ sei $\frac{a}{b} = \{ \langle x,y \rangle \; ; \; x,y \in H \wedge a \cdot y = b \cdot x \}$. Dann setzen wir $Q(H) = \{ \frac{a}{b} \; ; \; a \in H \wedge b \in H \}$. Dann ist $Q(H)$ in Bezug auf die Multiplikation $\frac{a}{b} \cdot \frac{c}{d} = \frac{a \cdot c}{b \cdot d}$ eine Gruppe (die sogenannte Gruppe der Quotienten). Wenn wir noch $\frac{a}{b} < \frac{c}{d} \Leftrightarrow a \cdot d < b \cdot c$ setzen,dann ist $\langle Q(H), \cdot, \leqq \rangle$ eine linear-geordnete abelsche Gruppe.Wir können H als Unterhalbgruppe von $Q(H)$ auffassen, denn H kann durch die Zuordnung $\eta : x \mapsto \frac{x^2}{x}$ in die Gruppe $Q(H)$ eingebettet werden. Sei $e = \frac{x}{x}$ das neutrale Element der Multiplikation von $Q(H)$. Wir behaupten, daß $H = \{ g \in Q(H) \; ; \; e < g \}$ gilt.

Sei $a \in H$ beliebig gewählt. Da H positiv-geordnet ist, gilt $a^2 \leqq a^3$, und weil H regulär ist, gilt dann notwendig $a^2 < a^3$ (man beachte,daß H wegen (S I) kein Eins-Element besitzen kann!) und folglich $e = \frac{a}{a} < \frac{a^2}{a} = a$. Also gilt $\{ g \in Q(H) \; ; \; e < g \} \supseteq H$.

Sei jetzt $g \in Q(H)$ mit $e < g$ beliebig gewählt. Dann existieren Elemente $a,b \in H$ mit $e = \frac{a}{a} < \frac{b}{a} = g$. Das heißt $a^2 < a \cdot b$, also $a < b$ weil H regulär ist. Da H natürlich geordnet ist, gibt es ein Element $v \in H$ mit $a \cdot v = b$. Dann gilt aber $\frac{b}{a} = \frac{a \cdot v}{a} = v$, also $g = \frac{b}{a} \in H$. Also ist H die Menge aller positiven Elemente von $Q(H)$.

Daraus folgt jetzt, daß auch $Q(H)$ die Bedingungen (S I) und (S II) von Souslin erfüllt. Nach Lemma 1 gilt daher $\langle Q(H),\cdot,\leqq\rangle \cong \langle \mathbb{R}^{+},+,\leqq\rangle$ und daraus folgt sofort die Behauptung des Lemmas, Q.E.D.

KOROLLAR : *Sei* $\mathcal{H} = \langle H,\cdot,\leqq\rangle$ *eine natürlich geordnete archimedische Halbgruppe. Falls* $\langle H,\leqq\rangle$ *die Bedingungen* (S I) *und* (S II) *von Souslin erfüllt, dann sind* $\langle H,\cdot,\leqq\rangle$ *und* $\langle \mathbb{R}^{+},+,\leqq\rangle$ *als geordnete Halbgruppen isomorph.*

Beweis. Die Bedingung (S I) impliziert, daß H kein maximales Element hat. Weil $\langle H,\cdot,\leqq\rangle$ außerdem archimedisch und regulär ist, ist H nach nach einem Lemma von Clifford regulär (siehe L.Fuchs [6] p.226-227). Natürlich-geordnete archimedische reguläre Halbgruppen sind nach einem Satz von O.Hölder kommutativ (siehe L.Fuchs [6] p.227). Nach Lemma 4 folgt jetzt die Behauptung, Q.E.D.

BEMERKUNG 1: Man kann das oben stehende Korollar auch ohne einen Verweis auf Lemma 4 direkt beweisen. Nach dem Satz von Hölder ist $\langle H,\cdot,\leqq\rangle$ eine Unterhalbgruppe von $\langle \mathbb{R}^{+},+,\leqq\rangle$. Daß sogar notwendig $H = \mathbb{R}^{+}$ gilt, folgt dann aus einem Satz von L.Kronecker (siehe Hardy-Wright [8] p. 423, Satz 439).

BEMERKUNG 2: Weder in Lemma 4 noch in seinem Korollar kann man die Bedingung der natürlichen Ordnung durch die schwächere Bedingung der positiven Ordnung ersetzen. Das folgt sofort aus der Existenz des folgenden Gegenbeispiels (dieses Gegenbeispiel verdanke ich Herrn Dr.J.Schulte-Mönting): sei

$$H = \{\langle 1,x\rangle\,;\ x\in\mathbb{R} \wedge x>0\}\cup\{\langle n,x\rangle\,;\ 2\leqq n\in\omega \wedge x\in\mathbb{R} \wedge x\geqq 0\}.$$

Mit $\lessapprox$ bezeichnen wir die lexikographische Ordnung auf der Menge H , also: $\langle n,x\rangle \prec \langle m,y\rangle \Longleftrightarrow [n<m \vee (n=m \wedge x<y)]$. Offenbar sind $\langle H,\lessapprox\rangle$ und $\langle \mathbb{R},\leqq\rangle$ ordnungs-isomorph. $\langle H,\lessapprox\rangle$ erfüllt also (S I),(S II) und (S III). In Bezug auf die folgende "Addition": $\langle n,x\rangle + \langle m,y\rangle = \langle n+m\,,\,x+y\rangle$ ist

H eine kommutative reguläre,archimedische positiv-geordnete Halbgruppe. H ist nicht natürlich geordnet, denn zu $\langle 1,x\rangle$ und $\langle 1,y\rangle$ mit $x<y$ gibt es kein $\langle n,z\rangle \in H$ mit $\langle 1,x\rangle + \langle n,z\rangle = \langle 1,y\rangle$. Die Halbgruppe H besitzt anomale Paare (vergl.Fuchs [6] p.225) und ist daher keine Unter-Halbgruppe von $\langle \mathbb{R}, +, \leqq\rangle$.

Das soeben geschilderte Gegenbeispiel berührt jedoch nicht die folgende Frage: *Existiert unter der Voraussetzung von* V = L *ein Souslin-Kontinuum* $\langle M, \leqq\rangle$ *derart, daß auf* M *eine binäre assoziative Operation "•" so erklärt werden kann, daß* $\langle M, \bullet, \leqq\rangle$ *eine kommutative reguläre positiv-geordnete archimedische ist?* Ich vermute eine positive Antwort auf diese Frage. Nach einem Satz von Hion (vergl.L.Fuchs [6]p.230-233) müßte $\langle M, \cdot, \leqq\rangle$ notwendig anomale Paare besitzen.

Die Frage kann auch wie folgt gestellt werden: *Besitzt eine reguläre kommutative, positiv-geordnete archimedische Halbgruppe eine dichte abzählbare Teilmenge, vorausgesetzt sie erfüllt die Bedingungen* (S I), (S II) *und* (S III) *von Souslin?*

Ein topologischer Raum heißt nach Frechet *separabel*, falls er eine abzählbare dichte Teilmenge enthält(vergl. C.Kuratowski, *Topologie I*, p.88). Das alte Souslin'sche Problem ist also äquivalent mit der Frage ob jede stetig-geordnete linear-geordnete Menge $\langle M, \leqq\rangle$,welche weder ein erstes noch ein letztes Element besitzt und die Bedingung (S III) erfüllt, in Bezug auf die Intervall-Topologie einen separablen Raum bildet. Wie wir heute wissen, geben die mengentheoretischen Axiome ZF+AC keine Antwort auf Souslin's Frage.

Die Souslin'sche Frage, ob für stetig-geordnete Mengen $\langle M, \leqq\rangle$ die Bedingung (S III) die Separabilität nach sich zieht,wollen wir jetzt wie folgt modifizieren: *Ist jede dicht-geordnete Gruppe* $\langle G, \cdot, \leqq\rangle$, *welche die Eigenschaft* (S III) *erfüllt, notwendig ein separabler Raum?*

Diese Frage werden wir im folgenden § 5 für abelsche Gruppen positiv beantworten. Bisher hatten wir nur die Stetigkeits-Bedingung (S II) untersucht. Jetzt diskutieren wir den algebraischen Gehalt von (S III) .

§ 5. Separabilität und Souslins Bedingung (S III)

Eine (multiplikativ geschriebene) Gruppe $\langle G,\cdot\rangle$ heißt *teilbar* (oder *dividierbar*), falls für jede positive ganze Zahl $n \geqq 2$ und jedes $g \in G$ ein $h \in G$ mit $g = h^n$ existiert. B.H.Neumann hat gezeigt, daß jede Gruppe in eine teilbare Gruppe einbettbar ist. Die Existenz einer teilbaren Hülle ist bisher allerdings nur für lokal-nilpotente Gruppen G bewiesen worden (A.I.Malzew) - siehe A.G.Kurosch [15] p.84 - 86)

LEMMA 5 : *Sei* $\langle G,+,\leqq\rangle$ *eine dicht-geordnete abelsche Gruppe und sei* D *die teilbare Hülle von* G. *Dann kann die lineare Ordnung* $\leqq$ *von* G *zu einer dichten linearen Ordnung* $\dot{\leqq}$ *auf* D *erweitert werden. Falls* $\langle G,\leqq\rangle$ *die Bedingung* (S III) *von Souslin erfüllt, dann erfüllt auch* $\langle D,\dot{\leqq}\rangle$ *die Bedingung* (S III).

Beweis. Sei $\mathbb{N}$ die Menge der von Null verschiedenen positiven ganzen Zahlen und $\mathbb{N}\times G = \{\langle n,g\rangle ;\ n\in\mathbb{N} \wedge g\in G\}$. Für $n\in\mathbb{N}$ und $g\in G$ setzen wir: $\frac{g}{n} = \{\langle m,h\rangle\in \mathbb{N}\times G ;\ mg = nh\}$, und damit $D = \{\frac{g}{n} ;\ g\in G \wedge n\in\mathbb{N}\}$.
Es sei : $\frac{g}{n} + \frac{h}{m} = \frac{mg+nh}{m\cdot n}$, dann ist $\langle D,+\rangle$ die teilbare Hülle von G, und $\langle D,+\rangle$ ist eine abelsche Gruppe. Eine lineare Ordnung $\dot{\leqq}$ können wir auf D wie folgt erklären: $\frac{g}{n} \dot{<} \frac{h}{m} \Longleftrightarrow mg < nh$. Wenn wir g mit $\frac{g}{1}$ identifizieren (für $g\in G$), dann ist G eine Untergruppe von D und $\dot{\leqq}$ ist eine Fortsetzung von $\leqq$.

Falls $0 \dot{\leqq} \frac{g}{n} \dot{<} \frac{h}{m}$, dann existiert in G ein Element x derart, daß $mg < x < nh$ gilt. Dann folgt $\frac{g}{n} \dot{<} \frac{x}{n\cdot m} \dot{<} \frac{h}{m}$. Also ist $\dot{\leqq}$ eine dichte Ordnung auf D.

Angenommen, die Bedingung (S III) gilt nicht für $\langle D,\dot{\leqq}\rangle$. Dann gibt es eine geordnete Menge $\langle I,\preceq\rangle$ und für jedes $i\in I$ ein offenes Intervall $A_i \subseteq D$, sodaß $\forall i,j\in I\ \forall a\in A_i\ \forall b\in A_j\ [\,i \prec j \Longrightarrow a \dot{<} b\,]$ und auch (wobei I überabzählbar ist!)

$\forall i,j \in I\,[i \neq j \Longrightarrow A_i \cap A_j = \emptyset]$ gilt. Da $\langle D, \dot{\leqq} \rangle$ dicht-geordnet ist, enthält jedes Intervall A_i mindestens zwei verschiedene Elemente. Wenn $\frac{a}{n}, \frac{b}{m} \in A_i$, dann ist offenbar $\frac{a}{n} = \frac{ma}{m \cdot n} \in A_i$, $\frac{b}{m} = \frac{nb}{m \cdot n} \in A_i$. In jedem Intervall A_i können wir daher zwei verschiedene Elemente mit demselben Nenner wählen. Wir können also für jedes $i \in I$ zwei Elemente $a_i \in G$, $b_i \in G$ und eine Zahl $z_i \in \mathbb{N}$ mit

$$\frac{a_i}{z_i} \in A_i \quad , \quad \frac{b_i}{z_i} \in A_i \quad \text{und} \quad \frac{a_i}{z_i} \dot{<} \frac{b_i}{z_i}$$

wählen. Weil $\mathbb{N}$ abzählbar ist, I jedoch überabzählbar, muß eine überabzählbare Teilmenge J von I existieren, derart daß $z_\xi = z_\nu$ für alle $\xi, \nu \in J$ gilt. Es folgt: $\forall \xi, \nu \in J\,[\nu \dot{<} \xi \Longrightarrow a_\nu < b_\nu < a_\xi < b_\xi]$. Für $\xi \in J$ definieren wir: $B_\xi = \{ g \in G \,;\, a_\xi < g < b_\xi \}$. Also ist $\{ B_\xi \,;\, \xi \in J \}$ eine überabzählbare Menge von paarweise disjunkten offenen Intervallen von $\langle G, \leqq \rangle$. Die Bedingung (S III) gilt also auch nicht für $\langle G, \leqq \rangle$, ein Widerspruch zur Voraussetzung! Q.E.D.

<u>LEMMA 6 :</u> *Sei $\langle G, +, \leqq \rangle$ eine dicht-geordnete teilbare abelsche Gruppe. Falls $\langle G, \leqq \rangle$ die Bedingung* (S III) *von Souslin erfüllt, dann besitzt $\langle G, \leqq \rangle$ eine abzählbare dichte Teilmenge.*

Beweis. Für $g \in G$ sei $|g| = Max\{ g, -g \}$. Zwei Elemente $g, h \in G$ heißen *archimedisch äquivalent*, falls entweder $|g| \leqq |h| \wedge \exists n \in \mathbb{N}\,(|h| \leqq |ng|)$ oder $|h| \leqq |g| \wedge \exists m \in \mathbb{N}\,(|g| \leqq |mh|)$ gilt. Mit $A(g)$ bezeichnen wir die Archimedische Klasse von g, also:

$$A(g) = \{ h \in G \,;\, g \text{ und } h \text{ sind archimedisch äquivalent} \}.$$

Sei $K = \{ A(g) \,;\, g \in G \}$. Für $\kappa, \lambda \in K$ mit $\kappa \neq \lambda$ sei $\kappa < \lambda$ genau dann wenn es $g \in \kappa$, $h \in \lambda$ mit $|g| < |h|$ gibt. Dann ist $\langle K, \leqq \rangle$ eine linear-geordnete Menge. Für $\kappa \in K$ setzen wir:

$$H_\kappa = \{ g \in G \,;\, A(g) \leqq \kappa \} \qquad \text{und} \qquad H^*_\kappa = \{ g \in G \,;\, A(g) < \kappa \}.$$

Aus $g, h \in H^*_\kappa$ und $g < x < h$ folgt $x \in H^*_\kappa$. Also ist H^*_κ eine konvexe Teilmenge von G. Analog gilt: H_κ ist eine konvexe Teilmenge von G. Ferner sind H^*_κ und H_κ Untergruppen von G.

Für jedes $g \neq 0$ ist $\{h \in G;\ h \in A(g) \wedge 0 < h\}$ ein offenes Intervall in $\langle G, \leqq \rangle$. Aus (S III) folgt daher: K ist abzählbar.

H_κ ist eine teilbare Untergruppe von G (aus $ny = x \in H_\kappa$ folgt nämlich $y \in A(x) \subseteq H_\kappa$). Analog folgt: H^*_κ ist eine teilbare Untergruppe von H_κ.

Interludium: Da H^*_κ teilbare Untergruppe von H_κ ist, ist H^*_κ ein direkter Summand von H_κ, also $H_\kappa = H^*_\kappa \oplus B_\kappa$. Es gilt $B_\kappa \subseteq \kappa$ und B_κ ist archimedisch geordnet, also nach O.Hölder eine Untergruppe der reellen Zahlen. Nach dem Einbettungs-Satz von Hahn ist G mit einer Untergruppe des anti-lexikographisch geordneten vollen Cartesischen Produktes $\prod B_\kappa$ ($\kappa \in K$) isomorph(vergl. A.H.Clifford, Proc.A.M.S.5(1954)pp.860-863). K ist abzählbar und aus $H^* \neq \{0\}$ folgt unter Verwendung von (S III), daß auch B_κ abzählbar ist, denn $H^*_\kappa \oplus B_\kappa$ ist anti-lexikographisch geordnet. Es ist jedoch schwer, daraus die Separabilität von $\langle G, \leqq \rangle$ zu folgern. Wir verfahren daher wie folgt:

Zunächst bemerken wir, daß $\langle K, \leqq \rangle$ ein kleinstes Element besitzt,und dieses ist $A(0) = \{0\}$. Sei daher $K^0 = \{A(g);\ 0 \neq g \in G\} = K - \{\{0\}\}$. Weil K^0 abzählbar ist, können wir für jedes $n \in \omega$ ein Element $\delta_n \in K^0$ so wählen, daß $\delta_0 \geqq \delta_1 \geqq \ldots \geqq \delta_n \geqq \delta_{n+1} \geqq \ldots$ für alle $n \in \omega$ gilt und außerdem diese Folge $\langle \delta_n \mid n \in \omega \rangle$ koninitial in K^0 liegt (das heißt: $\forall \kappa \in K^0\ \exists n \in \omega : \kappa \geqq \delta_n$).

Es sei $\delta_{n+1} < \delta_n$. Dann ist $H_{\delta_{n+1}}$ eine echte teilbare Unter-gruppe von H_{δ_n} und daher ein direkter Summand. Also $H_{\delta_n} = H_{\delta_{n+1}} \oplus D_n$ für eine gewisse Untergruppe D_n von H_{δ_n}. Um die Notation etwas zu vereinfachen, setzen wir: $H_n = H_{\delta_n}$. Aus $0 \neq d \in D_n$ folgt $d \notin H_{n+1}$ (da die Summe $H_{n+1} \oplus D_n$ direkt ist) und somit $\delta_{n+1} < A(d) \leqq \delta_n$. Aus $h \in H_{n+1}$ folgt andererseits sofort: $A(h) \leqq \delta_{n+1}$. Deshalb gilt also:

(+) $\forall h \in H_{n+1} \forall d \in D_n \left[d \neq 0 \Longrightarrow h < |d| \right]$.

Für $d_1, d_2 \in D_n$ mit $d_1 < d_2$ und für beliebige Elemente $h_1, h_2 \in H_{n+1}$ gilt $0 < d_2 - d_1 \in D_n$ und daher nach (+): $(h_2 + d_2) - (h_1 + d_1) = (h_2 - h_1) + (d_2 - d_1) > 0$. Deshalb gilt also auch:

(++) $\forall h_1, h_2 \in H_{n+1} \forall d_1, d_2 \in D_n \left[d_1 < d_2 \Longrightarrow h_1 + d_1 < h_2 + d_2 \right]$.

Als Konsequenz ergibt sich daraus: $H_n = H_{n+1} \oplus D_n$ ist anti-lexikographisch geordnet: $h_1 + d_1 < h_2 + d_2 \Longleftrightarrow \left[d_1 < d_2 \vee \left(d_1 = d_2 \wedge h_1 < h_2 \right) \right]$ für alle $h_1, h_2 \in H_{n+1}$, $d_1, d_2 \in D_n$.

Für $d \in D_n$ setzen wir $I_d = \{ h+d \,;\, h \in H_{n+1} \}$. Dann folgt aus (+ +): wenn $d_1, d_2 \in D_n$, $d_1 < d_2$, $x \in I_{d_1}$ und $y \in I_{d_2}$, dann $x < y$. Daher gilt $I_{d_1} \cap I_{d_2} = \emptyset$ für $d_1 \neq d_2$. Wenn $H_{n+1} \neq \{0\}$, dann ist H_{n+1} eine konvexe Untergruppe der dicht-geordneten Gruppe $\langle G, +, \leqq \rangle$. Weil H_{n+1} und I_d (für $d \in D_n$) ordnungs-isomorph sind, ist dann H_n die Union der paarweise disjunkten offenen Intervalle I_d (für $d \in D_n$). (S III) impliziert daher: $\forall n \in \omega \left[H_{n+1} \neq \{0\} \Longrightarrow D_n \textit{ ist abzählbar} \right]$.

Wir treffen jetzt die folgende Fallunterscheidung:

<u>1.Fall:</u> K^0 besitzt ein kleinstes Element. Sei μ das kleinste Element von K^0. Wir wählen die Folge der Archimedischen Klassen $\delta_n \in K^0$ so, daß $\delta_n = \mu$ für alle $n \in \omega$ gilt. Dann ist $H_0 = H_1 = \ldots = H_n = \ldots\ldots$ und $G = H_0 \oplus D$ für eine gewisse Untergruppe D von G und wie oben folgt, daß D abzählbar ist. Je zwei von 0 verschiedene Elemente von $H_0 = H_\mu$ sind archimedisch äquivalent, da μ minimal in K^0 ist. Daher ist H_0 archimedisch und also einer Untergruppe der reellen Zahlen isomorph. Insbesondere besitzt H_0 daher eine abzählbare dichte Teilmenge M. Da $G = H_0 \oplus D$ anti-lexikographisch geordnet ist,ist dann offenbar $\{x+d \,;\, x \in M \wedge d \in D\}$ eine abzählbare dichte Teilmenge von G.

<u>2.Fall:</u> K^0 besitzt kein kleinstes Element.

In diesem Falle können wir die Klassen $\delta_n \in K^0$ so wählen ,daß für alle $n \in \omega$ gilt: $\delta_{n+1} < \delta_n$, und $\langle \delta_n \mid n \in \omega \rangle$ koninitial in K^0 liegt.

Sei $H_n = H_{\delta_n}$. Dann gilt $H_n = H_{n+1} \oplus D_n$, $H_{n+1} \neq \{0\}$ und D_n ist abzählbar. Wenn $H_o = H_{\delta_o} \neq G$, dann ist auch H_o ein direkter Summand von G. Also $G = H_o \oplus E$ für eine gewisse Untergruppe E von G. Wie vorher folgt, daß auch E abzählbar ist. Wir setzen jetzt:

$$T = E \cup \bigcup_{n \in \omega} D_n \quad .$$

Zunächst ist klar, daß T abzählbar ist. Wir behaupten, daß $\hat{T}$ dicht in $\langle G, \leqq \rangle$ liegt, wenn $\hat{T}$ die von T erzeugte Untergruppe ist.

Es sei $f, g \in G$ mit $0 \leqq g < f$. Wir müssen die Existenz eines $t \in \hat{T}$ mit $g \leq t \leq f$ nachweisen. Sei $\theta = A(f-g) \in K^o$ die archimedische Klasse von $f - g > 0$. Da die Folge der δ_n koninitial in K^o liegt, gibt es eine Zahl $m \in \omega$ mit $\delta_m < \theta$. Dann gilt:

$$G = H_m \oplus D_{m-1} \oplus D_{m-2} \oplus \ldots . \oplus D_1 \oplus D_o \oplus E$$

(anti-lexikographisch geordnete direkte Summe). Aus $f, g \in G$ folgt daher

$$f = h + d_{m-1} + d_{m-2} + \ldots . + d_1 + d_o + e \quad ,$$
$$g = h' + d'_{m-1} + d'_{m-2} + \ldots . + d'_1 + d'_o + e'$$

für gewisse $h, h' \in H_m$, $d_{m-1}, d'_{m-1} \in D_{m-1}$, ..., $d_1, d'_1 \in D_1$, $d_o, d'_o \in D_o$ und $e, e' \in E$. Laut Definition von $H_m = H_{\delta_m} = \{ x \in G \, ; \, A(x) \leqq \delta_m \}$ haben wir: $f - g \notin H_m$, also $f - g \neq h - h'$. Daher gilt aufgrund der anti-lexikographischen Ordnung (vergl. (††)):

$$d'_{m-1} + \ldots \quad d'_1 + d'_o + e' < d_{m-1} + \ldots . + d_1 + d_o + e \quad .$$

Aus $h \in H_m = H_{m+1} \oplus D_m$ folgt $h = a + d_m$ für gewisse $a \in H_{m+1}$ und $d_m \in D_m$. Wähle jetzt ein $q \in D_m$ mit $q < d_m$, und setze:

$$t = q + d_{m-1} + d_{m-2} + \ldots . + d_1 + d_2 + e \quad .$$

Es ist klar, daß $g < t < f$ gilt, und daß $t \in \hat{T}$ gilt. $\langle G, +, \leqq \rangle$ besitzt also eine abzählbare dichte Teilmenge, nämlich $\hat{T}$.

Lemma 6 ist damit endlich bewiesen! Q.E.D.

Die Lemmata 5 und 6 haben zur Konsequenz:

SATZ 3 : *Sei* $\langle G,+,\leqq\rangle$ *eine dicht-geordnete abelsche Gruppe. Wenn* $\langle G,\leqq\rangle$ *die Bedingung* (S III) *von Souslin erfüllt, dann besitzt* $\langle G,\leqq\rangle$ *eine abzählbare dichte Teilmenge.*

Beweis. Sei D die teilbare Hülle von G. Nach Lemma 5 ist D auch dicht geordnet und erfüllt (S III). Nach Lemma 6 besitzt D eine abzählbare Teilmenge, welche dicht in D liegt. Die lineare Ordnung von D ist eine Erweiterung der linearen Ordnung von G (siehe Lemma 5). Also hat auch $\langle G,\leqq\rangle$ eine abzählbare Teilmenge (wenn $X=\{x_i\ ;\ i\in\omega\}$ in D dicht liegt, dann wähle man aus jedem Intervall $\{z\in G\ ;\ x_i\leqq z\leqq x_j\}$ welches nicht leer ist, ein Objekt $z_{i,j}$ aus. Die Menge dieser $z_{i,j}$ ist eine abzählbare dichte Teilmenge von G), Q.E.D.

Es ist mir nicht bekannt, ob auch für nicht-abelsche dicht-geordnete Gruppen aus (S III) die Existenz einer abzählbaren dichten Teilmenge folgt. Aus Satz 3 folgt aber unmittelbar das folgende Korollar:

KOROLLAR : *Es sei* $\mathfrak{A}=\langle A,+,\cdot,\leqq\rangle$ *entweder ein kommutativer geordneter Körper, ein geordneter Schiefkörper, ein geordneter Ring, oder ein rechts-geordneter (oder links-geordneter) Fastkörper. Falls* $\langle A,\leqq\rangle$ *dicht-geordnet ist und die Bedingung* (S III) *von Souslin erfüllt, dann besitzt* $\langle A,\leqq\rangle$ *eine dichte abzählbare Teilmenge.*

Beweis: die additive Gruppe $\langle A,+,\leqq\rangle$ ist in allen Fällen eine dicht-geordnete abelsche Gruppe, sodaß sich die Behauptung sofort aus Satz 3 ergibt. Q.E.D.

Bemerkung: Archimedizität kann aus (S III) nicht gefolgert werden, da mit Hilfe des Satzes von Löwenheim-Skolem die Existenz abzählbarer nicht-archimedischer Strukturen gefolgert werden kann.

L I T E R A T U R

[1] R.BAER: *Dichte, Archimedizität und Starrheit geordneter Körper;* Mathematische Annalen 188 (1970) pp.165-205.

[2] C.C.CHANG - H.J.KEISLER: *Model Theory.* (Studies in Logic, vol.73) Amsterdam-London-New York 1973.

[3] P.CONRAD: *Right-ordered Groups.* The Michigan Math.Journal 6(1959) pp.267-275.

[4] K.J.DEVLIN - H.JOHNSBRÅTEN: *The Souslin Problem.* Springer-Lecture-Notes in Mathematics, vol.405. Berlin-Heidelberg-New York 1974.

[5] F.R.DRAKE: *Set Theory - an introduction to large Cardinals.* (Studies in Logic, vol.76) Amsterdam-London-New York 1974.

[6] L.FUCHS: *Teilweise geordnete algebraische Strukturen.* Göttingen 1966.

[7] P.HÁJEK - P.VOPĚNKA: *Some permutation submodels of the model* ∇. Bull.Acad.Polon.Sci.14(1966)pp.1-7.

[8] G.H.HARDY - E.M.WRIGHT: *Einführung in die Zahlentheorie.* München 1958.

[9] D.HILBERT: *Grundlagen der Geometrie.* Festschrift zur Feier der Enthüllung des Gauss-Weber Denkmals in Göttingen. Teubner-Verlag in Leipzig 1899 (10.Auflage Teubner-Stuttgart 1968).

[10] O.HÖLDER: *Die Axiome der Quantität und die Lehre vom Maß.* Berichte der Verh.d.Sächsischen Ges.d.Wissenschaften zu Leipzig, Math.Phys. Classe, 53(1901)pp.1-64.

[11] K.ISEKI: *On Simply ordered Groups.* Portugaliae Mathematica 10, (1951)pp.85-88.

[12] Th.J.JECH: *Non-Provability of Souslin's Hypothesis.* Commentationes Mathematicae Universitatis Carolinae (Prag) 8(1967)pp.291-305.

[13] A.S.JESENIN-VOLPIN: *Unprovability of Souslin's Hypothesis without the aid of the axiom of choice in the Bernays-Mostowski axiom - system.* In: Amer.Math.Soc.Translations (Series 2), vol.23 (1963) pp.83-87.

[14] H.KARZEL: *Bericht über projektive Inzidenzgruppen.* Jahresberichte der Deutschen Math.Vereinigung 67 (1965) pp.58-92.

[15] A.G.KUROSCH: *Gruppentheorie II.* Berlin 1972.

[16] F.LOONSTRA: *Ordered Groups.* Indagationes Math. 6 (1946) pp.41-46.

[17] B.H.NEUMANN: *On the commutativity of addition.* The Journal of the London Math.Soc., 15 (1940) pp.203-208.

[18] M.E.RUDIN: *Souslin's conjecture.* Amer.Math.Monthly 76 (1969) pp. 1113-1119.

[19] W.SIERPINSKI: *Cardinal and Ordinal Numbers.* Monografie Matematyczne TOM.34. 2nd revised edition Warszawa 1965.

[20] R.M.SOLOVAY - S.TENNENBAUM: *Iterated Cohen Extensions and Souslin's Problem.* Annals of Math. 94 (1971) pp.201-245.

[21] M.SOUSLIN: *Problème 3.* Fundamenta Mathematicae 1 (1920) p.223.

[22] S.TENNENBAUM: *Souslin's Problem.* Proceedings of the National Acad. Sci.USA, vol. 59 (1968) pp.60-63.

[23] J.L.ZEMMER: *Near-fields, planar and non-planar.* The Mathematics Student (India) vol. 32 (1964) pp.145-150.

NACHTRAG (1.Juni 1976). Satz 3 kann von der Voraussetzung der Kommutativität befreit werden. Frau S.Koppelberg hat vor wenigen Tagen das hier gestellte Problem gelöst und gezeigt, daß *jede* geordnete Gruppe $\langle G, \cdot, \leqq \rangle$, welche (S III) erfüllt, separabel ist. Nachdem ich von diesem Ergebnis gehört hatte, realisierte ich, daß die hier gegebenen Beweise von Lemma 6 und Satz 3 nur unwesentlich geändert werden müssen, um auch im nicht-kommutativen Falle zu funktionieren. Im Beweis von Lemma 6 wurde die Kommutativität und Teilbarkeit von G benutzt, um die konvexe Untergruppe H_n in eine direkte Summe $H_{n+1} \oplus D_n$ zu zerlegen. Im Beweis wird dann aber im Wesentlichen nur benötigt, daß D_n ein Repräsentanten-System für die linken Nebenklassen von H_{n+1} ist. Falls G eine beliebige multiplikativ geschrieben Gruppe ist, dann definiere die H_n wie zuvor. Aus (S III) folgt $[H_n : H_{n+1}] \leq \aleph_0$ und als D_n wählen wir ein Repräsentanten-System für die linken Nebenklassen von H_{n+1} in H_n. Analog sei $G = \bigcup \{xH_0 \; ; \; x \in E\}$. Die von $E \cup \bigcup \{D_n \; ; \; n \in \omega\}$ erzeugte Untergruppe ist abzählbar und dicht in G.

U.Felgner, Math.Institut der Universität
74 TÜBINGEN, Fed.Rep.Germany

A CUMULATIVE SYSTEM OF FUZZY SETS

by Siegfried Gottwald (Leipzig)

1. Introduction.

The theory of fuzzy sets began with papers by L.A.Zadeh [13], [14] and D.Klaua [7], [8]. They independently realized the idea to generalize the usual notion of set in such a way that new "fuzzy" sets had "unsharp boundaries"; via characteristic functions this can be done by defining the fuzzy sts to be functions into a generalized set of truth-values. As set w of truth-values one considers the real interval $[0,1]$ or a finite subset of the form

$$\left\{ 0 , \frac{1}{m} , \frac{2}{m} , \dots , 1 \right\}$$

for natural $m \geq 2$ - the case $m = 1$ would give the classical sets.

Zadeh and his followers are mainly interested in applications of the notion of fuzzy set to system theory, pattern recognition, artificial intelligence, algorithms etc. Surveys of these results are given in [12] and [4].

A first cumulative system of fuzzy sets was given by Klaua in [8], a second used in [10]. In the second system one has classical and fuzzy sets interchangably, in the first system only fuzzy sets but no generalized identity. In the present note we will sketch a system with only fuzzy sets and a generalized identity; this system is a modification of my [3].

We will work in an impredicative set theory which admits proper classes, e.g. the system of Klaua [5], [6] or Morse/Kelley ; the class of ordinals we denote by On , ordinals by α, β, γ, ξ, λ , quantification by $\bigwedge$ (for every), $\bigvee$ (there exist), implication by $\Longrightarrow$.

For the formulations in connection with fuzzy sets let us use a special many - valued logic with propositional connectives defined by

$$\neg_w s =_{df} 1 - s ,$$

$$s \wedge_w t =_{df} \min\{s,t\} , \qquad s \wedge^w t =_{df} \max\{0, s+t-1\} ,$$

$$s \vee_w t =_{df} \max\{s,t\} , \qquad s \vee^w t =_{df} \min\{1, s+t\} ,$$

$$s \to_w t =_{df} \min\{1, 1-s+t\} , \qquad s <\!-\!>_w t =_{df} 1 - |s-t| .$$

for arbitrary truth-values $s,t \in w$. Later on special predicates $\in_w$, $=_w$, $\equiv_w$, $\sqsubset_w$ will be defined as functions into the truth-value set w . Bounded w-quantification over the elements of a class M we define by

$$\forall_M x\, H(x) =_{df} \inf_{x \in M} H(x) \quad , \qquad \exists_M x\, H(x) =_{df} \sup_{x \in M} H(x)$$

(with $\forall_\emptyset x\, H(x) = 1$ and $\exists_\emptyset x\, H(x) = 0$) for functions $H: M \to w$ which may be given as well-formed formulas H with free variable x in the language given by

$$\neg_w, \wedge^w, \wedge_w, \vee^w, \vee_w, \to_w, <\!-\!>_w \;;\; \forall_M, \exists_M \;;\; \in_w, =_w, \equiv_w, \sqsubset_w$$

and variables $x,y,z,\ldots$ (also with indices) for w-elements (as defined later), and constants for every element of w .

The connectives $\neg_w$, ${}_w$, ${}_w$, $\to_w$, $<\!-\!>_w$ are the same as used by Łukasiewicz [11] in the three-valued case. The universal algebra $[w, \wedge^w, \vee^w, \wedge_w, \vee_w, \neg_w, 0, 1]$ of type $(2,2,2,2,1,0,0)$ is a MV-algebra in the sense of Chang [1] (with $\wedge^w$ for $\cdot$, $\vee^w$ for $+$, $\wedge_w$ for $\wedge$, $\vee_w$ for $\vee$, and $\neg_w$ for $^-$) .

Since we identify for the sake of simplicity well-formed formulas (wff) with their truth-values, we consider a wff H with free variables as a function into w and a wff G without free variables as a truth-value.

If H is a wff then to state H let us mean to state: $H = 1$.

Sometimes we have to consider $\wedge^w, \wedge_w$ or also $\vee^w, \vee_w$ simultaneously, then instead to write down two formulas we use $\wedge.$ or $\vee.$, e.g.

$$s \wedge . \; t \to_w t$$

stands for both the formulas

$$s \wedge_w t \to_w t \,, \qquad\qquad s \wedge^w t \to_w t \,.$$

Two or more points within one formulas always have the same meaning.

2. Construction of the fuzzy sets.

We begin our construction with a set U ; the elements of U we call urelements and put

$$R(o) =_{df} U \,.$$

Two-place functions $\epsilon_o, \sqsubset_o$ and $=_o$ on $R(o)$ we define by

$$a \; \epsilon_o \; b = 0 \,,$$

$$a \sqsubset_o b = 1 \,,$$

$$a =_w b = \begin{cases} 1 & \text{for } a = b \\ 0 & \text{otherwise} \end{cases}$$

for all $a,b \in R(o)$.

Let be $\alpha \in On$ and for all $\beta \leqslant \alpha$ sets $R(\beta)$ and two-place functions $\epsilon_\beta, \sqsubset_\beta, =_\beta$ already defined, then now we define $R(\alpha+1)$, $\epsilon_{\alpha+1}, \sqsubset_{\alpha+1}, =_{\alpha+1}$. For this let be $S(\alpha)$ the set of all functions $f : R(\alpha) \to w$ for which the following two conditions hold:

(D1) $$\bigwedge_{\beta<\alpha} \bigvee_{\gamma} \bigvee_{x} (\beta < \gamma \leqslant \alpha \wedge x \in R(\gamma) \setminus \bigcup_{\xi<\gamma} R(\xi) \wedge f(x) \neq 0) \,,$$

(D2) $$\bigwedge_{x,y \in R(\alpha)} (f(x) \wedge^w y =_\alpha x \wedge^w y \sqsubset_\alpha x \wedge^w x \sqsubset_\alpha y \leqslant f(y)) \,,$$

and let be

$$R(\alpha + 1) =_{df} R(\alpha) \cup S(\alpha) \,.$$

The two-place functions $\epsilon_{\alpha+1}, \sqsubset_{\alpha+1}, =_{\alpha+1}$ from $R(\alpha + 1)$ into w shall be defined for all $a,b \in R(\alpha + 1)$ by:

$$a \in_{\alpha+1} b =_{df} \begin{cases} a \in_\alpha b \ , & \text{for } a,b \in R(\alpha) \\ b(a) \ , & \text{for } b \in S(\alpha) \ , \ a \in \mathrm{dom}(b) \\ 0 & \text{otherwise} \end{cases}$$

$$a \sqsubset_{\alpha+1} b =_{df} \begin{cases} a \sqsubset_\alpha b \ , & \text{for } a,b \in R(\alpha) \\ 1 \ , & \text{for } b \in S(\alpha) \\ 0 & \text{otherwise ,} \end{cases}$$

$$a =_{\alpha+1} b =_{df} \begin{cases} a =_\alpha b \ , & \text{for } a,b \in R(\alpha) \\ \forall_{R(\alpha)} x(x \in_\alpha a \leftrightarrow_w x \in_\alpha b) & \text{otherwise .} \end{cases}$$

Finally for $\lim(\lambda)$ we put

$$R(\lambda) =_{df} \bigcup_{\xi<\lambda} R(\xi)$$

and define $\in_\lambda, \sqsubset_\lambda, =_\lambda$ similar as join on all $\xi < \lambda$.

This is a correct definition by transfinite recursion.

Now we are able to introduce the following five proper classes which are basic for our further treatment:

(1) the class $\underline{E}_w$ of all w-elements:

$$\underline{E}_w =_{df} \bigcup_{\alpha\in On} R(\alpha) \ ,$$

(2) the class $\underline{M}_w$ of all fuzzy sets:

$$\underline{M}_w =_{df} \bigcup_{\alpha\in On} S(\alpha) = \underline{E}_w \setminus U \ ,$$

(3) the w-membership-function $\in_w$:

$$\in_w =_{df} \bigcup_{\alpha\in On} \in_\alpha$$

(4) the w-rank-function $\sqsubset_w$:

$$\sqsubset_w =_{df} \bigcup_{\alpha\in On} \sqsubset_\alpha \ ,$$

(5) the w-equality $=_w$:

$$=_w \;=_{df}\; \bigcup_{\alpha \in On} =_\alpha \quad .$$

With their help we define for $a,b \in \underline{E}_w$ also

$$a \sqsupset\!\sqsubset_w b =_{df} a \sqsubset_w b \wedge^w b \sqsubset_w a \quad ,$$

$$a \sqsubset_w b =_{df} a \sqsubset_w b \wedge^w a \not\sqsupset\!\sqsubset_w b \quad ,$$

with as usual $a \not\sqsupset\!\sqsubset_w b = \neg_w (a \sqsupset\!\sqsubset_w b)$, and the w-identity

$$a \equiv_w b =_{df} a =_w b \wedge^w a \sqsupset\!\sqsubset_w b \quad .$$

If $a,b \in R(o)$ or if $a,b \in S(\alpha)$ it immediately follows that

$$a \equiv_w b <\!\!-\!\!>_w a =_w b \quad .$$

Then the above condition (D2) gives that for all $a,b \in \underline{E}_w$ and fuzzy sets A :

$$a \equiv_w b \wedge^w b \in_w A \rightarrow_w a \in_w A \quad .$$

As here we will use capital letters A,B,C for fuzzy sets; furthermore w-quantification over $\underline{E}_w$ we will denote by $\forall , \exists$ without subscript.

3. Fundamental properties of fuzzy sets.

From the last definition it follows for all w-elements a,b that

$$a \in_w b \rightarrow_w a \sqsubset_w b \quad ,$$

$$a \sqsubset_w b \rightarrow_w b \notin_w a \quad ,$$

and therefore also

$$a \notin_w a$$

$$a \in_w b \to_w b \notin_w a \quad .$$

Together with for all $\emptyset \neq M \subseteq \underline{E}_w$:

$$\exists_M x \forall_M y\ (x \sqsubset_w y)$$

this gives a kind of regularity. For the formulation we need as further notion for every fuzzy set A the support $|A|$ defined by

$$|A| =_{df} \{x \in \underline{E}_w \mid (x \in_w A) \neq 0\} \ .$$

COROLLARY 1 (regularity): For every fuzzy set A there hold

$$\exists x(x \in_w A) \to_w \exists_{|A|} u \forall_{|A|} v(u \sqsubset_w v) \quad .$$

THEOREM 1. (a) For all fuzzy sets A,B

$$A =_w B \leftrightarrow_w \forall x(x \in_w A \leftrightarrow_w x \in_w B) \quad .$$

(b) For all w-elements a,b

$$a \equiv_w b \leftrightarrow_w \forall x(a \in_w x \leftrightarrow_w b \in_w x) \quad .$$

COROLLARY 2. For all $a,b,c \in \underline{E}_w$:

$$a =_w a \quad ,$$

$$a =_w b \leftrightarrow_w b =_w a \quad ,$$

$$a =_w b \wedge^w b =_w c \to_w a =_w c \quad .$$

THEOREM 2. Let be $a,b,c \in \underline{E}_w$ and H a wff with one free variable, then

$$a \equiv_w a \quad ,$$

$$a \equiv_w b \to_w (a \equiv_w c \to_w b \equiv_w c) \quad ,$$

$$a \equiv_w b \to_w (H(a) \to_w H(b)) \quad .$$

THEOREM 3. (weak extensionality for fuzzy sets):

$$A =_w B \Longrightarrow A \equiv_w B .$$

This last theorem doesn't hold with $\rightarrow_w$ instead of $\Rightarrow$. One only can show that always

$$(A =_w B \rightarrow_w A \equiv_w B) \neq 0 .$$

THEOREM 4. (replacement): For every wff F which is supposed not to contain any free occurence of the variable b there hold

$$\forall x \forall y \forall z\ (F(x,y) \wedge_w F(x,z) \rightarrow_w y \equiv_w z) \rightarrow_w$$
$$\forall a \exists b \forall y\ (y \in_w b <\!\!-\!\!>_w \exists x\ (x \in_w a \wedge .\ F(x,y))) .$$

COROLLARY 3. (separation): Let H be a wff which doesn't contain any free occurence of the variable "b" then

$$\forall a \exists b \forall x\ (x \in_w b <\!\!-\!\!>_w x \in_w a \wedge .\ H(x)) .$$

A characterization of all wff H with exactly one free variable comprehend fuzzy sets is possible if one uses not only w-notions.

THEOREM 5. (comprehension): Let H be a wff with exactly one free variable which doesn't contain any free occurence of the variable "a" , or a function from $\underline{E}_w$ into w ; then holds

$$\exists a \forall x\ (x \in_w a <\!\!-\!\!>_w H(x))$$

if and only if

$$(*) \qquad \bigvee_{\alpha} \bigwedge_{x \in \underline{E}_w \setminus R(\alpha)} (H(x) = 0) .$$

Because of theorem 1 every H with property (*) defines exactly one $A \in \underline{M}_w$ with

$$\forall x\ (x \in_w A <\!\!-\!\!>_w H(x)) ;$$

this A will be denoted by $\{x \,\|\, H(x)\}$. Therefore

$$y \in_w \{x \parallel H(x)\} \leftrightarrow_w H(y) \quad .$$

COROLLARY 4. (empty set):

$$\exists A \forall x \, (x \notin_w A) \; .$$

<u>Proof</u>. Take $A = \emptyset_w$ with

$$\emptyset_w =_{df} \{x \parallel x \neq_w x\} \quad .$$

COROLLARY 5. (pairing):

$$\forall a \forall b \exists A \forall x \, (x \in_w A \leftrightarrow_w (x \equiv_w a \vee. \; x \equiv_w b)) \; .$$

<u>Proof</u>. For given $a, b \in \underline{E}_w$ the wff $x \equiv_w a \veebar x \equiv_w b$ has property (*) .

COROLLARY 6. (sum-set):

$$\forall a \exists A \forall x \, (x \in_w A \leftrightarrow_w \exists z \, (x \in_w z \wedge. \; z \in_w a)) \; .$$

<u>Proof</u>. For given $a \in \underline{E}_w$ the wff $z(x \in_w z \wedge. \; z \in_w a)$ has property (*) .

For setting up a notion of fuzzy power set we first need a generalized inclusion. Corresponding with $=_w$, and $\equiv_w$, and theorem 1 (a) we define for fuzzy sets A, B

$$A \subseteq_w B =_{df} \forall x \, (x \in_w A \rightarrow_w x \in_w B) \quad ,$$

$$A \subseteq^*_w B =_{df} A \subseteq_w B \wedge^w A \sqsubset_w B \quad .$$

Straightforward compution gives

$$A \subseteq_w B \wedge_w B \subseteq_w A \leftrightarrow_w A =_w B \quad ,$$

$$A \subseteq^*_w B \wedge_w B \subseteq^*_w A \leftrightarrow_w A \equiv_w B \quad ,$$

and also

$$A \subseteq_w A \quad ,$$

$$A \subseteq_w B \wedge^w B \subseteq_w C \to_w A \subseteq_w C \quad ,$$

$$A \subseteq_w^* A \quad ,$$

$$A \subseteq_w^* B \wedge^w B \subseteq_w^* C \to_w A \subseteq_w^* C \quad .$$

But for given fuzzy set B the wff $\forall x(x \in_w A \to_w x \in_w B)$ hasn't property (*) because the class of all fuzzy sets A with

$$(A \subseteq_w B) \neq 0$$

is a proper class. For let be r a (constant for a) truth-value different from 0 and 1 , let $\alpha \in On$ and $B \in R(\alpha)$, $a \in S(\alpha)$, then the wff

$$x \in_w B \vee_w (x \equiv_w a \wedge^w r)$$

has property (*) and therefore comprehends a fuzzy set X_a with

$$(X_a \subseteq_w B) = 1 - r \neq 0 \quad .$$

However the wff

$$\forall x\, (x \in_w A \to_w x \in_w B) \;\wedge^w\; A \sqsubset_w B$$

has property (*) for every fuzzy set B . This enables us to define

$$P_w B \;=_{df}\; \{x \,\|\, x \subseteq^* B\}$$

for every $B \in \underline{M}_w$; if x is an urelement the value $x \subseteq_w^* B$ shall be taken as zero -that can be reached by addition of "$\wedge_w \exists y(y \sqsubset_w x)$" to the definition of $\subseteq_w^*$. So we have proved

COROLLARY 7. (power-set): If $R(o) = \emptyset$ then

$$\forall a \exists b \forall x(x \in_w b \leftrightarrow_w \forall y(y \in_w x \to_w y \in_w a) \wedge^w x \sqsubset_w a) \quad ;$$

if $R(o) \neq \emptyset$ then

$$\forall a \exists b \forall x(x \in_w b \leftrightarrow_w \forall y(y \in_w x \to_w y \in_w a) \wedge^w x \sqsubset_w a \wedge^w \exists z(z \sqsubset_w x)).$$

It is clear from the construction of the sets $R(\alpha)$ that for $\alpha \geq \omega$ always $R(\alpha)$ is infinite. Therefore if we consider the function $A_\omega : R(\omega) \to w$ with $A_\omega(x) = 1$ for all $x \in R(\omega)$ there hold $A_\omega \in S(\omega)$. The fuzzy set A_ω proves the existence statements of the following corollary.

COROLLARY 8. (infinity):

(a) $\exists a(\exists x(x \in_w a) \wedge . \forall x(x \in_w a \to_w \exists y(y \in_w a \wedge .$

$$\forall z(z \in_w y <\!\!\to_w z \equiv_w x)))) \quad ,$$

(b) $\exists a(\exists x(x \in_w a) \wedge . \forall x(x \in_w a \to_w \exists y(y \in_w a \wedge .$

$$\forall z(z \in_w y <\!\!\to_w z \in_w x \vee . \; z \equiv_w x)))) \quad ,$$

(c) $\exists a(\exists x(x \in_w a) \wedge . \forall x(x \in_w a \to_w \exists y(y \in_w a \wedge . \; x \sqsubset_w y)))$.

The first two parts of corollary 8 formulate infinity as usual in ZF (cf. [2], p.46), the 3rd part formulates it corresponding to [9] by stating the existence of a "Grenzbereich".

THEOREM 6 (choice): For every fuzzy set A let be fc(A) the class of all functions with domain $|A|$. Then it hold

$$\forall x(x \in_w A \to_w \exists y(y \sqsubset_w x)) \to_w$$

$$\exists_{fc(A)} f \, \forall x(x \in_w A \wedge_w \exists y(y \in_w x) \to_w f(x) \in_w x) \; .$$

The proof is by choice in the underlaying classical set theory.

Because we have used fc(A) this isn't a formulation which is given fully in our many-valued framework. But for other possible translations of classical formulations of the axiom of choice into our many-valued logic the question is open if they will hold. But in the case

$$\forall b[\forall x(x \in_w b \to_w \exists y(y \in_w x) \wedge^w$$

$$\forall y(y \in_w b \wedge^w \exists z(z \in_w x \wedge_w z \in_w y) \to_w x =_w y)) \to_w$$

$$\exists a \forall x(x \in_w b \to_w \exists z \forall u(u =_w z <\!\!\to_w u \in_w a \wedge^w u \in_w x))]$$

and $m = 2$, i.e. $w = \{0, 1/2, 1\}$, the example

$$b = \left\{\{c\}_{1/2} , \{d\}_{1/2} , \{c,d\}_1\right\}_{1/2}$$

with $\{x,y,z\}_\tau =_{df} \{u \| (u \equiv_w x \vee_w u \equiv_w y \vee_w u \equiv_w z) \wedge^w \tau\}$ for $\tau \in w$, and with c , d fuzzy sets, for which $(c \equiv_w d) = 0$ hold, shows that the given "translation" of the axiom of choice doesn't hold.

References

[1] Chang,C.C. , Algebraic analysis of may valued logics, Trans. Amer. Math. Soc. 88 (1958), 467-490.

[2] Fraenkel,A.A., Y.Bar-Hillel, A.Levy ,Foundations of Set Theory, Amsterdam 1973.

[3] Gottwald,S., Untersuchungen zur Mehrwertigen Mengenlehre I, II, III , Math. Nachr. (in print).

[4] Kaufmann,A. , Introduction à la théorie des sous-ensembles flous. Tome 1 , Eléments théoriques de base, Paris 1973.

[5] Klaua,D. , Ein Aufbau der Mengenlehre mit transfiniten Typen , formalisiert im Prädikatenkalkül der ersten Stufe, Ztschr. math. Logik Grdl. Math. 3 (1957), 303-316.

[6] Klaua,D. , Allgemeine Mengenlehre, Berlin 1964.

[7] Klaua,D. , Über einen Ansatz zur mehrwertigen Mengenlehre, Monatsber. Dt. Akad. Wiss., 7 (1965), 859-876.

[8] Klaua,D. , Über einen zweiten Ansatz zur mehrwertigen Mengenlehre , Monatsber. Dt. Akad. Wiss., 8 (1966), 161-177.

[9] Klaua,D. , Zur Mengenlehre mit Stufenrelation, Monatsber. Dt. Akad. Wiss., 11 (1969), 265-268,

[10] Klaua,D. , Stetige Gleichmächtigkeiten kontinuierlichwertiger Mengen, Monatsber. Dt. Akad. Wiss. 12 (1970), 749-758.

[11] Łukasiewicz,J. , O logice trójwartościowej, Ruch filozoficzny, 5 (1920), 169-171.

[12] Gusev,L.A. , I.M.Smirnova , Fuzzy sets. Theory and applications (A Survey), (russ.) Avtom. Telemeh. 1973, H.5, 66-85, engl. translation, Automat. Remote control 34 (1973), 739-755.

[13] Zadeh,L.A. , Fuzzy sets , Inf. and Control 8 (1965), 338--353.

[14] Zadeh,L.A. , Fuzzy sets and systems, Proc. Symp. on System Theory, New York 1965, pp. 29-39.

NON - STANDARD SATISFACTION CLASSES

by S. Krajewski

Introduction

The notion of satisfaction, introduced by Alfred Tarski in order to formalize the intuition of mathematical truth, proved to be one of the most fundamental in mathematical logic. Satisfaction is defined for formalized languages. Every such language, say L , can be arithmetized. Hence every non-standard model of arithmetic determines an extension of L containing non-standard formulas.

One may attempt to investigate the semantics of this non-standard language. The corresponding structures would be just the structures for L . The problem is how to define satisfaction (and truth) for non-standard formulae. This question has no clear meaning in full generality. One major difficulty is with the definition of appropriate non-standard finite sequences of individuals. We shall restrict our attention to the case when a given model contains non-standard formulae and determines in a natural way what non-standard finite sequences are. In this case we say that a satisfaction of non-standard formulae (from the model) in the model itself is defined, when an appropriate subset of the universe of the model, called satisfaction class, is given. A satisfaction class should consist of ordered pairs (in the sense of the model) $\langle e,x \rangle$, where e is a (possibly non-standard) formula and x is a corresponding (possibly non-standard) finite sequence. Our main requirement concerning this class is that suitably formulated Tarski's inductive conditions should hold true.

There exists another approach leading to the concept of a satisfaction class. Let us look at the notion of satisfaction as an object to be investigated in an axiomatic way. Thus we formalize the <u>meta</u>-theory of a given theory T in a first order language L_T. We add to L_T a new primitive predicate S which will be interpreted over models of T as a satisfaction class. Axioms concerning S will be just suitable expressions of Tarski's conditions. In order to formulate them it is sufficient to have the possibility of going down from a formula to its constituent immediate subformulas. Our definition will thus allow partial satisfaction classes, among them rather trivial ones. However, a series of stronger definitions will be considered, e.g. that of a closed satisfaction class deciding with a formula any of its subformulae or that of a full satisfaction class deciding all (non-standard) formulae.

To fulfil the above programme we shall consider theories T with two properties:

1^0 Aritmetic is relatively interpretable in T

2^0 In T the concept of a finite (in the sense of this interpretation) sequence of individuals is definable.

Arithmetic and different set theories (more generally-theories semantically closed in the sense of Montague [61]) and also theories like GB can serve as examples. Condition 2^0 will be made precise in an axiomatic way in § 0. We investigate satisfaction classes over models of such theories T.

A serious approach to the possibility of non-absoluteness of the finite (and so of the logical syntax too) was realised first by A. Robinson in [63]. To the present author the work of A.Mostowsi([50]) was also an inspiration.

A general result of the present paper is that those conditions of Tarski, admitted as the only axioms, are weak. There exists a rather rich variety of satisfaction classes which can be mutually inconsistent, also for sentences, not only for formulae with free variables. Non-standard languages have no uniquely determined semantics. This state of affairs was anticipated by Tarski in his famous work [33] and was shown explicitly by A. Robinson in the article mentioned above,

It should be emphasised that the study of satisfaction classes finds applications in at least two domains: investigations concerning non-standard models and, secondly, problems connected with second order expansions of theories (considered in first order logic-like GB relative to ZF) and their models. Some applications are indicated in § 6.

Basic definitions are given in § 1. Some general properties of satisfaction classes are proved.

In § 2 the theory T is assumed to be axiomatizable by a scheme. We investigate a special kind of satisfaction classes which can occur as parameters in instances of the scheme. Assuming the existence of such a class does not lead to a conservative extension of T . Satisfaction classes of this kind are most useful in applications (cf.§ 6).

In § 3 we take into consideration infinite generalisations of Tarski's conditions and give Robinson's definition of "external satisfaction" for some non-standard formulae.

In § 4 we show first that each non-ω-model of T of a power λ admits 2^{λ} pairwise inconsistent satisfaction classes which can, however, be very trivial. Models with indiscernibles have less trivial satisfaction classes, deciding all sentences, with the same properties.

§ 5 contains results to the effect that certain models admit many non-trivial full satisfaction classes mutually inconsistent either for sentences or for formulae with one free variable."Many" means either as many as the cardinality of the model in question or more than the cardinality of the model, respectively. Finally, Robinson's [63] result introduced in § 3 is proved.

I am greatly indebted to the late Professor A.Mostowski whose inspiration and suggestions were most important to me. I am also grateful to P. Hàjek and W. Marek for valuable suggestions as well as to H. Kotlarski, A. Krawczyk, Z. Ratajczyk and other colleagues for their useful remarks.

§ 0. Preliminaries.

Throughout the paper T will be a theory in a first order language L_T such that:

1^0 Peano aritmetic P is relatively interpretable in T . Let "N(·)" denote the domain of the interpretation. Given an arithmetical formula ϕ , let ϕ^N be the corresponding formula of L_T . We assume that free variables of ϕ remain unrestricted to N(·) in ϕ^N (but of course $T \vdash \phi^N(x) \to N(x))$. 0 and 1 will have the obvious meaning in L_T .

2^0a) In T ordered pair $\langle x,y\rangle$ is definable.

2^0b) In T finite sequences are definable in this sense that there exist formulae (and terms) of L_T

$$\mathrm{Seq}(x)\ ,\ x^\frown y\ ,\ \mathrm{Lh}(x)\ ,\ (x)_k\ ,\ \emptyset\ ,\ [u]$$

satisfying the following axioms, formulated with the use of convention that x,y,z satisfy Seq(•) (in particular "$\forall x$" stands for "$(\forall x)(\mathrm{Seq}(x) \rightarrow \ldots)$" etc.) and k,l,m,n satisfy N(•) in the same sense.

(1) $(\forall x,y)(\exists!\ z)\ x^\frown y = z$

(2) $(\forall x)\ N(\mathrm{Lh}(x))$

(3) $(\forall x)(\forall k)_{\mathrm{Lh}(x)}(\exists!\ u)\ (u = (x)_k)$,
where $(\forall k)_{\mathrm{Lh}(x)}$ abbreviates $(\forall k)((\exists l)(l = \mathrm{Lh}(x) \wedge (k < l)^N \rightarrow \ldots)$.

(4) $\mathrm{Seq}\ (\emptyset) \wedge \mathrm{Lh}(\emptyset) = 0$

(5) $(\forall x,y)(\mathrm{Lh}(x) = \mathrm{Lh}(y) \wedge (\forall l)_{\mathrm{Lh}(x)}((x)_l = (y)_l) \rightarrow x = y)$

(6) $(\forall u)(\exists! x)(x = [u] \wedge \mathrm{Lh}(x) = 1 \wedge (x)_o = u)$

(7) $(\forall x,y)(\forall k)_{\mathrm{Lh}(x)}(\forall l)_{\mathrm{Lh}(y)}((x^\frown y)_k = (x)_k \wedge (x^\frown y)_{\mathrm{Lh}(x)+l} = (y)_l)$

(8) $(\forall n,x)(\exists m)$ [("n is a sequence"$)^N \rightarrow$ ("m is an increasing sequence"$)^N \wedge (\forall l)$(("l is a member of m"$)^N \leftrightarrow (\exists t,u)((x)_l =$
$= \langle t,u\rangle \wedge N(t) \wedge$ ("t is a member of n"$)^N$))]

(9) $(\forall x,m)$[("m is an increasing sequence"$)^N \rightarrow (\exists! y)[\mathrm{Lh}(y) =$
$=$ ("the length of m"$)^N \wedge (\forall l)_{\mathrm{Lh}(y)}(\exists k)((y)_l = (x)_k \wedge$
$\wedge$ ("k = m(l)"$)^N$)]] .

(when m enumerates an initial segment of length n , we write $y = x \restriction n)$.

We shall often assume the full (for all L_T-formulae) induction scheme. This is the case of semantically closed theories (Montague [61]) as e.g. P, ZF, Z, A_2 (second order arithmetic), KM (Kelley--Morse theory), etc.

The above axioms are chosen ad hoc because they will suffice in the sequel. They are satisfies in the case of the theory GB , in its subtheory Prexp (ZF) (obtained by substitution of the replacement axiom with the usual ZF replacement scheme), and in the analoquous expansions of other theories than ZF (see Krajewski [74] for details).

We will use the following abbreviations:

- "[a,b]" for "[a]$^\frown$[b]" and similarly for longer standard sequences,
- "y subseq x" for $Seq(y) \wedge Seq(x) \wedge (\exists m)\ [N(m) \wedge (\text{"m is an increasing map from } Lh(y) \text{ to } Lh(x)\text{"})^N \wedge (\forall l)_{Lh(y)}(\exists k)((y)_l = (x)_k \wedge (\text{"}k = m(l)\text{"})^N)]$,
- "a ∈ x" (read: a is a member of the sequence x) for $Seq(x) \wedge (\exists k)(N(k) \wedge (k < Lh(x))^N \wedge a = (x)_k)$.

We fix an arithmetization of L_T and the following formulae (and terms) of L_T, being translations of some <u>arithmetical</u> formulae: "fm(e)" (e is a formula of L_T), "e sub f" (e is a subformula of f), "v_k" (k-th variable), "fr(e)" (the sequence [in P!] of free variables of e), "k ∈ fr(e)" (v_k is free in e), "¬ e", "e ∨ f", "e ∧ f", "($\exists v_k$)e", "($\forall v_k$)e".

By virtue of the axioms (1)-(9) we may also use the following abbreviations of L_T expressions:

- "x: fr(e)" (read: x is a valuation for e) for

  $Seq(x) \wedge fm(e) \wedge (\forall k)(k \in fr(e) \rightarrow (\exists! u)(\langle k,u\rangle \in x)) \wedge (\forall y)(y \in x \rightarrow$
  $\rightarrow (\exists k,u)(\langle k,u\rangle = y \wedge k \in fr(e)))$.

- "x | fr(e)" (read: restriction of x to free variables of e) for the unique y such that

  $y \text{ subseq } x \wedge (\forall k,u)(\langle k,u\rangle \in y \leftrightarrow \langle k,u\rangle \in x \wedge k \in fr(e))$

- "x$^\frown\langle k,a\rangle$" for "x$^\frown[\langle k,a\rangle]$" or the unique y such that

  $x = y \upharpoonright Lh(x) \wedge (\forall u)(u \in y \leftrightarrow u \in x \vee u = \langle k,a\rangle)$.

§ 1. <u>Definitions of satisfaction classes</u>.

Our definition of a satisfaction class will be as general as possible. We formulate it in principle in the language $L_T(S)$ (with a new unary predicate S added). However, it is convenient to treat S as a class and consider classes elementarily definable from S as e.g. $D(S) = \{x\colon (\exists y)\ S(\langle x,y\rangle)\}$. That is why we express the definition and why we reason in the language L_T' - the second order language over L_T in which new variables (denoted by X,Y,...) and new atomic formulas "x ∈ X" (where x is an old (individual) variable) are added.

The formalization of satisfaction in the language L_T' was first realized in Mostowski [50].

Definition 1.1. A class Φ of formulas is closed under immediate subformulae, iff $\neg e \in \Phi \to e \in \Phi$, $e \vee f \in \Phi \to e \in \Phi$ and $f \in \Phi$, $e \wedge f \in \Phi \to e \in \Phi$ and $f \in \Phi$, $(\exists v_k)e \in \Phi \to e \in \Phi$, $(\forall v_k)e \in \Phi \to e \in \Phi$.

Definition 1.2. S is a satisfaction class iff

(i) $x \in S \to (\exists e,y)(fm(e) \wedge y\colon fr(e) \wedge x = \langle e,y \rangle)$.

(ii) The class $\Phi(S) = D(S) \cup \{e\colon (\forall x)(x\colon fr(e) \to \langle \neg e,x\rangle \in S)\}$ is closed under immediate subformulae.

(iii) If R_i is i-th atomic relation and p_i its Gödel number, then

$(1)_i$ if $v_{k_0},\dots,v_{k_l}$ are all free variables of an atomic formula e of the shape $p_i\,(v_{k_0},\dots,v_{k_l})$ such that $e \in \Phi(S)$ and if $x\colon fr(e)$, then $\langle e,x\rangle \in S \iff R_i(x_{k_0},\dots,x_{k_l})$, where $\langle k_j, x_{k_j}\rangle \in x$ for $j = 0,\dots,l$.

(2) if $\neg e \in \Phi(S)$ and $x\colon fr(e)$, then $\langle \neg e,x\rangle \in S \iff \langle e,x\rangle \notin S$

(3) if $e \vee f \in \Phi(S)$ and $x\colon fr(e \vee f)$, then

$$\langle e \vee f,x\rangle \in S \iff \langle e,x|_{fr(e)}\rangle \in S \vee \langle f,x|_{fr(f)}\rangle \in S$$

(3') similarly for $e \wedge f$

(4) if $(\exists v_k)e \in \Phi(S)$, then

$$\langle (\exists v_k)e,x\rangle \in S \iff (k \in fr(e) \wedge (\exists u)(\langle e,x^\frown\langle k,u\rangle\rangle \in S)) \vee (k \notin fr(e) \wedge \langle e,x\rangle \in S)$$

(4') similarly for $(\forall v_k)e$.

Notice that if $\neg e \in \Phi(S)$, then S decides e i.e. $(\forall x)(x\colon fr(e) \to \langle e,x\rangle \in S \vee \langle \neg e,x\rangle \in S)$.

Over every model $\mathfrak{M}$ for T there exists a "true" satisfaction class:

$$S_{tr} = \{\langle e,x\rangle^{\mathfrak{M}} : e \text{ is the Gödel number of a formula } \varphi \text{ of } L_T, x \text{ is a sequence (in } \mathfrak{M}) \text{ with the members } \langle k_0,x_0\rangle^{\mathfrak{M}},\dots, \langle k_l,x_l\rangle^{\mathfrak{M}}, \text{ where } k_0,\dots,k_l \text{ are all free variables of } \varphi, \text{ and } \mathfrak{M} \models \varphi\,[x_0,\dots,x_l]\}.$$

The expression "satisfaction class" will be abbreviated as "sat. cl.".

Definition 1.3. A sat. cl. S is strong iff $\Phi(S)$ contains all the atomic formulae and if $e,f \in \Phi(S)$, then $\neg e$, $e \vee f$, $e \wedge f$, $(\exists v_k)e$, $(\forall v_k)e \in \Phi(S)$.

Lemma 1.1. Over every model for T the union of a chain of sat. classes is a sat. cl. again.

Lemma 1.2. Let S be a sat. cl. over a model $\mathfrak{M}$ for T . There exists a sat. cl. $\bar{S}$ such that $\bar{S}$ is strong, $S \subseteq \bar{S}$ and $\bar{S}$ is the least sat. cl. which is strong and includes S .

Proof. One adds what should be added. The construction is made straightforward but it is rather tedious to check, in detail, that it works.
In general $\Phi(S)$ need not be closed under taking subformulae because in general a non-standard (i.e. infinite) sequence can lead from a formula to its subformula. That is why we introduce the following

Definition 1.4. A sat. cl. S is closed iff

$$e \in \Phi(S) \wedge f \text{ sub } e \rightarrow f \in \Phi(S) .$$

Proposition 1.1. If the induction scheme for formulas of the language $L_T(S)$ holds, then the sat. cl. S is closed.

Proof. Let $\varphi(x)$ be the following formula:

$$fm(x) \wedge x \in \Phi(S) \wedge (\exists f)(f \text{ sub } x \wedge f \notin \Phi(S)) .$$

Assume that $(\exists e)\varphi(e)$. Let e_0 be the least such e. e_0 is not atomic. Its immediate subformulae have no subformulae which do not belong to $\Phi(S)$, whence an immediate subformula of e_0 does not belong to $\Phi(S)$, which is a contradiction with (ii) of definition 1.2.

Definition 1.5. A sat. cl. S is unlimited iff

$$(\forall e)(\exists f)\ (e \text{ sub } f \wedge f \in \Phi(S))$$

A sat. cl. S is full iff $\Phi(S)$ consists of all formulae.

Definition 1.6. Satisfaction classes S and S' are inconsistent iff

$$(\exists e,x)((\langle e,x\rangle \in S \wedge \langle \neg e,x\rangle \in S') \vee (\langle e,x\rangle \in S' \wedge \langle \neg e,x\rangle \in S))$$

S and S' are inconsistent for a sentence iff

$$(\exists e)((\langle e,\emptyset\rangle \in S \wedge \langle \neg e,\emptyset\rangle \in S') \vee (\langle e,\emptyset\rangle \in S' \wedge \langle \neg e,\emptyset\rangle \in S)) .$$

Lemma 1.3. (a) If S and S' are sat. classes, if both are strong, and if they are not inconsistent, then $S \cup S'$ is a sat.cl.
(b) Sat. classes S, S' (over a fixed model for T) are inconsistent if and only if $S \cup S'$ is not included in any sat. cl. (over the model).

Proof. It is easy to check (a). To prove (b) we have to show that given consistent S and S', $\bar{S}$ and $\bar{S}'$ from lemma 1.2 are consistent, which will be enough. The proof that $\bar{S}$ and $\bar{S}'$ are consistent is quite natural and uses a kind of induction on the stages of the specific construction carried out in proof of lemma 1.2.

Proposition 1.2. If the induction scheme for formulae of the language $L_T(S,S')$ holds, then sat. classes S and S' are not inconsistent.

Proof. To show that

$$e \in \Phi(S) \cap \Phi(S') \rightarrow (\forall x)(\langle e,x\rangle \in S \leftrightarrow \langle e,x\rangle \in S') ,$$

it is enough to apply induction to the formula

$$(\forall f,x)(f \in \Phi(S) \cap \Phi(S') \wedge lh(f) \leq k \wedge f \text{ sub } e \rightarrow (\langle f,x\rangle \in S \leftrightarrow \langle f,x\rangle \in S'))$$

The above proof shows that over any model inconsistent sat. classes determine a class of natural numbers (in the sense of the model) with no least element, namely the class of formulae for which they differ.

Theorem 1.1. Each sat. cl. (over a given model for T) extends to a maximal (w. r. t. inclusion) sat. cl., which happens to be closed and unlimited.

Proof. The existence of the sat. cl. S follows from lemma 1.1 and Zorn Lemma. S is maximal hence $S = \bar{S}$ (cf. lemma 1.2), i.e S is strong.

In the proof that S is unlimited we will use an anomaly that is allowed by the apparently smart definition of a satisfaction class. We may assume that our model is not ω-standard.

Let e be a formula (in the given model) and k be a non-standard natural number majorising the indices of free variables of e. Let us define for $i = 0,1,2,\ldots$

$$f_i : (\exists v_i) \ldots (\exists v_k)(e \wedge (v_0 = v_0 \wedge (\ldots \wedge v_k = v_k) \ldots)) .$$

Each f_i is a formula, because it is provable in P that for any k the string of symbols of the above shape is a well formed formula. Now let U be $\bigcup_{i=0}^{\infty}\{\langle f_i,x\rangle : x \colon fr(f_i)\}$. U is a sat. cl. (very trivial indeed).

If U and S are inconsistent, then $\langle \neg f_i,x\rangle \in S$ for some i and x , so e sub f_i and $f_i \in \Phi(S)$.

If U and S are consistent, then by virtue of lemma 1.3 (b) $U \cup S \subseteq S'$ for a sat. cl. S' . But by maximality $S = S'$, so $U \subseteq S$ i.e. e sub f_0 , $f_0 \in \Phi(S)$.

§ 2. Satisfaction classes substitutable in axiom scheme of the theory T .

If a satisfaction class can be used as a parameter in (an instance of) the induction scheme, then it is closed (see proposition 1.1) and has all the properties proof of which requires induction. Using a classical idea, we can characterise those substitutable classes by the following lemma (we fix notation to be used in the sequel: given a model $\mathfrak{M}, \mathfrak{A}, \ldots, M, A, \ldots$ is (respectively) its universe; given a model $\mathfrak{M}$ and $S \subseteq M$, Def $(\mathfrak{M},S)$ is the expansion of $\mathfrak{M}$ obtained by adding all classes (subsets of M) definable with parameters over $(\mathfrak{M},S)$):

Lemma 2.1. Let $\mathfrak{M}$ be a model for T and S a sat. cl. over $\mathfrak{M}$. Then $(\mathfrak{M},S) \models$ "Induction for formulae of $L_T(S)$" iff $\mathrm{Def}(\mathfrak{M},S) \models (\forall X)(0 \in X \wedge (\forall x)(N(x) \wedge x \in X \rightarrow x+1 \in X) \rightarrow (\forall x)(N(x) \rightarrow x \in X))$.

For the rest of the present section let us assume that T is axiomatizable by a scheme (by virtue of Vaught [67] it suffices that T is recursively axiomatizable). The familiar examples are: P (with induction), Z (with comprehension), ZF (with replacement), A_2 and KM (with class existence). We will distinguish the sat. classes that can be used as a parameter in the scheme. The above lemma suggests a definition which requires generalising of the building GB from ZF to other theories.

Given any theory T with a definable ordered pair and axiomatisable by a scheme, we form a theory T_{pr} in the second order language L'_T (cf. the beginning of § 1) by adding axioms assuring the existence of predicatively defined classes and by replacing the scheme of T with an appropriate Π-sentence (implying all the instances of the scheme). Details are given in Krajewski [74].

<u>Definition</u> 2.1 A sat. cl. S over a model $\mathfrak{M}$ for T is <u>substitutable</u> iff $\mathrm{Def}(\mathfrak{M},S) \models T_{pr}$. In the case of ZF S is called a GB-satisfaction class.

If we assume that L_T has a finite signature, then a kind of syntactical counterpart of definition 2.1 is given in

<u>Definition</u> 2.2. $\overline{T_{pr}}$ is the theory $T_{pr} + (\exists S)$("S is a full. sat. cl.") ; $\overline{T}$ is the theory $\overline{T_{pr}} \cap L_T$. We write $\overline{GB}$ for $\overline{ZF_{pr}}$

We have a classical.

<u>Proposition</u> 2.1. (a) $\overline{T_{pr}} \subseteq T_{pr} + \Sigma^1_1$-scheme of class existence
(b) $\overline{T_{pr}} \subseteq T_{pr} + \Delta^1_1$-scheme of class existence + Σ^1_1 - induction scheme.

On the other hand $T_{pr} + \Delta^1_1$ - scheme of class existence is not sufficient for a proof of existence of a full sat. cl. This follows from Barwise and Schlipf [75], where it is proved that there are models $\mathfrak{A}$ for P such that $\mathrm{Def}(\mathfrak{A}) \models P_{pr} + \Delta^1_1$ - scheme of class existence. And the famous Tarski theorem states that no full sat. cl. is definable over any model. In connection with this, let us show, that over some models of ZF, say, there are no sat. classes S with $\Phi(S) \supseteq \{f: f \text{ sub } e\}$ (for some e) and such that S is definable with parameters (this proves a conjecture from Mostowski [50]). For a proof consider an ultrapower of a standard model for ZF (modulo an ultrafilter on ω) and take as e (the equivalence class of) the sequence $(\phi_i)_{i<\omega}$, where ϕ_i is the Lévy's formula defining satisfaction for Σ_i-formulae in the Lévy hierarchy.

It is important to realize that $\mathbb{T}$ is stronger than T. The following variant of some results by Tarski and Mostowski holds true.

Proposition 2.2. If T contains the induction scheme, then $\mathbb{T} \vdash \mathrm{Cons}(T)$.

The idea of the proof is to show by induction that all the theorems of T (as considered inside T) are true in the sense of a given substitutable satisfaction class.

I would like to stress that $\overline{T}_{pr}$ seems to be an interesting example of an impredicative expansion of T. E.g. the Ehrenfeucht [66] proof that finitely generated models of P have no expansions to models of A_2 refers in fact to $\overline{P}_{pr}$ only and not to the whole A_2 (by the way: Barwise and Schlipf [75] obtained stronger results using other methods). Moreover $\overline{P}_{pr}$ is a relatively weak natural theory strong in the sense of Wilkie [73].

Let us recall now that a well known result of Montague and Vaught strenghtens proposition 2.2. for ZF. Namely $\overline{GB} \vdash \forall_\beta \exists_\alpha (\beta < \alpha \wedge \text{"}R_\alpha \prec V\text{"})$, where R_α is the set of sets of rank $< \alpha$ and "$R_\alpha \prec V$" abbreviates "$(\forall e)(\forall x)_{R_\alpha} (\langle e,x\rangle \in S \leftrightarrow R_\alpha \models e[x])$", where S is the (unique, cf. proposition 1.2) full sat. cl. It follows, as an important corollary, that the full sat. cl. S_{tr} over the minimal (standard) model for ZF in not a GB-sat. cl.

Anticipating further sections, we show now an example of a model for ZF with a full GB-sat. cl. and a full non-GB-sat. cl. (inconsistent with the former). Let θ be an inaccesible cardinal. Let ξ be the least η such that $R_\eta \prec R_\theta$. Clearly: $R_\theta \models \overline{ZF}$ and S_0 $(= S_{tr})$ over R_ξ is not a GB-sat. cl., while S_1 $(= S_{tr})$ over R_θ is a GB-sat. cl. By virtue of the Keisler-Shelah isomorphism theorem there exists a cardinal λ and ultrafilter $\mathcal{D}$ such that

$$R_\xi^\lambda/\mathcal{D} \cong R_\theta^\lambda/\mathcal{D} \quad .$$

S_0 and S_1 determine in the obvious way the required sat. classes over the ultrapower.

We have seen that for some models of $\mathbb{T}$ no full substitutable sat. cl. exists. However, each model of $\mathbb{T}$ is elementarily extendable to a model with such a class. This follows from a general lemma which will be used again in § 5.

Lemma 2.2. Let $Q \subseteq L_T'$ be consistent.
(a) If $\mathfrak{M} \models Q \cap L_T$, then there exists $\mathfrak{N}$ such that $\mathfrak{M} \prec \mathfrak{N}$ and $\mathfrak{N}$ is a Q-model, i.e. some model $\mathfrak{A}$ for Q restricted to the language L_T is isomorphic to $\mathfrak{N}$.
(b) If $\mathfrak{M} \models Q \cap L_T$ and $\mathfrak{M}$ is special of power $|\mathfrak{M}| = 2^{|\mathfrak{M}|}$, then $\mathfrak{M}$ is a Q-model (see Chang-Keisler [73] for properties of special models).
Analogous results hold for $Q \subseteq L_T(P_1,\ldots,P_k)$, where $P_1,\ldots,P_k$ are new predicate symbols.

Proof. (a) The set $Q \cup \mathrm{Th}((\mathfrak{M},c)_{c\in M})$ is consistent by virtue of Robinson consistency lemma because $(Q \cup \mathrm{Th}(\mathfrak{M})) \cap \mathrm{Th}((\mathfrak{M},c)_{c\in M}) =$ $= \mathrm{Th}(\mathfrak{M})$ is complete.
(b) Let $\mathfrak{M}_0$ be a countable elementary submodel of $\mathfrak{M}$. It follows from (a) that there is a countable model $\mathfrak{M}_1$ and a countable family $\mathcal{F} \subseteq P(M_1)$ such that $(\mathcal{F},\mathfrak{M}_1,\epsilon) \models Q$. Let $X \subseteq M_1$ codes $\mathcal{F}$, i.e. $\mathcal{F} = \{X^{(a)}: a \in M_1\}$, where for $a \in M_1$: $x \in X^{(a)} \iff \langle a,X\rangle^{\mathfrak{M}_1} \in X$. It is easy to see that there exists an effective map $o : L_T' \to L_T(P)$ (P is a new predicate symbol interpreted in $\mathfrak{M}_1$ as X) such that for all $\varphi \in L_T'$ $(\mathcal{F},\mathfrak{M}_1,\epsilon) \models \varphi \iff (\mathfrak{M}_1,X) \models \varphi^o$. Take now special model $(\mathfrak{N},V) \equiv (\mathfrak{M}_1,X)$ of power $|\mathfrak{M}|$. Hence $\mathfrak{N} \cong \mathfrak{M}$, so for some $Z \subseteq M$ it holds that $(\mathfrak{M},Z) \models \varphi^o$ for $\varphi \in Q$, whence $(\{Z^{(a)}: a \in M\}, \mathfrak{M},\epsilon) \models Q$.

§ 3. Complete satisfaction classes and external satisfaction.

Nonstandard finite is infinite in the external world. That is why it seems reasonable to consider infinite (i.e. nonstandard finite) generalisations of Tarski's conditions. The following definition is formulated in the language $L_T(S)$.

Definition 3.1. (a) A sat. cl. S is $\wedge$-complete if for any formula e of the shape $e_0 \wedge (e_1 \wedge (\ldots \wedge e_{p-1}) \ldots)$ and any x: fr(e)

$$\langle e,x\rangle \in S \iff (\forall 1)_p \langle e_1, x|_{fr(e_1)}\rangle \in S$$

and the same holds for every other distribution of parenthesis in e.
(b) <u>$\vee$-completeness</u> is defined analogously.
(c) S is <u>E-complete</u> iff for any formula e of the shape $(\exists v_0) \ldots (\exists v_{p-1}) f$ and every $x: fr(e)$, $\langle e,x\rangle \in S \Leftrightarrow (\exists s)$ ("s is a sequence with members $\langle l, s_l\rangle$, where $0 \leq l \leq p-1$ and $v_l \in fr(f)$" $\wedge \langle f, x^\frown s\rangle \in S)$, and the same holds for any other string of variables, not only $v_0,\ldots,v_{p-1}$, provided that no collision of variables occurs.
(d) S is <u>weakly complete</u> iff S is $\wedge$-complete, $\vee$-complete and E-complete.

<u>Lemma</u> 3.1. If S is a sat. cl. and induction for formulae of $L_T(S)$ holds, then S is weakly complete.

<u>Proof</u>. Fix $e = e_0 \wedge (e_1 \wedge (\ldots \wedge e_{p-1}) \ldots)$ and $x: fr(e)$. Let $\varphi(i)$ be

$$\langle e_i \wedge (e_{i+1} \wedge (\ldots \wedge e_{p-1})\ldots),\; x|\ldots\rangle \in S\,.$$

If $\langle e,x\rangle \in S$, then by induction $(\forall i)_p\, \varphi(i)$, whence $(\forall i)_p \langle e_i, x|_{fr(e_i)}\rangle \in S$.
If $(\forall i)_p \langle e_i, x|_{fr(e_i)}\rangle \in S$, then by induction $(\forall i)_{p+1}\, \varphi(p-i)$, whence $\varphi(0)$.

Other parts of the proof are similar.

Examples of non-trivial non-weakly complete sat. classes are given in § 4. To show a trivial one, let p be a non-standard natural number (for a fixed $\mathfrak{M} \models T$).
Let e_i be "$v_i = v_i$" for $i = 0,\ldots,p-1$.
Let e^j be $e_j \wedge (e_{j+1} \wedge (\ldots \wedge e_{p-1})\ldots)$ for <u>standard</u> j.
Then

$$U = \{\langle e_i,x\rangle: x: fr(e_i),\ i<p\} \cup \{\langle \neg\, e^j,x\rangle: x: fr(e^j),\ j=0,1,2,\ldots\}$$

is a non-$\wedge$-complete sat. cl. Note that U may be extended to a strong (so including S_{tr}), unlimited sat. cl. (thm. 1.1).

To treat an arbitrary sequence of quantifiers $\forall$ and $\exists$ A.Robinson (in [63]) used the idea, now well known, borrowed from game theory.

It is provable in P that each formula e is of the shape qe_0, where q is a string of quantifiers and e_0 does not begin with a quantifier, and both q and e_0 are uniquely determined. For such a sequence q let q_A be the sequence whose members are indices of variables quantified by $\forall$ in q and similarly q_E (we assume that no collision of variables occurs).

The following definition is in metatheory (in "the world").

Definition 3.2. A function Ψ is a Skolem operator for a formula e of the shape qe_0 iff $\Psi : \forall(q) \to \exists(q)$, where

$$\forall(q) = \{x\colon x \text{ is a sequence with members } \langle l, x_l\rangle \text{ for } l \in q_A\}$$

$$\exists(q) = \{x\colon x \text{ is a sequence with members } \langle l, x_l\rangle \text{ for } l \in q_E\},$$

and

for every $x, x' \in \forall(q)$ and each $m \in q_E$ the following holds: if $(\forall l)(l \in q_A \wedge$ "v_l occurs in q before v_m" $\to x_l = x'_l$, where $\langle l, x_l\rangle \in x$, $\langle l, x'_l\rangle \in x'$), then $y_m = y'_m$, where $\langle m, y_m\rangle \in \Psi(x)$, $\langle m, y'_m\rangle \in \Psi(x')$.

The name "Skolem operator" was used in Robinson [63]. Formulas and sequence are taken from a fixed model $\mathfrak{M}$ for T. This is also the case in the next definition that generalizes the definition of weak completeness.

Definition 3.3. (a) A sat. cl. S (over a given model for T) is complete w.r.t. a collection $\mathfrak{K}$ of Skolem operators iff S is $\wedge$-complete $\vee$-complete and for every formula e of the shape qe_0 and every $x\colon \mathrm{fr}(e)$

$$\langle e, x\rangle \in S \Longleftrightarrow \text{there exists } \Psi \in \mathfrak{K} \text{ such that for every } y \in \forall(q)$$
$$\langle e_0, x \frown y \frown \Psi(y)\rangle \in S .$$

(b) S is complete iff S is complete w.r.t the family of all Skolem operators.

It is possible to code small collections of Skolem operators as a class of the model. The next definition is formulated in $L_T(S,X)$ (or L'_T, cf. § 1).

Definition 3.4. A sat. cl. S is complete w.r.t. X iff S is $\wedge$-complete, $\vee$-complete and for every $e = qe_0$ and $x\colon \mathrm{fr}(e)$

$$\langle e, x\rangle \in S \Longleftrightarrow (\exists a)(\text{"}X^{(a)} \text{ is a Skolem operator for } e\text{"} \wedge$$
$$\wedge (\forall y)(y \in \forall(q) \to \langle e_0, x \frown y \frown X^{(a)}(y)\rangle \in S))$$

The classical method of skolemization proves that S_{tr} is always complete. Skolem operator is a sequence of Skolem functions. Moreover $Def(\mathfrak{M}, S_{tr}, \prec) \models$ "S_{tr} is complete", if $\mathfrak{M} \models T$ and $\prec$ is a well ordering of $\mathfrak{M}$.

The following definition (by induction in "the world") is due to A. Robinson.

<u>Definition</u> 3.5. Let be a model for T . The class of <u>simple</u> formulae (over $\mathfrak{M}$) is the least class of formulae containing all atomic formulae and satisfying the following conditions.

(i) if e is simple, then so is $\neg e$

(ii) if x is a finite sequence (in $\mathfrak{M}$) of simple formulae, then both

$$(x)_0 \wedge (x)_1 \wedge \ldots \wedge (x)_{Lh(x)-1} \quad \text{and} \quad (x)_0 \vee (x)_1 \vee \ldots \vee (x)_{Lh(x)-1}$$

are simple for all admissible distributions of parenthesis (i.e. yielding a formula in the sense of $\mathfrak{M}$) .

(iii) if e is simple and q is a finite sequence (in $\mathfrak{M}$) of quantifiers $\forall$ or $\exists$ then the formula qe is simple.

Simultaneously <u>Robinson</u> <u>rank</u> (abbreviated $Rr(\cdot)$) of a simple formula is defined: $Rr(e) = 0$ for atomic e , $Rr(\neg e) = Rr(qe) = Rr(e) + 1$, where writing "qe" we mean that e does not begin with quantifier, and finally $Rr((x)_0 \wedge \ldots \wedge (x)_{k-1}) = \max_{0 \leq i < k} Rr((x)_i) + 1$ and the same for $\vee$.

It is noteworthy that every formula in prenex normal form has Robinson rank ≤ 4 .

Robinson defined by induction so called external satistaction of simple formulae of finite Robinson rank.

<u>Definition</u> 3.6. Let $\mathfrak{M}$ be a model for T . Let e be a simple formula in $\mathfrak{M}$, let $Rr(e)$ be (really) finite and let $x \in M$ and $\mathfrak{M} \models x\colon fr(e)$.

Then: e is <u>externally satisfied by x</u> (in $\mathfrak{M}$) iff either e is atomic and the clause (1) of definition 1.2 (iii) holds ,

or e is $\neg f$ and f is not externally satisfied by x ,

or e is a conjuction and each conjunct is externally satisfied by the suitable restriction of x ,

or e is a disjunction and some disjunct is externally satisfied by the suitable restriction of x ,

or e is qe_0 (and e_0 does not begin with quantifier) and there exists (in "the world") a Skolem operator Ψ such that for each

$y \in \forall(q)\ e_0$ is externally satisfied by $x \frown y \frown \Psi(y)$.

It may seem that one can extend the definition, putting for any formula e .

e is satisfied by x iff $(e)_n$ is satisfied externally by x, where $(e)_n$ is(the least,say)formula in prenex normal form logically equivalent to e . However, Robinson proved that this is not true, which will be shown in § 5.

§ 4. Automorphisms and mutually inconsistent satisfaction classes.

We shall use the following formulas d^k_j , where k,j are natural numbers and $j < k$:

$$d^k_j(v_0): (\exists v_j)\ldots(\exists v_k)(v_1=0 \wedge (v_2=v_1+1 \wedge (\ldots \wedge (v_k=v_{k-1}+1 \wedge v_0=v_k+1)\ldots)$$

It is easy to prove in P (by induction) that a formula of this shape exists for all k,j $(j<k)$. d^k_j is an arithmetical formula and so (according to the assumptions of § 0) may be treated as a formula of L_T . In non-ω-models of T d^k_j is non-standard for non-standard k.

The following lemma is easy to prove.

Lemma 4.1. Let T contain the induction scheme and $\mathfrak{M}$ be a model for T . If S is a weakly complete sat. cl. over $\mathfrak{M}$, then for all standard j , all k and any $a_0,\ldots,a_j$

$$\langle d^k_{j+1},[\langle 0,a_0\rangle,\ldots,\langle j,a_j\rangle]\rangle \in S \leftrightarrow a_0=k \wedge a_1=0 \wedge \ldots\ a_j=j-1 .$$

Our first general result on inconsistent sat. classes is as follows.

Theorem 4.1. Let $\mathfrak{M}$ be a non-ω-model for T and let $\varkappa$ be the power of $\mathfrak{M}$. Then

(a) There are $2^{\varkappa}$ pairwise inconsistent strong unlimited sat. classes over $\mathfrak{M}$.

(b) There are $2^{\varkappa}$ pairwise inconsistent sat. classes over $\mathfrak{M}$ such that from each $\varkappa$ other classes are definable (with parameters) over $\mathfrak{M}$ and, moreover, there exist among them $\varkappa$ classes definable (with parameters) from any other one.

Proof. Fix a non-standard integer k. For any $A \subseteq M$ (M is the universe of $\mathfrak{M}$) let S^A be the class

$$\{\langle d_j^k,[\langle 0,t\rangle,\langle 1,0\rangle,\dots,\langle j-1,j-2\rangle]\rangle : t \in A,\ j=1,2,3,\dots\} \cup$$
$$\cup \{\langle \neg d_1^k,[\langle 0,t'\rangle]\rangle : t' \notin A\} .$$

S^A is a sat. cl. and S^A is inconsistent with $S^{A'}$ for $A \neq A'$. Applying theorem 1.1 we get (a).

Now, notice that S^A is definable predicatively over $\mathfrak{M}$ from k, A and ω (the class of standard natural numbers). At the same time for any (nonempty) A ω is definable from S^A as the class of the indices of free variables of formulae in $D(S^A)$. Hence if A is definable from A' with parameters $p_0,\dots,p_n$, then S^A is definable from $S^{A'}$, k, $p_0,\dots,p_n$. Also, if $A = \{a\}$, then S^A is definable from any $S^{A'}$ (with parameters a, k).

The above theorem is rather trivial. We are going to prove a less trivial result concerning classes deciding all sentences (so all formulas with a standard number of free variables).

Theorem 4.2. Let T contain the induction scheme, $\mathfrak{M}_0$ be a model for T and S_0 a weakly complete sat. cl. over $\mathfrak{M}_0$ deciding all sentences (e.g. $\mathfrak{M}_0$ is a ω-model, $S_0 = S_{tr}$). Then for any cardinal λ there exists a model $\mathfrak{M}$ of power λ and $S \subseteq M$ such that $(\mathfrak{M},S) \equiv (\mathfrak{M}_0,S_0)$ and there exists 2^λ pairwise inconsistent sat. classes over $\mathfrak{M}$, which decide all sentences and are consistent for sentences.

Proof. By the Ehrenfeucht-Mostowski theorem there exists a model $(\mathfrak{M},S)$ such that $(\mathfrak{M}_0,S_0) \equiv (\mathfrak{M},S)$ (or even $(\mathfrak{M}_0,S_0) \prec (\mathfrak{M},S)$ in the case when $\lambda \geq |\mathfrak{M}_0|$). in $\mathfrak{M}$ there is a set of indiscernibles of order type $(\omega^* + \omega)\cdot\lambda$, each element of the set is a natural number (in $\mathfrak{M}$) and finally $\mathfrak{M}$ is the closure of the set under Skolem functions. Therefore there are 2^λ automorphisms of $\mathfrak{M}$ generated by automorphisms of $(\omega^*+\omega)\cdot\lambda$ and for each such automorphism a we can define a class S_a as follows:

$$\langle e,x\rangle \in S_a \iff \langle e,\ a * x\rangle \in S \wedge e \in \Phi_a ,$$

where

$$\Phi_a = \{e:\ 1 < Lh(fr(e)) \to a(1) = 1\}$$

(i.e. the number of free variables of e is hereditarily a-invariant) and $a * x$ is the sequence with members $\langle i, a(x_i)\rangle$ for $\langle i, x_i\rangle$ being members of the sequence x. It is not difficult but somewhat long to prove that such a sequence always exists and that S_a is a sat, cl. (some details are given in Krajewski [74 a], where the present theorem is proved.

To finish the proof it is sufficient to note that

$$\langle d_1^k,\ [\langle 0,u\rangle]\rangle \in S_a <-> a(u) = k$$

by lemma 4.1.

It is worthwhile to note that the proof works when we start with the class S_{ext} of external satisfaction (instead of S_0) over any model $\mathfrak{M}_0$. We get then the satisfaction classes deciding all simple sentences (of finite Robinson rank), inconsistent for a simple formula d_1^k.

<u>Corollary</u> 4.1. Let us add to the hypothesis of theorem 4.2 the assumption that $\mathfrak{M}_0$ is pointwise definable. Then from each pair of the constructed sat. classes infinitely many others are definable (without parameters) over $\mathfrak{M}$.

<u>Proof</u>. According to the construction the class S_a is definable from (the graph of) the automorphism a. We can show that given any two distinct S_a and S_b, the authomorphism $a^{-1}b$ is definable from them. Then all (finite) iterations of $a^{-1}b$ are definable too. The definition of "$a^{-1}b(x) = y$" is as follows:

$$(*)\quad (\exists e)(\langle(\exists! v_0)e,\ \emptyset\rangle \in S_a \wedge \langle e,[\langle 0,x\rangle]\rangle \in S_b \wedge \langle e,[\langle 0,y\rangle]\rangle \in S_a)$$

If $a^{-1}b(x) = y$, then $a(y) = b(x)$ and for some e $\langle e \wedge (\exists! v_0)e,\ [\langle 0,a(y)\rangle]\rangle \in S$, because $\mathfrak{M}_0$ was pointwise definable. It means that (*) holds.
Conversely, if (*) holds, then $a(y)$ should equal $b(x)$.

Finally, one more remark. Let T contain induction $\mathfrak{M}$ be a model for T and a an automorphism of $\mathfrak{M}$ such that $a(m) \neq m$ for some integer m (in the sense of $\mathfrak{M}$). Then $\langle d_1^m,[\langle 0,m\rangle]\rangle \in S_{ext}$

and $\langle \neg\, d_1^m, [\langle 0, a(m)\rangle]\rangle \in S_{ext}$. Therefore we can say that such an automorphism does not preserve a certain "non-standard property" corresponding to the formula d_1^m .

§ 5. Mutually inconsistent full satisfaction classes

We shall deal first with sat. classes inconsistent for sentences. Recall that M is the universe of $\mathfrak{M}$, etc.

Theorem 5.1. (a) Let $\mathfrak{M}$ be a model for T , S be a full sat. cl. over $\mathfrak{M}$, X be an arbitrary subset of M (which will be specified in applications). For every cardinal λ there exists a model $\mathfrak{N}$ of power λ , sets X_α, $S_\alpha \subseteq N$ for $\alpha < \lambda$ such that $(\mathfrak{M}, S, X) \equiv$ $\equiv (\mathfrak{N}, S_\alpha, X_\alpha)$ for all $\alpha < \lambda$, and S_α is inconsistent with S_β for a sentence (in prenex normal form, if desired) for all distinct $\alpha, \beta < \lambda$.
(b) One can weaken the assumption in (a) to $\mathfrak{M} \models L_T \wedge$ (T + "S is a full sat. cl.") and simultaneously strenghten the conclusion requiring the existence of an automorphism $a_{\alpha\beta}$ of $\mathfrak{N}$, such that $a_{\alpha\beta}(S_\alpha) = S_\beta$ for $\alpha, \beta < \lambda$.

Proof. (a) Only a slight modification of the proof given in Krajewski [75] is necessary. Let $Th((\mathfrak{M}, S, X))$ in the language $L_T(U,V)$ be denoted by $A(U,V)$.

It is sufficient to prove the consistency of the set

$$B_\lambda : \bigcup_{\alpha<\lambda} A(S_\alpha, X_\alpha) \cup \bigcup_{\alpha<\beta<\lambda} \{(\exists x)(\text{"x is in prenex normal form"} \wedge \langle x, \emptyset\rangle \in S_\alpha \wedge \langle \neg\, x, \emptyset\rangle \in S_\beta)\}$$

By compactness, it is sufficient to prove the consistency of the sets B_n , $n = 1, 2, \ldots$, defined as B_λ but with n instead of λ. We poceed by induction.

If B_n is consistent, then so is the set

$$\Sigma' : \bigcup_{i<n} A(S_i', X_i') \cup \{(R'(x) \leftrightarrow \text{"x is in prenex normal form"} \wedge \bigwedge_{i<n} \langle x, \emptyset\rangle \in S_i'\} \cup$$

$$\cup \bigcup_{i<j<n} \{(\exists x)(\text{"x is in prenex normal form"} \wedge \langle x, \emptyset\rangle \in S_i \wedge \langle \neg x, \emptyset\rangle \in S_j)\}$$

let Σ'' arises from Σ' by replacing the predicates $S_0', \ldots, S_{n-1}', R'$ with new predicates $S_0'', \ldots, S_{n-1}'', R''$.

We claim that the following set is consistent:

$$(*) \qquad \Sigma' \cup \Sigma'' \cup \{(\exists x)(R'(x) \wedge \neg R''(x))\} .$$

Otherwise the beth theorem gives a formula $\varphi \in L_T$ such that

$$\Sigma' \vdash (\forall x)(R'(x) \leftrightarrow \varphi(x)) .$$

But then for any model $\mathfrak{A}$ for Σ' and $\Psi \in L_T$ in prenex normal form

$$\mathfrak{A} \models \Psi \leftrightarrow \langle \ulcorner \Psi \urcorner , \emptyset \rangle \in S_i \text{ for all } i < n \leftrightarrow R(\ulcorner \Psi \urcorner) ,$$

where $\ulcorner \Psi \urcorner$ is the Gödel number of Ψ . This contradicts the Tarski theorem.

Now, any model for (*) determines a model for B_{n+1} .

(b) The first claim follows from the lemma 2.2. For a proof of the second claim consider an elementary extension of the model constructed in (a) to a special model of a power $\mu > \lambda$

$$(\mathfrak{M}^+ , S^+_\alpha , X^+_\alpha , c)_{\alpha<\lambda, c\in N}$$

Because $(\mathfrak{M}^+, S^+_\alpha) \equiv (\mathfrak{M}^+, S^+_\beta)$ (for all $\alpha,\beta < \lambda$) , there exists an authomorphism $a^+_{\alpha\beta}$ of $\mathfrak{M}^+$ such that $a^+_{\alpha\beta}(S^+_\alpha) = S^+_\beta$. Applying the downward Löwenheim-Skolem-Tarski theorem to the structure

$$(\mathfrak{M}^+ , S^+_\alpha , X^+_\alpha , a_{\alpha\beta}, c)_{\alpha,\beta<\lambda,\ c\in N} ,$$

we get the required model.

To illustrate the point (b) let us prove.

<u>Proposition</u> 5.1. Let Δ be the theory ZF + "S is a full sat. cl." . Then $L_{ZF} \cap \Delta = ZF$.

<u>Proof</u>. In the hope of finding a contradiction, assume that for some sentence $\varphi \in L_{ZF} \cap \Delta$ the theory $ZF + \neg\varphi$ is consistent. Let $\models ZF + \neg\varphi$. Let ZF_k $(k = 1,2,3,\ldots)$ be the theory arising from ZF by restricting the replacement scheme to formulas with at the most k quantifiers. By virtue of the reflection principle $\mathfrak{M} \models (\exists_\alpha)(\text{"}R_\alpha \models ZF_k + \neg\varphi)$, $k = 1,2,\ldots$. Therefore for each k

there exists $\alpha_k \in M$ such that the submodel $\mathfrak{M}_k$ of $\mathfrak{M}$ with the universe $R_{\alpha_k}^{\mathfrak{M}}$ satisfies $ZF_k + \neg\varphi$. Moreover,

$$\mathfrak{M} \models (\forall_\alpha)(\exists x)(x \subseteq R_\alpha \wedge \text{"}x \text{ is a full sat. cl. over } (R_\alpha, \in)\text{"}) ,$$

whence for each k there exists $x_k \in M$ such that $x_k^* = \{a \in M ; \mathfrak{M} \models a \in x_k\}$ is a full sat. cl. over $\mathfrak{M}_k$ (we have to use suitable absoluteness results).

Let $\mathcal{D}$ be an ultrafilter on ω. The model $\mathfrak{N} = \prod_{k\in\omega} \mathfrak{M}_k/\mathcal{D}$ admits a full sat. cl., so it satisfies $\Delta \cap L_{ZF}$. On the other hand $\mathfrak{N} \models \neg\varphi$. But $\varphi \in \Delta \cap L_{ZF}$, a contradiction.

If in theorem 5.1 one drops the requirement of the inconsistency for sentences, demanding simple inconsistency, then one can strenghten the number of classes significantly. As a matter of fact this follows from generalized definability theory (see Chang-Keisler [73]), in particular Shelah [71]. Repeating Shelah's proof with a minor modification one arrives at the following. For any cardinal λ let $\varkappa(\lambda)$ be $\min\{\nu: 2^\nu > \lambda\}$ (e.g. $\varkappa(\aleph_0) = \aleph_0$, $\varkappa(\aleph_1) = \aleph_0 \Leftrightarrow \neg$ CH).

<u>Theorem</u> 5.2. (a) Let be a model for T, S be a full sat. cl. over $\mathfrak{M}$ and $X \subseteq M$. For every cardinal $\lambda \geq |\mathfrak{M}|$ there exists a model $\mathfrak{N}$ of power λ and $S_\alpha, X_\alpha \subseteq N$ for $\alpha < 2^{\varkappa(\lambda)}$ such that $(\mathfrak{M}, S, X) \prec (\mathfrak{N}, S_\alpha, X_\alpha)$ for each $\alpha < 2^{\varkappa(\lambda)}$ and S_α is inconsistent with S_β for a formula with one free variable (for all distinct α, β).
(b) as in theorem 5.1.

The proof is based on the existence of a tree with $\leq \lambda$ models and $2^{\varkappa(\lambda)}$ maximal branches (see Shelah [71]). Assuming this, one applies a lemma analogous to theorem 5.1 to get an elementary chain of models indexed by the ordinals less than the height of the tree, in such a way that the α-th model admits as many pairwise inconsistent for a formula with one free variable satisfaction classes as there are elements on the α-th level of the tree. Now, it is sufficient to take the union of the models and unions of these classes along maximal branches.

Finally, we show the result on external satisfaction mentioned in § 3. Recall that $(e)_n$ is a formula in prenex normal form logically equivalent to e.

Corollary 5.1. (A.Robinson). There exists a model $\mathfrak{N}$ for T such that a (non-standard) tautology of the shape "$\neg e \Leftrightarrow (\neg e)_n$" is externally false.

Proof. Let $\mathfrak{M}$ be an ω-model and let $X \subseteq M$ be such that S_{tr} is complete w.r.t. X (see § 3). Applying theorem 5.1, we get a model $\mathfrak{N}$ and $S_0, S_1, X_0, X_1 \subseteq N$ such that $(\mathfrak{N}, S_i, X_i) \equiv (\mathfrak{M}, S_{tr}, X)$, $i = 0,1$, and $\langle e, \emptyset \rangle^{\mathfrak{N}} \in S_0$, $\langle \neg e, \emptyset \rangle^{\mathfrak{N}} \in S_1$, for a sentence e in prenex normal form. By virtue of the elementary equivalence S_i is complete w.r.t. X_i $(i = 0,1)$ and $\langle (\neg e)_n, \emptyset \rangle^{\mathfrak{N}} \in S_1$ (because $\langle (\forall e)(e \Leftrightarrow (e)_n), \emptyset \rangle^{\mathfrak{M}} \in S_{tr}$). We can conclude that e and $(\neg e)_n$ are both externally true over $\mathfrak{N}$, applying the following:

Lemma 5.1. Let S be a sat cl. complete w.r.t. x over a model $\mathfrak{M}$ for T. Let e be a formula in prenex normal form. If $\langle e, x \rangle^{\mathfrak{M}} \in S$, then s is externally satisfied by x over $\mathfrak{M}$.

Proof. First for quantifier-free e using $\wedge$ - and $\vee$ - completeness. Then the rest follows from definitions.

The above proof is essentially the Robinson [63] proof. Our formulations seem to be more convenient and are more general. In particular, theorem 5.1 has probably not appeared in literature in this generality.

§ 6. Indicating the applications.

1. Our first application concerns T_{pr} - expansions of models of T, i.e. models of T_{pr} (see § 2) obtained by adding classes to a given model of T. Assume that T contains the full induction scheme. Recall that definition of $\mathbb{T}$ is given in § 2.

Theorem 6.1. (cf. Krajewski [75]). If $\mathfrak{M} \models \mathbb{T}$, then for any cardinal $\lambda \geq |\mathfrak{M}|$ there exists a model $\mathfrak{N}$ of power λ such that $\mathfrak{M} \prec \mathfrak{N}$ and $\mathfrak{N}$ has $2^{\varkappa(\lambda)}$ pairwise incompatible T_{pr}-expansions (i.e. no two of then are included in a common T_{pr} - expansion of $\mathfrak{N}$).

The theorem follows from theorem 5.2(b). We can take $\mathrm{Def}(\mathfrak{N}, S_\alpha)$, $\alpha < 2^{\varkappa(\lambda)}$, where S_α is substitutable for each α (because of a lemma analoguous to lemma 2.1). Incompatibility follows from proposition 1.2. W. Marek observed that one can use Zorn Lemma to obtain at least $2^{\varkappa(\lambda)}$ maximal (w.r.t. inclusion) T_{pr} - expansions of $\mathfrak{N}$.

Assume now, that L_T has a finite signature (and that T contains full induction). Below "$\omega^{\mathfrak{M}}$" denotes $\{x \in M: \mathfrak{M} \models N[x]\}$.

Theorem 6.2. Let $\mathfrak{M}$ be a non-ω-model for T and S a substitutable full sat. cl. over $\mathfrak{M}$. There exists a family $(\mathfrak{A}_n)_{n \in \omega^{\mathfrak{M}}}$ of T_{pr} - expansions of $\mathfrak{M}$ such that for all $n,m \in \omega^{\mathfrak{M}}$

$$\mathfrak{A}_n \subsetneq \mathfrak{A}_m \iff \mathfrak{M} \models (n < m)^N .$$

The proof uses familiar ideas but is rather long . It follows from the theorem that any linear ordering is embeddable into a family of T_{pr} - expansions (ordered by inclusion) of some models for T .

2. The second application is to "very" non-standard models of ZF . The observation that there are models of ZF isomorphic to some of their R_α's can be strenghtened as follows.

Theorem 6.3. Let $\mathfrak{M}$ be a model of ZF and $\lambda \geq |\mathfrak{M}|$. There exists a model $\mathfrak{N}$ of power λ such that $\mathfrak{M} \prec \mathfrak{N}$ and for some cardinal τ $\mathfrak{N} = \bigcup_{\nu<\tau} \mathfrak{N}_\nu$, where $\mathfrak{N}_\nu \prec \mathfrak{N}_\mu$ for $\nu < \mu < \tau$, $\mathfrak{N}_\nu \cong \mathfrak{N}$ for $\nu < \tau$ and moreover $\mathfrak{N}_\nu$ has as its universe $\{x \in N: \mathfrak{N} \models x \in R_{\alpha_\nu}\}$ for an ordinal (in the sense of $\mathfrak{N}$) α_ν and finally

$$\mathfrak{N} \models (R_{\alpha_\nu} \neq R_{\alpha_\mu}) \qquad \text{for} \qquad \nu \neq \mu .$$

In particular every special model of ZF has this property.

A proof of this result (as well as of the preceding theorem) appears in the author's PhD thesis (in Polish).

3. Sat. classes can be also applied to some syntactical problems concerning the theory T . For example the theory: GB without the replacement axiom plus the axiom: $(\forall X)(\mathrm{Def}(X) \to (\forall y)(\exists z)(X''y = z))$, where $\mathrm{Def}(X) \iff (\exists S)(\exists e,k,a)(\text{"S is a sat.cl."} \wedge$

$$\wedge (\forall x)(x \in X \iff \langle e, a^\frown\langle k,x\rangle\rangle \in S)) ,$$

is a finitely axiomatisable theory strictly included in GB (and including ZF). Let us mention also the paper of Mostowski ([50]) , where it is shown that induction fails in GB and a more recent paper of Vopenka and Hàjek [73], where interpretations of GB in GB are considered.

References

I. Barwise, J.Schlipf [1975] , On Recursively Saturated Models of Arithmetic, preprint.

C.C. Chang, H.J. Keisler [1973] , Theory of Models, North Holland.

A. Ehrenfeucht, G. Kreisel [1966] , Strong Models of Arithmetic, Bull. Acad. Pol. Sci., 14, 107-110.

S. Krajewski [1974] , Predicative Expansions of Axiomatic Theries, Zeitsch. Math. Log. Gr. Math. 20, 435-452.

S. Krajewski [1974a] , Mutually Inconsistent Satisfaction Classes, Bull. Acad. Pol. Sci. 22, 883-887.

S. Krajewski [1975] , A Note on Expansions of Models for Set Theories, Proceedings of Karpacz 1974 Logic Conference.

R. Montague [1961] , Semantic Closure and Non-finite Axiomatizability, in: Infinistic Methods, PWN and Pergamon Press.

A. Mostowski [1950] , Some Impredicative Definitions in Axiomatic Set Theory, Fund. Math. 37, 111-124.

A. Robinson [1963] , On Languages Based on Non-standard Arithmetic, Nagoya Math. J. 22, 83-107.

S. Shelah [1971] , Remark to Local Definability Theory of Reyes, Ann. Math. Log. 2, 441-448.

A. Tarski [1933] , Pojęcie Prawdy w Językach Nauk Dedukcyjnych, nakładem Towarzystwa Naukowego Warszawskiego.

R.L. Vaught [1967] , Axiomatizability by a Schema, J. Symb. Log. 32, 473-479.

P. Vopenka, P. Hàjek [1973] , Existence of a Generalised Semantic Model of GB , Bull. Acad. Pol. Sci. 21, 1079-1086.

A. Wilkie [1973] , Arithmetical Parts of Strong Theories, preprint.

EHRENFEUCHT GAMES FOR GENERALIZED QUANTIFIERS

by A. Krawczyk and M. Krynicki

§ 0. Introduction

In [4] Ehrenfeucht gave a game theoretical characterization of elementary equivalence with respect to a first order language. A similar characterizations for the language $L(Q_\alpha)$ are due independently to Lipner [7], Brown [2], and Vinner [9], for the language with Henkin quantifier to Krynicki [5], for the language with Malitz quantifier to Badger [1] (as we read in [8]). The idea of all these games is rather uniform. Our paper grew out this observation. This paper was born during the 1975 Conference in Helsinki, where the authors could not get accustomed to polar night. We wish to thank our Finnish Friends for the inwitation to this conference. The authors wish to thank prof. A. Lachlan for fruitful discussions.

Structure $\langle A, A_1, \ldots, A_k \rangle$ s.t. $A_i \subseteq A^{n_i}$ we call structure of type $\langle n_1, \ldots, n_k \rangle$. Let L denote a first order language with equality allowing constant symbols but without functions and quantifier symbols. By a quantifier Q of type $\langle n_1, \ldots, n_k \rangle$ we mean some class of structures of this type. Formulae of language $L(Q)$ are defined by induction:

1. $L \subseteq L(Q)$

2. If $\varphi_0, \ldots, \varphi_{k+2}$ $L(Q)$ then $\varphi_{k+1} \vee \varphi_{k+2}$, $\neg \varphi_{k+1}$, $(Q\bar{x}_1, \ldots, \bar{x}_k)\ (\varphi_1, \ldots, \varphi_k) \in L(Q)$ provided no free variable of φ_i occurs in $\bar{x}_j$ for $i \neq j$.

We use Lindström [6] definition of satisfaction:

$$\underline{\underline{A}} \models (Q\bar{x}_1,\dots,\bar{x}_k)(\varphi_1,\dots,\varphi_k) \iff$$

$$\langle A, \{\bar{a} \in A^{n_1} : \underline{\underline{A}} \models \varphi_1[\bar{a}]\}, \dots, \{\bar{a} \in A^{n_k} : \underline{\underline{A}} \models \varphi_k[\bar{a}]\}\rangle \in Q$$

By Q_c we denote the Chang quantifier, by Q_I the Hartig quantifier, by Q_α^n the Malitz one. For definitions see [8]. By Q_H we denote the Henkin quantifier defined in [3] as follows:

$$(Qxyuv)\ \varphi(x,y,u,v) \iff \begin{pmatrix} \forall x \ \exists u \\ \forall y \ \exists v \end{pmatrix} \varphi(x,y,u,v)$$

Suppose the formulae and semantics of the language $L(Q_1,\dots,Q_n)$ have been defined and Q_{n+1} is a quantifier of some finite type we obtain $L(Q_1,\dots,Q_{n+1})$ from $L(Q_1,\dots,Q_n)$ in the same way as $L(Q)$ from L.

We use standard model theoretical notation.

§ 1. Quantifiers of the type $\langle q\rangle$

Definition 1.1. The quantifier Q of type $\langle q\rangle$ is called monotone if for every $\langle A,R\rangle \in Q$ and every $R \subseteq R' \subseteq A^q$ $\langle A,R'\rangle \in Q$

For example $\exists, \forall, Q_\alpha, Q_H, Q_\alpha^n$ are monotone.

Definition 1.2. The class $K \subseteq Q$ is a support for Q if for arbitrary $\langle A,R\rangle \in Q$ exists R' s.t. $R' \subseteq R$ and $\langle A,R'\rangle \in K$.

Examples: The quantifier "there existe exactly $\aleph_\alpha$" is a support for Q_α. The Machover quantifier $Q_M = \{\langle A,R\rangle : \bar{R} = \overline{A-R}\}$ is a support for Q_c. The class K s.t. $\langle A,R\rangle \in K$ iff $R \subseteq A^4$ and there are $f,g \in A^A$ s.t. $\langle a,b,c,d\rangle \in R$ $f(a) = c$ and $g(b) = d$ is a support for Q_H.

Let $\underline{\underline{A}}, \underline{\underline{B}}$ be structures with the same signature. By partial isomorphism we mean a partial function from the universe of $\underline{\underline{A}}$ into the universe of $\underline{\underline{B}}$ s.t. the domain of that function contain constans of $\underline{\underline{A}}$ and function preserve relations.

Assume now that Q is fixed monotonic quantifier of type $\langle q\rangle$ and K its support. The Ehrenfeucht game appropriate to such a quantifier is given by the following:

Definition 1.3. Consider the game $G_n(\underline{\underline{A}},\underline{\underline{B}})$ played by two players and having n moves. In each move player 1 chooses a structure (e.g. $\underline{\underline{A}}$) and its subset $(R \subseteq A^q)$ s.t. $\langle A,R\rangle \in K$, then player 2 chooses a subset $R' \subseteq B^q$ (if $R \subseteq B^q$ then $R' \subseteq A^q$). Later player 1 chooses a q-tuple $\langle b_1,\dots,b_q\rangle \in R'$ and then player 2 chooses a q-tuple $\langle a_1,\dots,a_q\rangle \in R$. So, we may, that after the n moves we have a partial mapping $\{\langle a_i,b_i\rangle : i=1,\dots,qn\}$ from the universe of $\underline{\underline{A}}$ into the universe of $\underline{\underline{B}}$. If this mapping is extendable to a partial isomorphism player 2 has won. Otherwise player 1 won.

Definition 1.4. $\underline{\underline{A}} \sim_n \underline{\underline{B}}$ iff player 2 has a winning strategy in the game $G_n(\underline{\underline{A}},\underline{\underline{B}})$.

By the rank of the formula $\varphi \in L(Q)$ $(\mathrm{rk}(\varphi))$ we mean the number of quantifiers occuring in φ. By $\underline{\underline{A}} \equiv_n \underline{\underline{B}}$ we denote that $\mathrm{Th}_n(\underline{\underline{A}}) = \mathrm{Th}_n(\underline{\underline{B}})$, where $\mathrm{Th}_n\, \underline{\underline{A}} = \{\varphi \in L(Q) :\ \varphi\text{-sentence},\ \mathrm{rk}(\varphi) \leq n,\ \underline{\underline{A}} \models \varphi\}$.

Theorem 1.1. If $\underline{\underline{A}} \sim_n \underline{\underline{B}}$ then $\underline{\underline{A}} \equiv_n \underline{\underline{B}}$.

Proof. By induction with respect to n. The case $n=0$ is obvious. Assume that $\underline{\underline{A}} \sim_{n+1} \underline{\underline{B}}$ and there is a sentence φ of rank $n+1$ s.t. $\underline{\underline{A}} \models \varphi$, $\underline{\underline{B}} \models \neg\varphi$. We may assume that φ is of the form $(Q\overline{x})\,\psi(\overline{x})$. Let $R_1 = \{\overline{a} \in A^q: \underline{\underline{A}} \models \psi[\overline{a}]\}$. Hence $\langle A,R_1\rangle \in Q$ and there is $R \subseteq R_1$ s.t. $\langle A,R\rangle \in K$. In the first move of the game $G_{n+1}(\underline{\underline{A}},\underline{\underline{B}})$ let player 1 choose this R. So player 2 using winning strategy choose R'. Claim that there is $\overline{b} \in R'$ s.t. $\underline{\underline{B}} \models \neg\,\psi[\overline{b}]$. Indeed otherwise $R' \subseteq \{\overline{b}: \underline{\underline{B}} \models \psi[\overline{b}]\}$ and by monotonicity of Q $\underline{\underline{B}} \models (Q\overline{x})\,\psi(\overline{x})$. Let player 1 choose such $\overline{b}$ and player 2 applaying his winning strategy returb $\overline{a}$. Hence $\langle \underline{\underline{A}},\overline{a}\rangle \sim_n \langle \underline{\underline{B}},\overline{b}\rangle$ and by induction hypothesis $\langle \underline{\underline{A}},\overline{a}\rangle \equiv_n \langle \underline{\underline{B}},\overline{b}\rangle$. This contradicts the fact that $\underline{\underline{A}} \models \psi[\overline{a}]$ and $\underline{\underline{B}} \models \neg\,\psi[\overline{b}]$.

Corollary. If for every n $\underline{\underline{A}} \sim_n \underline{\underline{B}}$ then $\underline{\underline{A}} \equiv_{L(Q)} \underline{\underline{B}}$.

It is well known that converse implication doesn't hold without additional assumption.

Now let τ denote a fixed finite signature. We define two finite sequences $\{l_p^{n,\tau}\}_{p=1,..,n}$, $\{m_p^{n,\tau}\}_{p=1,..,n}$ for each n.

$l_0^{n,\tau} = 0$

$m_{p+1}^{n,\tau}$ = "the number of formulas of rank $l_p^{n,\tau}$ in which occur at most the first $l_p^{n,\tau} + (n-p)\,q$ variables".

$l_{p+1}^{n,\tau} = 2^{m_{p+1}^{p,\tau}} \cdot l_p^{p,\tau} \cdot m_{p+1}^{n,\tau} + 1$

Claim: If τ' is an extension of τ and contains at most q new constants then for $p = 0,\ldots,n-1$

(1) $\qquad l_p^{n,\tau} \geqslant l_p^{n-1,\tau'}$

Theorem 1.2. Let $\underline{\underline{A}},\underline{\underline{B}}$ be structures of type τ. If $\underline{\underline{A}} \equiv_{l_p^{n,\tau}} \underline{\underline{B}}$ then $\underline{\underline{A}} \sim_n \underline{\underline{B}}$.

Proof. By induction with respect to n. For $n = 0$ it is obvious. Let $\underline{\underline{A}} \equiv_{l_{n+1}^{n+1,\tau}} \underline{\underline{B}}$ and $\varphi_0,\ldots,\varphi_{s-1}$ be formulas of rank $l_n^{n+1,\tau}$ in which occur at most the first $l_n^{n+1,\tau} + q$ variables and exactly q free variables. Naturally $s \leqslant m_{n+1}^{n+1,\tau}$.

Let for $\varepsilon \in 2^s$ $\quad \theta_\varepsilon = \bigwedge_{i<s} \varphi_i^{\varepsilon(i)}$ where $\varphi_i^0 = \varphi_i$, $\varphi_i^1 = \neg\varphi_i$

Note that:

(2) $\qquad \varepsilon_1 \neq \varepsilon_2 \Rightarrow$ for all $\bar{a} \in A^q$ $\quad \underline{\underline{A}} \models \neg\,\theta_{\varepsilon_1} \wedge \theta_{\varepsilon_2}[\bar{a}]$

and $\underline{\underline{A}} \models \bigvee_{\varepsilon\in 2^s} \theta_\varepsilon[\bar{a}]$

(3) $\qquad rk(\theta_\varepsilon) \leqslant s \cdot l_n^{n+1,\tau} \leqslant m_{n+1}^{n+1,\tau} \cdot l_n^{n+1,\tau}$

Assume that player 1 chooses $A_1 \subseteq A^q$ s.t. $\langle A,A_1\rangle \in K$. Let $J = \{\varepsilon \in 2^s : (\exists\bar{a})_{A_1}\ \underline{\underline{A}} \models \theta_\varepsilon[\bar{a}]\}$. Hence $A_1 \subseteq \{\bar{a} \in A^q : \underline{\underline{A}} \models \bigvee_{\varepsilon\in J} \theta_\varepsilon[\bar{a}]\}$

But Q is monotonic so $\underline{\underline{A}} \models (Q\bar{x}) \bigvee_{\varepsilon\in J} \theta_\varepsilon$. On the other hand

$rk((Q\bar{x}) \bigvee_{\varepsilon\in J}\theta_\varepsilon) \leqslant 2^s \cdot rk(\theta_\varepsilon) + 1 \leqslant 2^{m_{n+1}^{n+1,\tau}} \cdot m_{n+1}^{n+1,\tau} \cdot l_n^{n+1,\tau} + 1 = l_{n+1}^{n+1,\tau}$

and $\underline{\underline{B}} \models (Q\bar{x}) \bigvee_{\varepsilon\in J}\theta_\varepsilon$. By choice of the class K there exists $B_1 \subseteq B^q$ s.t. $B_1 \subseteq \{\bar{b} \in B^q : \underline{\underline{B}} \models \bigvee_{\varepsilon\in J}\theta_\varepsilon[\bar{b}]\}$ and $\langle B,B_1\rangle \in K$. This B_1 is adequate for player 2 reply. Now let $\bar{b} \in B_1$ be chosen by player 1, by (2) and definition B_1 there is exactly one $\varepsilon \in J$ such that $\underline{\underline{B}} \models \theta_\varepsilon[\bar{b}]$ so player 2 should choose $\bar{a} \in A_1$ s.t.. $\underline{\underline{A}} \models \theta_\varepsilon[\bar{a}]$. Then $\langle \underline{\underline{A}}, a\rangle \equiv_{1_n^{n+1},\tau} \langle \underline{\underline{B}}, \bar{b}\rangle$ which by (1) implice $\langle \underline{\underline{A}}, \bar{a}\rangle \equiv_{1_n^{n},\tau'} \langle \underline{\underline{B}}, \bar{b}\rangle$ where τ' is an expansion of τ and contains q new constans. By inductive assumption we obtain that $\langle \underline{\underline{A}}, \bar{a}\rangle \sim_n \langle \underline{\underline{B}}, \bar{b}\rangle$ which completes the proof.

Note that the relation $\sim_n$ for a monotonic quantifier does not depend on the choice of support.

Definition 1.5. Let $Q_1, \dots, Q_k$ be monotonic quantifiers, $\underline{\underline{A}}, \underline{\underline{B}}$ structures of the same type. The game $G^n_{Q_1,..,Q_k}(\underline{\underline{A}}, \underline{\underline{B}})$ is the following game for two gamblers. In every move player 1 chooses $i \leqslant k$ and both players play according to Definition 1.3 with Q_i in place of Q .

Theorem 1.3. Let $Q_1, \dots, Q_k$ be monotonic quantifiers, $\underline{\underline{A}}, \underline{\underline{B}}$ structures of the same finite signature. Then $\underline{\underline{A}} \equiv_{L(Q_1,..,Q_k)} \underline{\underline{B}}$ iff for each n player 2 has winning strategy in the game $G^n_{Q_1,..,Q_k}(\underline{\underline{A}}, \underline{\underline{B}})$.

The proof is a slight elaboration of the preceding one.

Until now we have assumed that Q is a monotonic quantifier. Now we prove that defined games doesn't work for arbitrary quantifier. More precisely we prove the following theorem,

Theorem 1.4. Let Q be a quantifier of type $\langle q\rangle$, $Q' = \{\langle A,R\rangle : \text{exists } R' \subseteq R \text{ and } \langle A,R'\rangle \in Q\}$. Then either $\underline{\underline{A}} \equiv_{L(Q)} \underline{\underline{B}}$ iff $\underline{\underline{A}} \equiv_{L(Q')} \underline{\underline{B}}$ or the games $G^n(\underline{\underline{A}}, \underline{\underline{B}})$ are not adequate for the quantifier Q .

Proof. It is enough note that Q is support for Q' .

Question. Is it possible to find a "natural" game theoretical characterization of elementary equivalence for an arbitrary quantifier of type $\langle q\rangle$ (e.g. for quantifier "there exists exactly $\aleph_\alpha$")?

§ 2. Quantifier of the type $\langle q_1, q_2 \rangle$

We restrict our discussion to the case of a quantifier of type $\langle 1,1 \rangle$, but all results hold for general case as well.

Definition 2.1. Quantifier Q of type $\langle 1,1 \rangle$ is monotonic if for arbitrary $\langle A, A_1, A_2 \rangle \in Q$ and arbitrary $A_1' \subseteq A_1$, $A \supseteq A_2' \supseteq A_2$ $\langle A, A_1', A_2' \rangle \in Q$

Note that by this definition Q_I is not monotonic but the following quantifier is monotonic: $(Qx)(\varphi,\psi) \Longleftrightarrow \overline{\{x: \varphi(x)\}} \leqslant \overline{\{x: \psi(x)\}}$. The language with this quantifier is a little stonger then the language with Q_I .

Definition 2.2. Now we define the games $G^n(\underline{\underline{A}},\underline{\underline{B}})$ as in Definition 1.3 with the following changes: the first step player 1 chooses one of the two structures (e.g. $\underline{\underline{A}}$) and A_1, A_2 s.t. $\langle A, A_1, A_2 \rangle \in Q$, player 2 replye choosing B_1, B_2 s.t. $\langle B, B_1, B_2 \rangle \in Q$ and player 1 chooses $i \in \{1,2,-1,-2\}$ and $b \in B_i$ where $B_{-j} = B - B_j$. Finally player 2 chooses $a \in A_i$.

Theorem 2.1. Let Q be a monotonic quantifier of type $\langle 1,1 \rangle$. Then $\underline{\underline{A}} \sim_n \underline{\underline{B}}$ implies $\underline{\underline{A}} \equiv_n \underline{\underline{B}}$.

Proof. As in Theorem 1.1. Let $rk(\varphi) = n+1$, $\underline{\underline{A}} \models \varphi$, $\underline{\underline{B}} \models \neg\varphi$, $\varphi = (Qx)(\varphi_1,\varphi_2)$ and player 1 choose $A_i = \{a \in A: \underline{\underline{A}} \models \varphi_i[a]\}$, player 2 using winning straregy reply B_1, B_2 . Let $B_i' = \{b \in B: \underline{\underline{B}} \not\models \varphi_i[b]\}$ hence $B_1' \neq B_1$ or $B_2' \neq B_2$. If for example $B_1' - B_1 \neq \emptyset$ then player 1 choose -1 and $x \in B_1' - B_1$. One can prove that this procedure followed the contradiction.

A converse theorem analogous to Thm. 1.2 holds as well.

Let τ be a fixed finite signature. Similarly as above we define sequences $\{l_p^{n,\tau}\}_{p=0,..,n}$, $\{m_p^{n,\tau}\}_{p=1,..,n}$

$$l_0^{n,\tau} = 0$$

$m_{p+1}^{n,\tau}$ = "the number of formulas of the rank $l_p^{n,\tau}$ in which occur at most the first $l_p^{n,\tau} + n - p$ variables".

$$l_{p+1}^{m,} = (3 \cdot l_p^{n,\tau} \cdot m_{p+1}^{n,\tau} + 2) \cdot 2^{m_{p+1}^{n,\tau^p}} + 1$$

Theorem 2.2. Let $\underline{\underline{A}},\underline{\underline{B}}$ be a structures of the signature τ. Then for each monotonic quantifier Q of type $\langle 1,1\rangle$ we have:

If $\underline{\underline{A}} \equiv_{l_n^{n,\tau}} \underline{\underline{B}}$ then $\underline{\underline{A}} \sim_n \underline{\underline{B}}$.

Proof. By induction. For $n=0$ it is obvious.
Assume $\underline{\underline{A}} \equiv_{l_{n+1}^{n+1,\tau}} \underline{\underline{B}}$. Let $\varphi_0,\varphi_1,\dots,\varphi_{s-1}$ be all formulas of the rank $l_n^{n+1,\tau}$ with one free variable and which occur at most first $l_n^{n+1,\tau}+1$ variables.
As above, for $\varepsilon \in 2^s$ let $\theta_\varepsilon = \bigwedge_{i<s} \varphi_i^{\varepsilon(i)}$. Note, that for $l_p^{n,\tau}$ $m_p^{n,\tau}$ and θ_ε defined in such a manner the conditions (1),(2),(3) from section 1 hold.
We will give a winning stategy for player 2 in the game $G^{m+1}(\underline{\underline{A}},\underline{\underline{B}})$.
In the first move let player 1 choose $A_1,A_2 \subseteq A$ s.t.

(4) $\langle A, A_1, A_2\rangle \in Q$

For $i=1,2$ let

$$J_0^i = \{\varepsilon \in 2^s\colon \text{ for all } a\in A \text{ if } \underline{\underline{A}} \models \theta_\varepsilon[a] \text{ then } a \in A_i\}$$

$$J_1^i = \{\varepsilon \in 2^s\colon \text{ there are } a\in A_i \text{ and } b\notin A_i \text{ s.t. } \underline{\underline{A}} \models \theta_\varepsilon[a]\wedge\theta_\varepsilon[b]\}$$

Let r (t) be a number of elements of J_1^1 (J_1^2).
By (4) and monotonicity of Q we have:

$$\underline{\underline{A}} \models (\exists y_1\dots y_r)(\exists z_1\dots z_t)\ \bigwedge_{\substack{i\leq r\\ \varepsilon^i\in J_1^1}} \theta_{\varepsilon^i}(y_i) \wedge \bigwedge_{\substack{i\leq t\\ \varepsilon^i\in J_1^2}} \theta_{\varepsilon^i}(z_i) \wedge$$

$$\wedge (Qx)\Big[\bigvee_{\varepsilon\in J_0^1} \theta_\varepsilon(x) \vee \bigvee_{i\leq r} x=y_i\ ,\ \bigvee_{\varepsilon\in J_0^2\cup J_1^2} \theta_\varepsilon(x)\wedge \bigwedge_{i\leq t} x\neq z_i\Big]$$

It is easy to see that $rk(\theta) \leq l_{n+1}^{n+1,\tau}$. Hence $\underline{\underline{B}} \models \theta$ and there are $b_1,\dots,b_r$, $c_1,\dots,c_t \in B$ s.t.

$$\underline{\underline{B}} \models (Qx)\Big[\bigvee_{\varepsilon\in J_0^1} \theta_\varepsilon(x) \vee \bigvee_{i\leq r} x=b_i\ ,\ \bigvee_{\varepsilon\in J_0^2\cup J_1^2} \theta_\varepsilon(x)\wedge \bigwedge_{j\leq t} x\neq c_j\Big]$$

and

$$\underline{\underline{B}} \models \bigwedge_{i \leq r} \theta_{\varepsilon^i}[b_i] \wedge \bigwedge_{j \leq t} \theta_{\varepsilon^j}[c_j]$$

Let player 2 choose the following subsets

$$B_1 = \{b : \underline{\underline{B}} \models \bigvee_{\varepsilon \in J_0^1} \theta_\varepsilon[b] \vee \bigvee_{i \leq r} b = b_i\}$$

$$B_2 = \{b : \underline{\underline{B}} \models \bigvee_{\varepsilon \in J_0^2 \cup J_1^2} \theta_\varepsilon[b] \wedge \bigwedge_{j \leq t} b \neq c_j\}$$

Let player 1 choose $i \in \{1,2,-1,-2\}$ and $b \in B_i$. There exists exactly one ε s.t. $\underline{\underline{B}} \not\models \theta_\varepsilon[b]$. From definition of B_1 follows that we may find $a \in A_i$ s.t. $\underline{\underline{A}} \models \theta_\varepsilon[a]$. Hence $\langle A, a\rangle \equiv_{1_n^{n+1},\tau} \langle \underline{\underline{B}}, b\rangle$ and $\langle \underline{\underline{A}}, a\rangle \equiv_{1_n^n,\tau'} \langle \underline{\underline{B}}, b\rangle$ where $\tau' = \tau \cup \{c\}$.

So, from the inductive assumption $\langle \underline{\underline{A}}, a\rangle \sim_n \langle \underline{\underline{B}}, b\rangle$.

References

[1] B.W. Badger, The Malitz quantifier meets its Ehrenfeucht game, Ph. D. Thesis, University of Colorado, (1975).

[2] W. Brown, Infinitary languages, generalized quantifier and generalized products, Ph. D. Thesis, Dartmouth (1971),

[3] L. Henkin, Some remarks on infinitely long formulas, in Infinitic Methods, p.167-183, (1961).

[4] A. Ehrenfeucht, An application of games to the completeness problem for formalized theories, Fund. Math. 49 (1961), p.129-141.

[5] M. Krynicki, Henkin quantifier and decidability, to appear in Proceed. of the Symposium in Helsinki (1975).

[6] P. Lindstrom, First order logic and generalized quantifier, Theoria 32 (1966), p. 186-195.

[7] L.D. Lipner, Some aspects of generalized quantifiers, Ph. D. Thesis, Berkeley (1970).

[8] A. Slomson, Decision Problems for Generalized Quantifiers A Survey, This volume.

[9] S. Vinner, A generalization of Ehrenfeucht's game and some applications, Israel J. Math. 12 (1972), p.279-298.

DIMENSION AND TOTALLY TRANSCENDENTAL THEORIES OF RANK 2 .

by A. H. Lachlan
Simon Fraser University
and
Instytut Matematyczny, PAN, Warszawa

This paper is concerned with the totally transcendental first--order theories and the notions of rank and degree introduced by Morley [10]. In the terminology of Shelah such theories are just those which are $\aleph_0$-stable. From Marsh [9] and later work of Baldwin and the author [3] it is clear that the idea of the dimension of a model or of a formula in a model is one that is useful for analysing the structure of models of totally transcendental theories. A set A in a model M is called <u>independent</u> if for any distinct members $a, \bar{b}$ of A and formula $\varphi(x,\bar{y})$, $M \models \varphi(a,\bar{b})$ implies that the formula $\varphi(x,\bar{b})$ has the same rank as the universe, i.e. as $x = x$. A model can be said to have dimension if all maximal independent sets have the same cardinality. Our main result is that every model of rank 1 or of rank 2 and degree 1 has dimension but that for models of higher rank or of rank 2 and degree > 1 there may be maximal independent sets of different cardinalities.

As a byproduct of our investigation of dimension we provide an analysis of the structure of $\aleph_0$-stable theories for which the rank of the universe is 2 and the degree 1. Roughly speaking we can divide such theories into four groups distinguished as follows:

(i) there exists a sequence $\langle \varphi_i(x) : i < \omega \rangle$ of formulas of rank 1 such that for every rank 1 formula $\psi(x,\bar{a})$ there exists n such that $\psi(x,\bar{a}) \wedge \neg \varphi_n(x)$ is finite

(ii) there exists a formula $\varphi(x)$ of rank 2 such that the theory restricted to $\varphi(x)$ is $\aleph_1$-categorieal ; in this case we call $\varphi(x)$ an $\aleph_1$-categorical formula

(iii) there exists no $\aleph_1$-categorical formula of rank 2 but there exists a definable equivalence relation which has infinitely many infinite equivalence classes

(iv) there is no $\aleph_1$-categorical formula of rank 2 but there exists a definable relation which is expressible uniquely as the disjunction of two equivalence relations, each having infinitely many infinite equivalence classes, and in some model is an automorphism taking one equivalence relation onto the other.

Examples of these various kinds of theories are as follows. For (i) take the theory of countably many infinite disjoint unary relations. For (ii) we can take any disjoint union of a theory having rank 1 and a theory which is $\aleph_1$-categorical of rank 2 and degree 1. Specifically we could take $Th(M)$ where $|M| = \omega \cup (\omega \times \omega)$ and there are two unary function symbols F and G defined on M by

$$F(i) = G(i) = i \qquad F(\langle i,j\rangle) = G(\langle j,i\rangle) = \langle i,i\rangle$$

for all $i,j < \omega$. For (iii) we can take the theory of an equivalence relation having infinitely many infinite equivalence classes. For (iv) we can take $Th(\langle \omega \times \omega ; R\rangle)$ where R is the binary relation defined by

$$\langle i,j\rangle \, R \, \langle m,n \rangle \iff i = m \vee j = n$$

for all $i,j,m,n < \omega$.

We shall also be able to indicate what spectra can be associated with the theories under consideration, For instance a theory falling under (i) either has $|\alpha|$ models in every power $\aleph_\alpha \geq \aleph_\omega$ or $|\alpha|^\omega$. A theory falling under (ii) is either $\aleph_1$-categorical or has $|\alpha|$ models in every power $\aleph_\alpha \geq \aleph_\omega$. A theory satisfying (iv) also has $|\alpha|$ models in $\aleph_\alpha$ for $\alpha \geq \omega$. The situation for theories of type (iii) is more complicated but here again we shall be able to delineate all the possibilities.

The notions of dimension, rank, and degree may seem at first glance extremely technical and not worthy of investigation in their own right. However, in the right context their role is extremely natural. For instance if we choose the right examples our dimension can become either the dimension of a vector space over the rationals or the cardinality of a transcendence basis for an algebraically closed field. The naturalness of the notion of rank is witnessed by the finiteness of the rank of $\aleph_1$-categorical theories (Baldwin [2]) and also by the finite equivalence relation theorem of Shelah, see 1.7 below.

The plan of the paper is as follows. In § 1 are various definitions and technical lemmas which will be required in the later part of the paper. In § 2 we will establish that dimension is well-defined for theories of rank 1. We also present counterexamples showing that in general dimension is not well defined and make some general observations about dimension. For example, when there is an infinite independent set then dimension is uniquely defined. In § 3 we show that dimension is well-defined for all models when the universe has rank 2 and degree 1. In § 4 profiting from the analysis in § 3 we describe the possibilities for the structure of the models of theories having rank 2 and degree 1. At the same time we point out what spectra are possible for such theories. In § 4 we have had to leave many details to the reader to prevent the paper from becoming unconscionably long.

§ 1. Definitions and some useful lemmas.

In general we shall follow the notational conventions of Shelah [12]. In particular we shall suppose that all models considered are elementary submodels of some large and very saturated model $\overline{M}$ of whatever theory T we may be considering. Here all theories considered, in the absence of some specific remark to the contrary, will be complete, countable and $\aleph_0$-stable. If $\varphi(x,\overline{a})$ is a formula and M a model then $\varphi(M,\overline{a})$ denotes the set of solutions of $\varphi(x,\overline{a})$ in M. We shall sometimes use the quantifiers $\exists^{\geq\omega}x$, "there exist infinitely many x", and $\forall^{a}x$, "for all but a finite number of x". These are always to be interpreted in the large model referred to above. $I(\varkappa)$ denotes the number of nonisomorphic models of power $\varkappa$ of whatever theory we are considering.

1.1. Rank and degree. We define a mapping r of the set of all 1-types into On and a mapping d of the same set into ω. Here a satisfiable formula $\varphi(x,\overline{a})$ is seen as a 1-type with cardinality 1, thus r and d are also defined on all formulas having just x free. Now r is defined implicitly by:

$r(p) \geq \alpha + 1$ if and only if there exist disjoint formulas $\varphi_0(x,\overline{a}_0)$, $\varphi_1(x,\overline{a}_1)$, ... such that $r(p \cup \{\varphi_i(x,\overline{a}_i)\}) \geq \alpha$ for all

$i < \omega$.
and

$d(p)$ = the greatest $i < \omega$ such that there are disjoint formulas $\varphi_0(x,\bar{a}_0), \ldots, \varphi_{i-1}(x,\bar{a}_{i-1})$ such that $r(p \cup \{\varphi_j(x,\bar{a}_j)\}) = r(p)$ for all $j < i$.

The rank of Morley [10] will be denoted r^M and is given by: $r^M(p)$ = least α such that $r(\varphi(x,\bar{a})) = \alpha$ for some $\varphi(x,\bar{a})$ in p . The degree d^M defined by Morley turns out to be the same as d . To see the difference between r and r^M consider the theory of $M = \langle |M| ; R_i^M , i\langle\omega\rangle$ where $R_0^M, R_1^M, \ldots$ are infinite disjoint unary relations. Let $p = \{\neg R_i(x) : i < \omega\}$ then $r(p) = 1$ but $r^M(p) = 2$. Since r and r^M agree on formulas the notion of independent set is the same whether we work with r or r^M . By $r^M(a,A)$ we denote the Morley rank of the 1-type over A realised by a .

Although we have given the definition of rank only for 1-types and formulas $\varphi(x,\bar{a})$ it should be clear that r and d can be defined for n-types and formulas $\varphi(\bar{x},\bar{a})$ in an entirely analagous way.

Independent sets of arbitrarily large cardinality can easily be constructed. It is not clear to whom to attribute this result but it follows easily from considerations in Morley [10]. Choose $\psi(x,\bar{a})$ such that $r(\psi(x,\bar{a})) = r(x = x)$ and $d(\psi(x,\bar{a})) = 1$. Choose $\langle b_i : i < \varkappa\rangle$ such that for all $i < \varkappa, \models \psi(b_i,\bar{a})$ and

$$r^M(b_i, \{b_j : j < i\} \cup \mathrm{Rng}\ \bar{a}) = r(x = x) .$$

Then $\{b_i : i < \varkappa\}$ is an independent set of power $\varkappa$.

Below we shall sometimes write $r(\varphi(x,\bar{a}))$ and $d(\varphi(x,\bar{a}))$ where $\varphi(x,\bar{a})$ is a formula but not first-order. In every case the formula $\varphi(x,\bar{a})$ can be regarded as a 1-type p over $\mathrm{Rng}\ \bar{a}$ and by $r(\varphi(x,\bar{a}))$, $d(\varphi(x,\bar{a}))$ we mean $r(p)$, $d(p)$ respectively, For example, $r(\exists^{\geq\omega} y\, \varphi(y,x))$ means $r(p)$ where $p = \{\exists^{\geq n} y\, \varphi(y,x) : n < \omega\}$.

1.2. Strongly minimal types and formulas. A formula $\varphi(x,\bar{a})$ is called <u>strongly minimal (s.m.)</u> if $r(\varphi(x,\bar{a})) = d(\varphi(x,\bar{a})) = 1$. Such formulas were first considered by Marsh [9] and later exploited in [3], [7]. Here we find it convenient to extend the definition to types, thus p will be called s.m. if $r(p) = d(p) = 1$.

Let $\varphi(x,\bar{a})$, $\psi(x,\bar{b})$ be s.m., they are said to be linked if for all M , M'

$$\operatorname{Rng}\bar{a} \cup \operatorname{Rng}\bar{b} \subset M \subset M' \Rightarrow \varphi(M,\bar{a}) = \varphi(M',\bar{a}) \Leftrightarrow \psi(M,\bar{b}) = \psi(M',\bar{b}).$$

Linked s.m. formulas were used in [7]. It is easy to show that the s.m. formulas $\varphi(x,\bar{a})$ and $\psi(x,\bar{b})$ are linked if and only if there exists a formula $\chi(x,y,\bar{c})$ such that the following sentences are true

$$\forall x \forall y\ (\chi(x,y,\bar{c}) \rightarrow .\ \varphi(x,\bar{a}) \wedge \psi(y,\bar{b}))$$

$$\forall x \forall y\ (\exists^{<\omega} y\ \chi(x,y,\bar{c}) \wedge \exists^{<\omega} x\ \chi(x,y,\bar{c}))$$

$$\forall^{a} x\ \forall^{a} y (\varphi(x,\bar{a}) \rightarrow \exists y\ \chi(x,y,\bar{c})\ .\wedge .\ \psi(y,\bar{b}) \rightarrow \exists x\ \chi(x,y,\bar{c}))\ .$$

Further in the above definition instead of saying "for all M, M' " we could say "for some M and all M' " without altering the nation of linked s.m. sets. Also instead of $\exists^{<\omega}$ we can write $\exists^{<n}$ for sufficiently large n and $\forall^{a}$ may be similarly replaced. Generally, when restricted to formulas of rank 1 the quantifiers $\exists^{<\omega}$, $\forall^{a}$, $\exists^{\geq\omega}$ behave like first-order quantifiers because of:

1.3. LEMMA. Let $r(\varphi(x,\bar{a})) = 1$. For every formula $\vartheta(x,\bar{y})$ there exists m such that

$$\models \forall \bar{y}\ (\exists^{\geq m} x\ (\varphi(x,\bar{a}) \wedge \vartheta(x,\bar{y})) \rightarrow \exists^{\geq\omega} x\ (\varphi(x,\bar{a}) \wedge \vartheta(x,\bar{y})))\ .$$

Similarly when $\varphi(x,\bar{a})$ is replaced by a 1-type p such that $r(p)=1$.

<u>Proof</u>. Otherwise by compactness we can find $\bar{b}_i$, $i<\omega$, such that $\varphi(x,\bar{a}) \wedge \vartheta(x,\bar{b}_i) \wedge \bigwedge_{j<i} \neg\, \vartheta(x,\bar{b}^j)$ is infinite for each $i<\omega$. This would contradict $r(\varphi(x,\bar{a})) = 1$.

This lemma is very similar to Lemma 9 of [3]. Another easy but useful observation is that when $\varphi(x,\bar{a})$ is s.m. the quantifiers $\exists^{\geq\omega}$ and $\forall^{a}$ restricted to $\varphi(x,\bar{a})$ are the same

1.4. Restricting a theory to a formula. Let $\varphi(x,\bar{a})$ be an infinite formula, i.e. one with infinitely many solutions, we can form a new theory T' by restricting T to $\varphi(x,\bar{a})$ as follows. Let M be a model of T , $\bar{a} \in M$. Form a structure M' letting $|M'|=\varphi(M,\bar{a})$ and taking a relation symbol R_ϑ for each formula $\vartheta(\bar{x},\bar{a})$ of T having $\bar{a}$ as parameters. For $\bar{b} \in M'$ define $R^{M}_{\vartheta}(\bar{b})$ to hold if and only if $\models \vartheta(\bar{b},\bar{a})$. Then T' is $Th(M')$ and of course is independent of the choice of M . Of course, this construction does not require that T be totally transcendental. However for a totally trascendental theory every model M' of T' is obtanied in the above manner from a model of T . For an arbitrary countable theory this may fail for some uncountable M' .

The value of this idea is that notions pertaining to T or to a model M may be restricted to $\varphi(x,\bar{a})$. Thus we may speak of $\varphi(x,\bar{a})$ being $\aleph_1$-<u>categorical</u> meaning that T' is $\aleph_1$-categorical. Also if $A \subset \varphi(M,a)$ we may speak of A being <u>independent</u> <u>in</u> $\varphi(x,\bar{a})$ meaning that A is independent considered as a subset of M' . The operation of restricting the theory to a formula is particularly natural in the context of stable theories, because from Shelah [12 , Theorem 3.1] or Baldwin [1] we know that every relation on M' using parameters from M is already definable using parameters from M' .

It is useful to recall from [3] that if $\psi(x)$ is s.m. then T is $\aleph_1$-categorical if and only if for all M , M'

$$M \subset M' \wedge \psi(M) = \psi(M') \Longrightarrow M = M' .$$

In the next lemma it is important to bear in mind that b runs through $\mathbf{M}$ the very saturated model of which all models considered are elementary subsystems.

1.5. LEMMA. Let $r(\exists y\, \varphi(y,x)) = d(\exists y\, \varphi(y,x)) = 1$. Suppose that $r(\varphi(x,b)) = 1 = d(\varphi(x,b))$ and that $\exists y\, \varphi(y,x)$ and $\varphi(x,b)$ are linked for every b such that $\models \exists y\, \varphi(y,b)$. Then $\exists y\, \varphi(x,y)$ is $\aleph_1$-categorical.

Proof. Observe that $\exists y\, \varphi(x,y) \vee \exists y\, \varphi(y,x)$ is $\aleph_1$-categorical because if $M \subset M'$ and $\exists y\, \varphi(y,M) = \exists y\, \varphi(y,M')$,then $\varphi(M,b) = \varphi(M',b)$ for every b in M , whence $\exists y\, \varphi(M,y) = \exists y\, \varphi(M',y)$. But whenever an infinite formula $\psi(x,\bar{a})$ implies an $\aleph_1$-categorical formula $\vartheta(x,\bar{b})$, $\psi(x,\bar{a})$ is also $\aleph_1$-categorical. Thus in the present case $\exists y\, \varphi(x,y)$ is $\aleph_1$-categorical.

There are other situations where we shall want to observe that a formula is $\aleph_1$-categorical. In each case the inference will be straightforward provided one recalls the criterion for $\aleph_1$-categoricity mentioned above.

The next lemma is similar to Lemma 2 of [6] and may be obtained by the same methods. We omit the proof. The key idea is to translate the problem into one concerning ψ-rank for a suitable formula $\psi(x,y)$, see [4] for the definition of ψ-rank.

1.6. Normalisation Lemma (NL) . Let $\varphi(x,\bar{y})$ be a formula, $l(\bar{y}) = m$ and p be a complete m-type over $\emptyset$. There exists a formula $\varphi'(x,\bar{y})$ such that for all $\bar{b}_0, \bar{b}_1$ realising p , if $\varphi(x,\bar{b}_0)$ and $\varphi(x,\bar{b}_0) \wedge \varphi(x,\bar{b}_1)$ have the same rank and degree then $\models \forall x(\varphi'(x,\bar{b}_0) \longleftrightarrow \varphi'(x,\bar{b}_1))$ and $\varphi(x,\bar{b}_0)$, $\varphi'(x,\bar{b}_0)$, and $\varphi(x,\bar{b}_0) \wedge \varphi'(x,\bar{b}_0)$ all have the same rank and degree. Further for each $\bar{b}'$ realising p , $\varphi'(x,\bar{b}')$ is equivalent to a positive Boolean combination of formulas of the form $\varphi(x,\bar{b})$ where $\bar{b}$ realises p .

The formula $\varphi'(x,\bar{y})$ in the conclusion of the lemma is said to be obtained by <u>normalizing</u> $\varphi(x,y)$ <u>with</u> <u>respect</u> <u>to</u> p .

The next lemma is due to Shelah and follows easily from 1.6. Again we shall omit the proof.

1.7. Finite equivalence relation lemma. Let p be a 1-type over A not necessarily complete and $d(p) = k$. There exists a formula $\varepsilon(x,y,\bar{z})$, $\bar{a} \in A$, and $b_0,\dots,b_{k-1}$ such that $\varepsilon(x,y,\bar{a})$ defines an equivalence relation with finitely many equivalence classes and $r(p \cup \{\varepsilon(x,b_i,\bar{a})\}) = r(p)$ for each $i < k$ and for $i < j < k$ $\varepsilon(x,b_i,\bar{a}) \wedge \varepsilon(x,b_j,\bar{a})$ is inconsistent.

<u>Remark</u>. Of course in the conclusion of the lemma we must have $d(p \cup \{\varepsilon(x,b_i,\bar{a})\}) = 1$ for each $i < k$. Otherwise $d(p) = k$ would be contradicted. We refer to the types $p \cup \{\varepsilon(x,b_i,\bar{a})\}$ as the <u>components</u> of p . If we are dealing with a formula $\psi(x,\bar{a})$ which we may identify with $\{\psi(x,a)\}$ then the components are written $\psi(x,\bar{a}) \wedge \varepsilon(x,b_i,\bar{a})$. The significance of the lemma is that it shows that p can almost be split into components over A . The formula $\varepsilon(x,y,\bar{a})$ is said to <u>split</u> p . The equivalence classes corresponding to extensions of p of lower rank will be ignored when we use this lemma below.

Another lemma in the same vein as 1.6 and 1.7 is the following which allows us to illustrate the technique, referred to above in connection with 1.6., of translating a problem concerning Morley rank into one concerning rank with respect to a single formula. Let ψ-rank and ψ-degree be denoted by r_ψ and d_ψ respectively and see [4] for definitions.

1.8. LEMMA. Let $\varphi(x,\bar{y})$ be a formula, $\bar{b}$ be a finite sequence of elements. For every formula $\psi(x,\bar{z})$ there exists a formula $\vartheta(\bar{y},\bar{z})$ such that for all $\bar{c}$, $r(\varphi(x,\bar{b}) \wedge \psi(x,\bar{c})) < r(\varphi(x,\bar{b}))$ if and only if $\models \vartheta(\bar{b},\bar{c})$.

Proof. Let $\varphi(x,\bar{y})$ be $x = x$ without loss of generality. Choose $\chi(x,\bar{y}')$ and $\bar{b}'$ such that $\chi(x,\bar{b}')$ is a Boolean combination of instances of $\psi(x,\bar{z})$, $\chi(x,\bar{b}')$ has the same rank and degree as $x = x$. Make the choice first so as to minimise $r_\psi(\chi(x,\bar{b}'))$ and then so as to minimise $d_\psi(\chi(x,b'))$. By normalisation - of course we need a different lemma than 1.6 because here we are dealing with ψ-rank, but the principle is exactly the same - we can eliminate the parameters $\bar{b}'$. Now $r(\psi(x,\bar{c})) < r(x = x)$ is the same as $r_\psi(\chi(x) \wedge \neg\, \psi(x,\bar{c})) = r_\psi(\chi(x))$. We can show that this is equivalent to $\models \vartheta(\bar{c})$ for some formula $\vartheta(\bar{z})$ by the method of proof of [4, Lemma 4].

The next lemma provides a simplification of the problem of showing that dimension is well-defined for every model of a theory satisfying $r(x = x) = 2$, $d(x = x) = 1$.

1.9. LEMMA. If $r(\neg\, \vartheta(x,\bar{a})) < r(x = x)$ and $\vartheta(x,\bar{a})$ is $\aleph_1$-categorical then there exists an $\aleph_1$-categorical formula $\vartheta(x)$ such that $r(\neg\, \vartheta(x)) < r(x = x)$.

Proof. From NL we obtain $\vartheta'(x,y)$ such that $r(\neg\, \vartheta'(x,\bar{a})) < r(x = x)$ and for any $\bar{a}'$ realising the same type as $\bar{a}$ $\models \forall x(\vartheta'(x,\bar{a}) \leftrightarrow \vartheta'(x,\bar{a}'))$. By compactness it follows that $\vartheta'(x,\bar{a})$ is equivalent to some formula $\vartheta(x)$ containing no parameters. Further from NL, $\vartheta(x)$ is equivalent to a positive Boolean combination of formulas of the form $\vartheta(x,\bar{a}')$ where $\bar{a}'$ realises the same type as $\bar{a}$. Since the conjunction of all the constituent formulas is infinite $\vartheta(x)$ will also be $\aleph_1$-categorical as required.

1.10. Remark. Suppose that $d(x = x) = 1$ then any n-tuple of elements in an independent set realises a unique n-type. In this case, if there exists $\vartheta(x)$ such that $r(\vartheta(x)) = r(x = x)$ and $\vartheta(x)$ is $\aleph_1$-categorical, then the uniqueness of dimension follows from the homogeneity of models of $\aleph_1$-categorical theories [3, §5]. Thus from the lemma when $r(x = x) = 2$ and $d(x = x) = 1$ for the uniqueness of dimension will suffice the existence of an $\aleph_1$-categorical formula $\vartheta(x,\bar{a})$ of rank 2.

1.11. Uniform decomposition. A formula $\varphi(x,y)$ is called a uniform decomposition of the universe, or of $x = x$, if $r(\varphi(x,\bar{b})) < r(x = x)$ for every $\bar{b}$, and for every $\psi(x,\bar{a})$ having the same rank as $x = x$, all $\alpha < r(x = x)$, and all $n < \omega$ there exist $\bar{b}_i$, $i < n$, such that

$$r\,(\psi(x,\bar{a}) \wedge \varphi(x,\bar{b}_j) \wedge \bigwedge_{i<j} \neg\,\varphi(x,\bar{b}_i)) \geq \alpha \quad .$$

We shall be exclusively concerned with the case in which $r(x=x) = 2$ and $d(x=x) = 1$; however we state the following lemma in general terms.

1.12. <u>LEMMA</u>. Let $d(x = x) = 1$ and there be no uniform decomposition of $x = x$. Let $r^M(a,\emptyset) = r(x = x)$ then for all A, $r(x = x) \neq r^M(a,A) + 1$.

<u>Proof</u>. Let $\models \varphi(a,\bar{b})$ and $r(x = x) = r^M(a,\emptyset) = \alpha + 1 = r(\varphi(x,\bar{b})) + 1$. Let $\varphi'(x,\bar{y})$ be obtained by normalising $\varphi(x,\bar{y})$ with respect to the type p realised by $\bar{b}$. From 1.8. we can suppose that $r(\varphi'(x,\bar{c})) \leq \alpha$ for every $\bar{c}$. Since there is no uniform decomposition of $x = x$, for $\bar{c}$ realising p there are only a finite number of possibilities for $\varphi'(x,\bar{c})$ modulo equivalence. Thus $\varphi'(x,\bar{y})$ may be chosen so that for every $\bar{c}'$ there exists $\bar{c}$ realising p such that

$$\models \neg \exists x\, \varphi'(x,\bar{c}') \vee \forall x(\varphi(x,\bar{c}') \leftrightarrow \varphi(x,\bar{c})) \quad .$$

Let $\psi(x)$ denote $\exists \bar{y}\, \varphi'(x,\bar{y})$ then $\psi(x)$, being equivalent to a finite disjunction of formulas of rank α, itself has rank α. From the normalisation $r(\varphi(x,\bar{b}) \wedge \neg\,\varphi'(x,\bar{b})) < \alpha$, whence $r(\varphi(x,\bar{b}) \wedge \neg\,\psi(x)) < \alpha$. By assumption $\models \varphi(a,\bar{b})$ and $r^M(a,\emptyset) = \alpha + 1$, whence $\models \varphi(a,\bar{b}) \wedge \neg\,\psi(a)$. Thus $r(a,A) < \alpha$ for any A containing $\bar{b}$, which completes the proof.

For $i < 2$ let

$$\beta_i = \omega^{\alpha_0} n_{i,0} + \omega^{\alpha_1} n_{i,1} + \dots + \omega^{\alpha_m} n_{i,m}$$

where $\alpha_0 > \alpha_1 > \dots > \alpha_m$ and $n_{i,j} < \omega$. We define $\beta_0 \oplus \beta_1$ to be

$$\omega^{\alpha_0}(n_{0,0} + n_{1,0}) + \omega^{\alpha_1}(n_{0,1} + n_{1,1}) + \dots + \omega^{\alpha_m}(n_{0,m} + n_{1,m}).$$

Lascar[8]and Wierzejewski[16]independently noticed the importance of this operation for the theory of rank. As we remarked above the whole theory of rank operates equally well for formulas $\varphi(\overline{x},\overline{a})$, having $\overline{x} = x_0,\dots,x_{k-1}$ as free variables, as for formulas $\varphi(x,\overline{a})$ having only x free. The next lemma will be useful in § 2 because it shows that if for some fixed k , $\overline{x} = \overline{x}$, that is $x_0 = x_0 \wedge \dots \wedge x_{k-1} = x_{k-1}$, has rank $\geqslant \omega$, then $r(x = x) \geqslant \omega$.

1.12. LEMMA. $r(\bigwedge_{i<k} \varphi_i(x_i,\overline{a}_i)) = \bigoplus_{i<k} r(\varphi_i(x,\overline{a}_i))$ and $d(\bigwedge_{i<k} \varphi_i(x_i,\overline{a}_i)) = \prod_{i<k} d(\varphi_i(x,\overline{a}_i))$.

<u>Proof</u>. We need only consider the case $k = 2$ and further it is easily seen that we may suppose $d(\varphi_0(x,\overline{a}_0)) = d(\varphi_1(x,\overline{a}_1)) = 1$. A straightforward argument by induction on $r(\varphi_0(x,\overline{a}_0)) \oplus r(\varphi_1(x,\overline{a}_1))$ establishes that

$$r(\varphi_0(x_0,\overline{a}_0) \wedge \varphi_1(x_1,\overline{a}_1)) \geqslant r(\varphi_0(x,\overline{a}_0)) \oplus r(\varphi_1(x,\overline{a}_1)) \quad .$$

To obtain the opposite inequality we must prove something more general namely that, if $\langle a_0,a_1\rangle$ realises p , a_0 realises p_0 and a_1 realises a type p_1 over $\{a_0\}$, then $r^M(p) \leqslant r^{M_0}(p_0) \oplus r^M(p_1)$. This is straightforward if we proceed by induction on the ordinal on the right-hand side.

§ 2. Counterexamples, the case $r(x = x) = 1$, an alternative definition of dimension.

We recall that A is <u>independent</u> if for every formula $\varphi(x,\overline{y})$ and sequence $a,\overline{b}$ of distinct elements of A

$$[\models \varphi(a,\overline{b})] \Longrightarrow r(\varphi(x,\overline{b})) = r(x = x) \quad .$$

Any maximal independent set in M is called a <u>basis of</u> M. Dimension is well defined for M if every basis has the same cardinality. We now present two examples which show that in general dimension is far from being well-defined.

2.1. <u>Example</u>. We will construct M' such that if $T = Th(M')$, then $r(x = x) = 2 = d(x = x)$ and M' has bases of cardinality 1 and 2. Let the language have two unary function symbols F_0, F_1 and unary predicate symbols $U, U_0, U_1, \ldots$. Let $|M| = \omega \cup (\omega \times \omega \times \{0,1\})$, $U^M = \omega$, $U_i^M = \{i\} \cup (\omega \times \{\langle i,1\rangle\})$, $F_0^M(\langle i,j,0\rangle) = F_0^M(\langle i,j,1\rangle) = i$, $F_1^M(\langle i,j,0\rangle) = j$, and otherwise F_0^M, F_1^M are the identity. The formulas $U(x)$, $F_1(x) \neq x$, $F_1(x) = x \wedge \neg U(x)$ partition the universe and have rank 1, 2, 2 respectively, each having degree 1. Replacing ω by $\omega + 2$ we get an elementary extension M' of M such that $|M'| = (\omega + 2) \cup ((\omega + 2) \times (\omega + 2) . \times \{0,1\})$. Now $\{\langle \omega, \omega+1, 0\rangle\}$ and $\{\langle \omega, \omega, 1\rangle, \langle \omega+1, \omega, 1\rangle\}$ are both bases of M'.

This example shows that when $d(x = x) > 1$ we cannot expect models to have well-defined dimension. If $d(x = x) = k$ it is more natural to consider the components of $x = x$ separately. Thus we can think of such a model as being eligible to have k possibly different dimensions, corresponding to the k components of $x = x$ having the same rank as $x = x$. The situation is unchanged even if we restrict attention to $\aleph_1$-categorical theories. Unfortunately as the next example shows, dimension of a model may fail to be well defined even when $d(x = x) = 1$.

2.2. <u>Example</u>. Let M be a structure for the language $+$, $\cdot$, U, U_0, U_1, $\ldots$, R where $+$, $\cdot$ are binary function symbols, U, U_0, U_1, $\ldots$ are unary predicate symbols and R is 4-ary. Let $|M| = C \cup (P \times \omega)$ where C is the set of complex numbers,

$$P = \{[c_0, c_1, c_2] \; ; \; c_0, c_1, c_2 \in C \text{ not all } 0\}$$

and $[c_0, c_1, c_2] = [b_0, b_1, b_2]$ if and only if there exists $a \in C$ such that $c_i = ab_i$ for $i < 3$. Let $U^M = C$, $U_i^M = P \times \{i\}$, $+$ and $\cdot$ have their usual meanings on C and take the value 0, i.e. the complex number, whenever at least one argument is not in C. Let

$$R^M(a_0, a_1, a_2, b) \iff : a_0, a_1, a_2 \in C \wedge b \in \{[a_0, a_1, a_2]\} \times \omega .$$

Now $r(U(x)) = 1$, $r(\neg U(x)) = 3$ and $d(U(x)) = d(\neg U(x)) = 1$. Consider an elementary extension M' of M such that $U^{M'} = C'$ has transcendental basis $\{e_i : i \leq 3\}$, $|M'| = C' \cup (P' \times \omega)$ where P' is defined from C' exactly as P from C , and the nonlogical symbols are interpreted as before but writing C', P' for C , P . Notice that M' has dimension 0 , because

$$|M'| = U^{M'} \cup \bigcup_{i<\omega} U_i^{M'} .$$

Let b_0 , b_1 be $[e_0,e_1,1]$ and $[e_2,e_3,1]$ and $b \in C'$ be the unique point orthogonal to both b_0 and b_1 . Let $M'' \geq M'$ satisfy $|M''| = |M'| \cup \{\langle b_0,\omega\rangle , \langle b_1,\omega\rangle , \langle b,\omega\rangle\}$, $U^{M''} = U^{M'}$,

$$R^{M''}(a_0,a_1,a_2, \langle b,\omega\rangle) \iff R^{M'}(a_0,a_1,a_2, \langle b,0\rangle)$$

and similarly for b_0 , b_1 . This fixes M'' uniquely. In M'' $\{\langle b,\omega\rangle\}$ and $\{\langle b_0,\omega\rangle, \langle b_1,\omega\rangle\}$ are both bases. To see that $\{\langle b,\omega\rangle\}$ is a basis we reason as follows. There is a formula $\psi(x,y)$ such that for $c \in P'$ and $\langle c,i\rangle \in M''$, $\models \psi(a,\langle c,i\rangle)$ for $a \in M''$ iff $a = \langle c',j\rangle$ where $c' \in P'$ is orthogonal to c . Now

$$|M''| = U^{M''} \cup \bigcup_{i<\omega} U_i^{M''} \cup \psi(M'',\langle b,\omega\rangle)$$

and $\psi(x, \langle b,\omega\rangle)$ has rank 2. This shows that $\{\langle b,\omega\rangle\}$ is a basis, and a similar argument shows that $\{\langle b_0,\omega\rangle, \langle b_1,\omega\rangle\}$ is also a basis.

2.3. The case $r(x = x) = 1$. In this case every model has a well-defined dimension. The argument is similar to that given by Marsh [9] for the case in which $r(x = x) = d(x = x) = 1$. For any set A in a model let $cl(A)$ denote the set of all b such that there exists a formula $\varphi(\overline{x},y)$, $\overline{a} \in A$, and $n < \omega$ such that $\models \varphi(\overline{a},b) \wedge \exists^{<n} y\, \varphi(\overline{a},y)$. To carry over the proof, see [3, §2] , from the case in which $x = x$ is s.m. to that in which we have only $r(x = x) = 1$ we require only:

<u>Exchange Lemma</u>. Let $r(x = x) = 1$, $\{b,c\} \cap cl(A) = \emptyset$, and $b \in cl(A \cup \{c\})$, then $c \in cl(A \cup \{b\})$.

<u>Proof</u>. There exists a formula $\varphi(y,z;\overline{x})$, $\overline{a} \in A$, and $n < \omega$ such that $\models \varphi(b,c;\overline{a})$ and $\models \forall\overline{x}\, \forall z\, \exists^{<n} y\, \varphi(y,z;\overline{x})$. If

$r(\varphi(b,x;\bar{a})) = 0$ we have the conclusion so suppose otherwise. From 1.3 there exists m such that

$$\models \forall \bar{x}\, \forall y (\exists^{\geq m} z\, \varphi(y,z,\bar{x}) \to \exists^{\geq \omega} z\, \varphi(y,z;\bar{x})) .$$

Let $\varphi'(y,z;\bar{x})$ denote $\varphi(y,z;\bar{x}) \wedge \exists^{\geq m} z\, \varphi(y,z;x)$, then $\models \varphi'(b,c;\bar{a})$. Suppose $\models \exists z\, \varphi'(b_i,z,\bar{a})$ for $i < i^* = kn$ where $k = d(x = x)$ and $i \neq j$ implies $b_i \neq b_j$. Let $\psi_i(x;\bar{a}_i)$, $i < k$, be formulas such that $\models \neg \exists x(\psi_i(x;\bar{a}_i) \wedge \psi_i(x;\bar{a}_j))$ whenever $i \neq j$, $\models \bigvee_{i<k} \psi_i(x;\bar{a}_i)$ and $r(\psi_i(x;\bar{a}_i)) = d(\psi_i(x;\bar{a}_i)) = 1$. Define $E \subset i^* \times k$ by

$$E(i,j) \iff r(\varphi(b_i,z;\bar{a}) \wedge \psi_j(z,\bar{a}_j)) = 1$$

Then $(\forall i \leq i^*)(\exists j < k)\, E(i,j)$ and thus $k' < k$ can be chosen such that $\exists^{\geq n} i\, E(i,k')$. Without loss assume $k' = 0$ and that $E(i,0)$ for each $i < n$. Since for $i < n$

$$r(\varphi'(b_i,x;\bar{a}) \wedge \psi_0(x;\bar{a}_0)) = r(\psi_0(x;\bar{a}_0)) = d(\psi_0(x;\bar{a}_0)) = 1$$

there exists c' such that $\models \varphi'(b_i,c';\bar{a})$ for $i < n$. This contradicts the choice of φ' since $\models \exists^{<n} y\, \varphi(y,c';\bar{a})$. Thus $\models \exists^{<i^*} y\, \exists z\, \varphi'(y,z;\bar{a})$ whence $b \in cl(A)$ contradiction.

2.4. The case of infinite dimension. It is well known that if a model has an infinite basis then all bases have the same cardinality. For instance Theorem 5.11 (B) of Shelah [12] has essentially this content. The fact we need here is that if I is independent and $r(\varphi(x,a)) < r(x = x)$ then $\varphi(x,a)$ has only a finite number of solutions in I. It suffices to treat the case in which $d(x = x) = 1$, because we can always work within one of the components of $x = x$. If $r(\varphi(x,a)) < r(x = x)$ then by the definition of independence we can enlarge I to an independent set I' such that $|I' - (M',\bar{a})| \geq \aleph_0$. If $|I' \cap \varphi(M',\bar{a})| \geq \aleph_0$ then clearly $\varphi(x,\bar{a})$ splits the infinite indiscernible set I' which contradicts stability, see Shelah[12, Theorem 5.9].

2.5. An alternative definition. The examples given above show that in many cases models and formulas fail to have a well defined dimension. However, we can remedy the situation by defining the upper (lower) dimension of M to be the least cardinal $\varkappa$ such that every basis has power $\leq \varkappa$ $(\geq \varkappa)$. The question naturally arises: if the lower dimension is finite, must the upper dimension also be finite? At present we have no definitive answer. However, if $r(x = x)$ is finite then the answer is affirmative. We can see this as follows. Let M be a model with lower dimension n and upper dimension $\geq \aleph_0$. From 2.4 the uper dimension must be exactly $\aleph_0$ but we shall not use this fact directly. Let $\bar{b}$ be an n-tuple such that Rng $\bar{b}$ is a basis of M. Choose $m < \omega$, There exists a basis of cardinality $\geq m \cdot d(x = x)$ and at least m members of this basis say $a_0, \ldots, a_{m-1}$ lie in the same component of $x = x$, see 1.7. Thus $\{a_0, \ldots, a_{m-1}\}$ can be extended to an infinite indiscernible independent set $\{a_i : i < \omega\}$ although of course $\{a_i : i < \omega\} \not\subset |M|$. Since Rng $\bar{b}$ is a basis there exists $\psi(x, \bar{y})$ such that $\models \psi(a_i; \bar{b})$ for $i < m$ and $r(\psi(x; \bar{b})) < r(x = x)$. There exists m^* such that if $\bar{b}' \in M$ realises the same type as $\bar{b}$ then $\models \neg \psi(a_i; \bar{b}')$ for some $i \leq m^*$, i.e. $\psi(M, \bar{b}')$ contains at most m^* members of $\{a_i : i < \omega\}$. Otherwise it would be consistent for $\models \psi(a_i; \bar{b}')$ to hold iff i is even, which would contradict stability, see Shelah [12, Theorem 5.9]. By compactness we can choose $\psi(x; \bar{y})$ such that $\models \forall \bar{y}[\bigvee_{i \leq m^*} \neg \psi(a_i; \bar{y})]$. Let $\psi_j(\bar{x}; \bar{a})$, $\psi_{j,i}(\bar{x}; \bar{a})$ denote the formulas

$$\bigwedge_{i<j} \psi(a_i; \bar{x}) \quad , \quad \psi_j(\bar{x}; \bar{a}) \wedge \psi(a_i; \bar{x})$$

respectively. Then $\psi_{j,i}(\bar{x}; \bar{a})$ implies $\psi_j(\bar{x}; \bar{a})$. Since $\langle a_0, \ldots, a_{j-1}, a_j \rangle$ and $\langle a_0, \ldots, a_{j-1}, a_i \rangle$ realise the same type for $i \geq j$, we have $r(\psi_{j,i}(\bar{x}, \bar{a})) = r(\psi_{j,j}(\bar{x}, \bar{a}))$ for $i \geq j$. At most m^* of $\psi_{j,j}(x, \bar{a}))$, $\psi_{j,j+1}(x, \bar{a})$, ... are simultaneously satisfiable. Hence $r(\psi_j(\bar{x}, \bar{a})) > r(\psi_{j,j}(\bar{x}, \bar{a}))$ provided $\models \exists \bar{x}\, \psi_{j,j}(\bar{x}, \bar{a})$. Also $\psi_{j,j}(\bar{x}, \bar{a})$ is the same as $\psi_{j+1}(\bar{x}, \bar{a})$. Since $\models \exists \bar{x}\, \psi_m(\bar{x}, \bar{a})$, $\bar{x} = \bar{x}$ has rank $\geq m$. From 1.12 since $\bar{x} = \bar{x}$ has rank $\geq \omega$ so does $x = x$.

§ 3. Dimension of models having rank 2 and degree 1.

Throughout this section we shall suppose we are dealing with a theory which satisfies $r(x = x) = 2$ and $d(x = x) = 1$. Our purpose is to show that any model of such a theory has a well-defined dimension, i.e. if I_0 and I_1 are maximal independent sets then $|I_0| = |I_1|$.

Consider first the case in which there is no uniform decomposition of $x = x$. From 1.12 if $\{a\}$ is independent for each $a \in A$ then either $r^M(a, A - \{a\}) = 2$ or a is algebraic over $A - \{a\}$. Thus we can show that $|I_0| = |I_1|$ exactly as in the case when $x = x$ is s.m., see [9] or [3].

Now suppose that there exists a uniform decomposition of $x = x$. Then we can find $\varphi(x,\bar{y})$ and a complete m-type p such that for all n there exist $\bar{b}_0,\ldots,\bar{b}_{n-1}$ all realising p such that for all $i,j < n$ with $i \neq j$

$$r(\varphi(x,\bar{b}_i)) = d(\varphi(x,\bar{b}_i)) = 1 \wedge r(\varphi(x,\bar{b}_i) \wedge \varphi(x,\bar{b}_j)) = 0 \quad .$$

From NL we obtain $\varphi'(x,\bar{y})$ such that for all $\bar{b}$, $\bar{b}'$ realising p $\varphi'(x,\bar{b})$ and $\varphi(x,\bar{b}) \wedge \varphi'(x,\bar{b})$ are s.m., and $\varphi'(x,\bar{b})$ and $\varphi'(x,\bar{b}')$ are equivalent if and only if $r(\varphi(x,\bar{b}) \wedge \varphi(x,\bar{b}')) = 1$. For $\bar{b}$, $\bar{b}'$ realising p there exists n such that in any model M

$$|\varphi'(M,\bar{b}') \cap \varphi'(M,\bar{b}')| \geq n \Rightarrow r(\varphi'(x,\bar{b}) \wedge \varphi'(x,\bar{b}')) = 1 \quad .$$

Otherwise by compactness we could contradict $\varphi'(x,\bar{b})$ being s.m. when $\bar{b}$ realises p . Applying compactness again we can choose $\varphi'(x,\bar{y})$ such that

$$\models \forall \bar{y} \forall \bar{y}' (\exists^{\geq n} x(\varphi'(x,\bar{y}) \wedge \varphi'(x,\bar{y}')) \rightarrow \forall x(\varphi'(x,\bar{y}) \leftrightarrow \varphi'(x,\bar{y}'))) \ .$$

Obviously $\forall x(\varphi'(x,\bar{y}) \leftrightarrow \varphi'(x,\bar{y}'))$ defines an equivalence relation on $l(\bar{y})$-tuples and it is notationally convenient to suppose that the equivalence classes are representable by elements of the model. Thus we shall write $\varphi(x,y)$ below instead of $\varphi'(x,\bar{y})$ and we now have the additional property (Φ_1)

$$\models \forall y \forall y' (\exists^{\geq n} x(\varphi(x,y) \wedge \varphi(x,y')) \rightarrow y = y') \ .$$

Also below p will denote the 1-type corresponding to the m-type p referred to above. This manoevre is perfectly legitimate since we could adjoin a new element for each equivalence class and a new (m+1)-ary relation $R(\bar{y},z)$ such that $\models R(\bar{b},c)$ if and only if $\bar{b}$ belongs to the equivalence class named by c . Instead of considering the dimension of the universe we could consider the dimension of the formula $\neg \exists \bar{y}\, R(\bar{y},x)$. Below we must only take care not to profit from the fact that $r(\exists x\, \varphi(x,y)) \leq 2$. We now proceed to the consideration of various cases and subcases. At each point we tacitly assume that the hypotheses of cases considered earlier fail.

Case 1. $r(\exists^{\geq\omega} y\, \varphi(y,x)) = 1$. Without loss of generality we may assume that p is the only nonalgebraic complete 1-type extending $\exists^{\geq\omega} y\, \varphi(y,x)$. There exists m such that if $\models \exists^{\geq m} y(\exists^{\geq\omega} x\, \varphi(x,y) \wedge \varphi(a,y))$ then $\models \exists^{\geq\omega} y(\exists^{\geq\omega} x\, \varphi(x,y) \wedge \varphi(a,y))$. Otherwise we can contradict the case hypothesis. It follows that

$$|\{a : \models \exists^{\geq m} y(\exists^{\geq\omega} x\, \varphi(x,y) \wedge \varphi(a,y))\}| < n.\ \deg\, (\exists^{\geq\omega} y\, \varphi(y,x)) \quad .$$

Otherwise we can find distinct $a_0,\dots,a_{n-1}$ and a component C of $\exists^{\geq\omega} y\, \varphi(y,x)$ such that for each $i < n$, $\varphi(a_i,x)$ almost contains C . Immediately we get distinct b_0, b_1 such that $\models \varphi(a_i,b_j)$ for all $i < n$ and $j < 2$ contradicting the property (Φ_1) established above. Applying compactness the set of elements on the left-hand side of the last inequality is first-order definable. Thus we may further refine the choice of $\varphi(x,y)$ such that $\models \forall x\, \exists^{<m} y\, \varphi(x,y)$. We shall call this property (Φ_2) . Let $\varepsilon(x,y)$ define an equivalence relation splitting $\exists^{\geq\omega} y\, \varphi(y,x)$. Observe that if $\models \varepsilon(b,b)$ then for each $i < \omega$, $\exists y(\varphi(x,y) \wedge \exists^{\geq i} x\, \varphi(x,y) \wedge \varepsilon(y,b))$ has rank 2. Thus if $k = d(\exists^{\geq\omega} y\, \varphi(y,x))$, for each $i < \omega$

$$r(\exists y_0 \dots y_{k-1}(\bigwedge_{j<k} \varphi(x,y_j) \wedge \bigwedge_{j<k} \exists^{\geq i} x\, \varphi(x,y_j) \wedge \bigwedge_{j<j'<k} \neg\, \varepsilon(y_j,y_{j'}))) = 2 \ .$$

From (Φ_2) it follows that if $\{a\}$ is independent then for every b' such that $\models \varepsilon(b',b')$ there exists b such that $\models \varepsilon(b,b')$, b realises p and $\models \varphi(a,b)$. We now have to consider three subcases.

Case 1.1. There exist models M, M' , a and b in M , $b' \in M'$, such that b and b' realise p , $\models \varepsilon(b,b')$, $b' \notin M$, M' is prime over $|M| \cup \{b'\}$ and $a \in \varphi(M', b) - \varphi(M,b)$. Then there exists a formula $\psi(x,y,\bar{z})$ and $\bar{c} \in M$ such that $b \in \mathrm{Rng}\, \bar{c}$ and $\psi(x,b',\bar{c})$ isolates the type of a over $|M| \cup \{b'\}$. Clearly we may suppose that

$$\models \forall x\, \forall y(\psi(x,y,\bar{c}) \rightarrow .\ \varepsilon(y,b) \wedge \varphi(x,b))$$

Further, since $r(\varphi(x,b)) = d(\varphi(x,b)) = 1$ we may suppose that $\models \forall y \forall \bar{z} \exists^{\leq i} x\, \psi(x,y,\bar{z})$ for suitable $i < \omega$. Again $\psi(a,x,\bar{c})$ is not satisfied by any element of M , whence $\psi(a,x,\bar{c})$ has only a finite number of solutions realizing p since

$$r(p \cup \{\varepsilon(x,b)\}) = d(p \cup \{\varepsilon(x,b)\}) = 1 \quad .$$

By compactness we may choose ψ such that all solutions of $\psi(x,a,\bar{c})$ realize p and hence such that $\models \forall y \forall \bar{z} \exists^{\leq i} x\, \psi(y,x,\bar{z})$ for suitable $i < \omega$. Let $\chi(x,\bar{z})$ be a formula such that $\chi(x,\bar{c})$ is $\exists y(\varphi(y,b) \wedge \psi(y,x,\bar{c}))$. Then $r(\chi(x,\bar{c})) = 1$ since $\models \chi(b',\bar{c})$, $r(\varphi(x,b)) = 1$ and $\models \forall y \exists^{\leq i} x\ \psi(y,x,\bar{c})$. Since we are admitting parameters from M we may further restrict the choice of ψ such that $d(\chi(x,\bar{c})) = 1$ also. Clearly $\models \chi(b',\bar{c})$. Since $b' \notin M$ it follows that

$$\varepsilon(x,b) \wedge \exists^{\geq\omega} y\ \ \varphi(y,x) \wedge \chi(x,\bar{c})$$

is infinite and hence that

$$\varepsilon(x,b) \wedge \exists^{\geq\omega} y\ \varphi(y,x) \wedge \neg\, \chi(x,\bar{c})$$

is finite. Also, since $a \notin M$ and $\models \varphi(a,b) \wedge \exists y\ \psi(a,y,\bar{c})$, $\varphi(x,b) \wedge \neg \exists y\ \psi(x,y,\bar{c})$ is finite. It follows that the s.m. formulas $\chi(x,c)$ and $\varphi(x,b)$ are linked. As we remarked after the definition of linked s.m. formulas any linkage is expressible by some first-order sentence. Thus since b' realizes the same type as b there exists $\bar{c}' \in M'$ such that $\chi(x,\bar{c}')$ is s.m., implies $\varepsilon(x,b')$, and is linked to $\varphi(x,b')$, and such that

$$\varepsilon(x,b') \wedge \exists^{\geq\omega}\ y\ \varphi(y,x) \wedge \chi(x,\bar{c}')$$

is infinite, while

$$\varepsilon(x,b') \wedge \exists^{\geq\omega}\ y\ \varphi(y,x) \wedge \neg\ \chi(x,\bar{c}')$$

is finite. Since $\varepsilon(x,b)$ and $\varepsilon(x,b')$ are equivalent, $\chi(x,\bar{c}) \wedge \chi(x,\bar{c}')$ is infinite. Thus $\chi(x,\bar{c})$ and $\chi(x,\bar{c}')$ differ only finitely, whence $\varphi(x,b')$ is linked to $\chi(x,\bar{c})$. Using once more the fact that linkages are expressible in a first-order way, there exists $\chi'(x,\bar{c})$ implying $\chi(x,\bar{c})$ such that $\chi(x,\bar{c}) \wedge \neg\, \chi'(x,\bar{c})$

is finite and $\models \chi'(b^o,\bar{c})$ implies that $\varphi(x,b^o)$ and $\chi'(x,\bar{c})$ are linked for every b^o. From 1.5 $\exists y(\varphi(x.y) \wedge \chi'(x,c))$ is $\aleph_1$-categorical. From 1.10 this suffices for the uniqueness of dimension.

Remark. The following observation will be useful in the next section. Suppose Case 1 holds but Case 1.1 fails. Let b, b' realise p, $b \in M$, $b' \notin M$ and $\models \varepsilon(b,b')$ then $\models \neg \exists x(\varphi(x,b) \wedge \varphi(x,b'))$. If not, let $\models \varphi(a',b) \wedge \varphi(a',b')$. Then since $\varphi(x,b) \wedge \varphi(x,b')$ is finite, $a' \in M'$ the model prime over $M \cup \{b'\}$. From the failure of Case 1.1 $a' \in M$. But this means that $\models \exists^{\geqslant\omega} y\,\varphi(a',y)$ contradicting $\models \forall x \exists^{<m} y\,\varphi(x,y)$. Thus $\varphi(x,b)$ and $\varphi(x,b')$ are indeed disjoint. From the strong minimality of $\varepsilon(x,b) \wedge \exists^{\geqslant y} \varphi(y,x)$ it follows that

$$\models \exists^{<\omega} y \exists x(\varphi(x,b) \wedge \exists^{\geqslant\omega} x\,\varphi(x,y) \wedge \varphi(x,y) \wedge \varepsilon(b,y)) .$$

By compactness we can replace $\varphi(x,y)$ by some formula $\varphi(x,y) \wedge \psi(y)$ where $\psi(x)$ is a suitable formula in p such that for some $i < \omega$

$$\models \forall y \exists^{<i} y' \exists x(\varphi(x,y) \wedge \varphi(x,y') \wedge \varepsilon(y,y')) .$$

Finally since there is a uniform bound on the cardinality of $\varphi(x,b) \wedge \varphi(x,b')$ for $b \neq b'$ we may choose φ such that

$$\models \forall y \forall y'(\varepsilon(y,y') \wedge y \neq y' . \rightarrow \neg \exists x(\varphi(x,y) \wedge \varphi(x,y'))) .$$

Case 1.2. There exist b, b' realising p such that $\models \varepsilon(b,b')$, b' is not algebraic over b and $\varphi(x,b)$, $\varphi(x,b')$ are linked. First note that 2-type of $\langle b,b'\rangle$ is fixed uniquely by the given conditions since $r(p \cup \{\varphi(x,b)\}) = d(p \cup \{\varphi(x,b)\}) = 1$. Thus in fact $\varphi(x,b)$, $\varphi(x,b')$ are linked for all such b and b'. Consider an arbitrary model M and bases $\{a_i : i < k\}$, $\{a'_i : i < k'\}$ of M. From the remark immediately preceding Case 1.1 there exist b_i, $i < k$, and b'_i, $i < k'$, realising p such that $\models \varphi(a_i,b_i)$, $\models \varphi(a'_{i'},b'_{i'})$ and $\models \varepsilon(b_i,b'_{i'})$ for all $i < k$ and $i' < k'$. Clearly $\{b_i : i < k\}$ and $\{b'_i : i < k'\}$ are independent sets in $\exists^{\geqslant\omega} y\,\varphi(y,x) \wedge \varepsilon(x,b_o)$. Suppose $k < k'$ then since $\exists^{\geqslant\omega} y\,\varphi(y,x) \wedge \varepsilon(x,b_o)$ is s.m. in the extended sense we can find $j < k'$ such that $\{b_i : i < k\} \cup \{b'_j\}$ is an independent set of power $k+1$ in $\exists^{\geqslant\omega} y\,\varphi(y,x) \wedge \varepsilon(x,b_o)$. Without loss $j = 0$. We shall show that there is an element $a''_o \in M$ such that $\models \varphi(a''_o, b'_o)$ and $\{a_i : i < k\} \cup \{a''_o\}$ is independent in the whole model. This will conclude the treatment of this case.

Choose b realising p such that $\models \varepsilon(b_0, b)$ and $b \notin M$. Let M^* be prime over $|M \cup \{b\}|$. Since Case 1.1 fails $\varphi(M, b_i) = \varphi(M^*, b_i)$, $i < k$, and $\varphi(M, b_i') = \varphi(M^*, b_i')$, $i < k'$. By choice of b, $\{b_i : i < k\} \cup \{b_0', b\}$ and $\{b_i' : i < k+1\} \cup \{b\}$ are independent sets of power $k+1$ in $\exists^{\geq\omega} y\,\varphi(y,x) \wedge \varepsilon(x, b_0)$. From the failure of Case 1.1 we can deduce further that $\{b_i : i < k\} \cup \{b_0'\}$ and $\{b_i' : i < k+1\}$ are independent in $\exists^{\geq\omega} y\,\varphi(y,x) \wedge \varepsilon(x, b_0)$ over $\{b\} \cup \varphi(M^*, b)$. Let $N, N' \subset M^*$ be prime over $\varphi(M^*, b) \cup \{b_i : i < k\} \cup \{b_0'\}$ and $\varphi(M^*, b) \cup \{b_i' : i < k+1\}$ respectively. From the hypothesis of this subcase $\varphi(N, b_i) = \varphi(M, b_i)$, $i < k$, $\varphi(N, b_0') = \varphi(M, b_0')$, and $\varphi(N', b_i') = \varphi(M, b_i')$ for $i < k+1$. Clearly there is an elementary embedding F of N' into N taking $b_0', \ldots, b_k'$ into $b_0, \ldots, b_{k-1}, b_0'$ respectively. Using the failure of Case 1.1 again we can see that $\langle a_0, \ldots, a_{k-1}, b_0, \ldots, b_{k-1}, b_0' \rangle$ and $\langle a_0', \ldots, a_{k-1}', b_0', \ldots, b_{k-1}', b_k' \rangle$ realise the same type. Thus $\langle F(a_0'), \ldots, F(a_{k-1}') \rangle$ realises the same type as $\langle a_0, \ldots, a_{k-i} \rangle$ over $\{b_0, \ldots, b_{k-1}, b_0'\}$. Now the theory of

$$\varphi(x, b_0) \vee \ldots \vee \varphi(x, b_{k-1}) \vee \varphi(x, b_0')$$

is $\aleph_1$-categorical because the disjunctands are s.m. and pairwise linked. Every model of an $\aleph_1$-categorical theory is homogeneous [3, §5]. Thus there exists a_0'' in $\varphi(M, b_0')$ such that $\langle a_0, \ldots, a_{k-1}, a_0'' \rangle$ realises the same type as $\langle F(a_0'), \ldots, F(a_{k-1}'), F(a_k') \rangle$.

Before considering the remaining subcase of Case 1 we observe that if $\{a\}$ is independent, $\models \varphi(a, b)$, and b realises p then a is not algebraic over b. Suppose the contrary then there is a formula $\psi(x,y)$ implying $\varphi(x,y)$ such that $\models \psi(a,b)$ and $\models \forall y \exists^{<i} x\,\psi(x,y)$ for some $i < \omega$. Over $\emptyset$ there is a unique 1-type p^* of Morley rank 2. From the definition of rank for arbitrary $k < \omega$ there exist disjoint formulas $\chi(x, \bar{c}_j)$, $j < k$, such that p^* is realised infinitely often in each $\chi(x, \bar{c}_j)$. Let $k = i.\, d(\exists^{\geq\omega} y\,\varphi(y,x))$. For each $j < k$

$$\models \exists^{\geq\omega} x (\exists^{\geq\omega} y\,\varphi(y,x) \wedge \exists y (\chi(y, \bar{c}_j) \wedge \psi(y,x))) .$$

Otherwise p^* could be realised only finitely often in $\chi(x, \bar{c}_j)$. Denoting $\exists y(\chi(y, \bar{c}_j) \wedge \psi(y,x))$ by $\vartheta_j(x, \bar{c}_j)$ we see that $\vartheta_j(x, \bar{c}_j)$ almost contains at least one of the components of $\exists^{\geq\omega} y\,\varphi(y,x)$. From the choice of k there is a component C of $\exists^{\geq\omega} y\,\varphi(y,x)$ such that at least i of the formulas $\vartheta_j(x, \bar{c}_j)$ almost contain C. Choosing b realising p in the intersection of there i formulas we have $\models \exists x(\chi(x,\bar{c}_j) \wedge \psi(x,b))$ for at least i values of j contradicting $\models \exists^{<i} x\, \phi(x,b)$.

Let q denote the unique 2-type such that if $\langle a,b\rangle$ realises p then $\models \varphi(a,b)$, $\{a\}$ is independent, and b realises p .

Case 1.3. Otherwise. Let b realise p . It suffices to establish: Claim. Let $\models \varepsilon(b_i,b)$ for $i < n$ and $\{b_0,\ldots,b_{n-1}\}$ be independent in $\exists^{\geqslant\omega} y\, \varphi(y,x)$. Let $\langle a_i,b_i\rangle$ realise q for $i < n$ then $\{a_0,\ldots,a_{n-1}\}$ is independent in the whole model.

The claim is enough because it allows us to infer the uniqueness of dimension in the whole model from the uniqueness of dimension in the formula $\exists^{\geqslant\omega} y\, \varphi(y,x) \wedge \varepsilon(x,b)$ which is s.m. in the extended sense.

To prove the claim, for a contradiction argument let $\langle a_i\colon i\leqslant n\rangle$ and $\langle b_i\colon i \leqslant n\rangle$ constitute a mimimal counter example. Thus $\{b_i\colon i \leqslant n\}$ is independent in $\exists^{\geqslant\omega} y\, \varphi(y,x) \wedge \varepsilon(x,b)$, $\langle a_i,b_i\rangle$ realises q for $i \leqslant n$, $\{a_i\colon i \leqslant n\}$ is not independent in the whole model, and the claim is true for n . We consider two subcases.

Case 1.3.1. b_n is algebraic over $\{a_i\colon i < n\} \cup \{b_i\colon i < n\}$. Choose the least $j < n$ such that for some formula ψ and $k < \omega$ we have

$$\models \psi(a_0,\ldots,a_j,\ b_0,\ldots,b_n) \wedge \exists^{<k} y\, \psi(a_0,\ldots,a_j,\ b_0,\ldots,b_{n-1},\ y)$$

We may suppose that $\{b_j,\ldots,b_n\}$ is independent in $\exists^{\geqslant\omega} y\, \varphi(x,y) \wedge \varepsilon(x,b)$ over $\{a_i\colon i < j\} \cup \{b_i\colon i < j\}$. Otherwise by reordering $b_j,\ldots,b_n$ we can diminish the value of j . Also there exists i such that

$$\models \exists^{<i} x\, \psi(a_0,\ldots,a_{j-1},\ x,\ b_0,\ldots,b_n) .$$

Otherwise we could again diminish the value of j by replacing ψ by a first-order formula ψ' such that $\psi'(a_0,\ldots,a_{j-1},b_0,\ldots,b_{n-1},y)$ is equivalent to

$$\exists^{\geqslant\omega} x(\varphi(x,b_j) \wedge \psi(a_0,\ldots,a_{j-1},\ x,\ b_0,\ldots,b_{n-1},\ y)) .$$

Such ψ' exists from 1.3.

Let M , M' be prime over $\{a_0,\ldots,a_{j-1}, b_0,\ldots,b_{n-1}\}$ and $|M| \cup \{b_n\}$ respectively. Since $\{b_j,\ldots,b_n\}$ is independent over $\{a_i: i < j\} \cup \cup \{b_i: i < j\}$, $b_n \notin M$ and hence $a_n \notin M$. But $a_n, b_n \in M'$ and so Case 1.1 holds.

Case 1.3.2. a_n is algebraic over $\{a_i: i< n\} \cup \{b_i: i \leq n\}$. Let j' be the least number such that a_n is algebraic over $\{a_i: i < j'\} \cup \{b_i: i \leq n\}$. If $j' = 0$, then we may suppose $\{b_i: i < n\}$ is independent in $\exists^{\geq\omega} y\, \varphi(y,x) \wedge \epsilon(x,b)$ over $\{a_n, b_n\}$. Otherwise reordering $b_0,\ldots,b_n$ we have the previous case. Thus $j' > 0$. Let $j = j' - 1$. Let ψ be a formula and $k < \omega$ be such that

$$\models \psi(b_0,\ldots,b_n,a_1,\ldots,a_j,a_n) \wedge \exists^{<k} x\, \psi(b_0,\ldots,b_n,a_1,\ldots,a_j,x) .$$

We may suppose

$$\models \forall \bar{y}\, \forall x_1 \ldots x_j\, \exists^{<k} x\, \psi(\bar{y},x_1,\ldots,x_j,x)$$

and

$$\models \forall \bar{y}\, \forall x_1 \ldots x_j\, x(\psi(\bar{y},x_1,\ldots,x_j,x) \rightarrow . \;\varphi(x_j,y_j) \wedge \varphi(x,y_n)) .$$

Abbreviate $\psi(b_0,\ldots,b_n,a_1,\ldots,a_{j-1},x,y)$ to $\psi(\bar{b},\bar{a}',x,y)$. Then $\models \exists y\, \psi(\bar{b},\bar{a}',a_j,y)$ and a_j is independent in $\varphi(x,b_j)$ over $\{b_i: i \leq n\} \cup \{a_i: i < j\}$ by the minimality of j . Thus

$$\varphi(x,b_j) \wedge \neg \exists y\, \psi(\bar{b},\bar{a}',x,y)$$

is finite. Also a_j is algebraic over $\{b_i: i \leq n\} \cup \{a_i: i < j\} \cup \{a_n\}$. Otherwise we can reduce j as in the last case. Thus we can choose k and ψ such that

$$\models \forall \bar{y}\, \forall x_1 \ldots x_{j-1}\, x_j\, \exists^{<k} x\, \psi(\bar{y},x_1,\ldots,x_{j-1},x,x_j) .$$

Repeating the argument used above

$$\varphi(x,b_n) \wedge \neg \exists y\, \psi(\bar{b},\bar{a}',y,x)$$

is finite. Clearly $\varphi(x,b_j)$ and $\varphi(x,b_n)$ are linked and so Case 1.2 obtains. This concludes the treatment of Case 1.

Case 2. $r(\exists^{\geq\omega} y\,\varphi(y,x)) > 1$. Choose disjoint formulas $\chi_i(x,\bar{c}_i)$, $i < 3$, such that $\chi_i(x,\bar{c}_i) \wedge \exists^{\geq\omega} y\,\varphi(y,x)$ which we denote by $\vartheta_i(x,\bar{c}_i)$ is s.m. in the extended sense. Recall that

$$\models \forall y \forall y'(\exists^{\geq n} x(\varphi(x,y) \wedge \varphi(x,y')) \to y = y') .$$

Using the strong minimality of $\vartheta_i(x,\bar{c}_i)$ we see that there is a first--order formula $\vartheta_i'(x,\bar{c}_i)$ satisfied exactly by those a's , and there are only a finite number of them, such that $\models \exists^{\geq\omega} y(\varphi(a,y) \wedge \vartheta_i(y,\bar{c}_i))$. Let $\vartheta'(x,\bar{c}')$ denote $\neg \bigvee\{\vartheta_i'(x,\bar{c}_i) : i < 3\}$.

Claim. For $i < j < 3$ we have

$$\models \forall^a y_i \forall^a y_j(\vartheta_i(y_i,\bar{c}_i) \wedge \vartheta_j(y_j,\bar{c}_j) .\to.$$

$$\exists x(\varphi(x,y_i) \wedge \varphi(x,y_j) \wedge \vartheta'(x,\bar{c}'))) .$$

If not, fix i,j so that the claim fails. Then by compactness and 1.3 there are disjoint first-order formulas $\vartheta_i'(x,\bar{c}_i)$, $\vartheta_j'(x,\bar{c}_j)$, implied by $\vartheta_i(x,\bar{c}_i)$, $\vartheta_j(x,\bar{c}_j)$ respectively, and $k < \omega$ such that

$$\models \exists^{\leq k} y_i \exists^{\geq k} y_j(\vartheta_i'(y_i,\bar{c}_i') \wedge \vartheta_j'(y_j,\bar{c}_j') .\to.$$

$$\exists x(\varphi(x,y_i) \wedge \varphi(x,y_j) \wedge \vartheta'(x,\bar{c}')) .$$

Now, clearly

$$r(\exists y(\vartheta_j'(y,\bar{c}_j') \wedge \varphi(x,y) \wedge \vartheta'(x,\bar{c}'))) \geq 2 .$$

There are infinitely many b such that $\models \vartheta_i(b,\bar{c}_i)$ and

$$\varphi(x,b) \wedge \exists y(\vartheta_j'(y,\bar{c}_j') \wedge \varphi(x,y)) \wedge \vartheta'(x,\bar{c}')$$

is finite. In fact this in true for all but at most k b's satisfying $\models \vartheta_i'(b,\bar{c}_i')$. Thus

$$r(\neg \exists y(\vartheta_j'(y,\bar{c}_j') \wedge \varphi(x,y) \wedge \vartheta'(x,\bar{c}')) \geq 2$$

which together with the last inequality contradicts $d(x = x) = 1$ and establishes the claim.

From the claim, compactness, and 1.3 we can choose $\chi_i(x,\bar{c}_i)$, $i < 3$, such that for $i < j < 3$

$$\models \forall y_i \forall y_j (\chi_i(y_i,\bar{c}_i) \wedge \chi_j(y_j,\bar{c}_j) \rightarrow.$$

$$\exists x(\varphi(x,y_i) \wedge \varphi(x,y_j) \wedge \vartheta'(x,\bar{c}')) .$$

This implies that for $i < 3$

$$\models \forall x(\chi_i(x,\bar{c}_i) \rightarrow \exists^{\geq\omega} y\, \varphi(y,x))$$

and thus $\chi_i(x,\bar{c}_i)$ is s.m. for $i < 3$. Choose b such that $\models \chi_2(b,\bar{c}_2)$. Let $\psi(x_0,x_1,\bar{c})$ denote

$$\chi_0(x_0,\bar{c}_0) \wedge \chi_1(x_1,\bar{c}_1) \wedge \exists x(\varphi(x,x_0) \wedge \varphi(x,x_1) \wedge \vartheta'(x,\bar{c}') \wedge \varphi(x,b)) .$$

The presence of $\vartheta'(x,\bar{c}')$ ensures

$$\models \forall x_0 \forall x_1 (\exists^{\leq\omega} x_1\, \psi(x_0,x_1,\bar{c}) \wedge \exists^{\leq\omega} x_0\, \psi(x_0,x_1,\bar{c})) .$$

Further $\varphi(x,b)$ intersects $\varphi(x,b_0) \wedge \vartheta'(x,\bar{c}')$ for every b_0 such that $\models \chi_0(b_0,\bar{c}_0)$. Thus

$$\models \exists^{\geq\omega} x(\vartheta'(x,\bar{c}') \wedge \varphi(x,\bar{b}) \wedge \exists x_0(\chi_0(x_0,\bar{c}_0) \wedge \varphi(x,x_0)))$$

and similarly with χ_1 instead of χ_0 . From strong minimality $\exists^{\geq\omega}$ and $\forall^a$ are the same restricted to $\varphi(x,b)$. Thus

$$\models \exists^{\geq\omega} x(\vartheta'(x,\bar{c}') \wedge \varphi(x,b) \wedge \exists x_0 \exists x_1(\chi_0(x_0,\bar{c}_0) \wedge \chi_1(x_1,\bar{c}_1) \wedge$$

$$\wedge \varphi(x,x_0) \wedge \varphi(x,x_1))) .$$

Since x is restricted to $\vartheta'(x,\bar{c}')$ we can deduce

$$\models \exists^{\geq\omega} x_0 \exists x_1\, \psi(x_0,x_1,\bar{c}) \wedge \exists^{\geq\omega} x_1 \exists x_0\, \psi(x_0,x_1,\bar{c}) .$$

Thus, since $\chi_0(x,\bar{c}_0)$ and $\chi_1(x,\bar{c}_1)$ are s.m.,

$$\models \forall^a x_0(\chi_0(x_0,\bar{c}_0) \rightarrow \exists x_1\, \psi(x_0,x_1,\bar{c})) \wedge \forall^a x_1(\chi_1(x_1,\bar{c}_1) \rightarrow$$

$$\rightarrow \exists x_0\, \psi(x_0,x_1,\bar{c})) .$$

This shows that $\chi_0(x,\bar{c}_0)$ and $\chi_1(x,\bar{c}_1)$ are linked. It is now easy to see that

$$\exists y_0 \exists y_1(\varphi(x,y_0) \wedge \varphi(x,y_1) \wedge \chi_0(y_0,\bar{c}_0) \wedge \chi_1(y_1,\bar{c}_1))$$

has rank 2 and is $\aleph_1$-categorical, the latter because it is included in the algebraic closure of the two linked s.m. sets. As was observed in 1.10 this is enough to give us a unique dimension.

§ 4. Structure and spectra of theories having rank 2 and degree 1.

The question as to what spectra are possible for $\aleph_0$-stable theories has apparently been solved by Shelah [14, p.190] . One can expect that his results in this direction will appear in [15]. Here we shall discuss what spectra are possible for theories of rank 2 and degree 1 and relate the spectra to the structure of the models. We shall follow the analysis of the last section.

Suppose that there is no uniform decomposition. Let $\varphi(x,\bar{y})$ be a formula and p be a complete type such that if $\bar{b}$ realises p then $r(\varphi(x,\bar{b})) = d(\varphi(x,\bar{b})) = 1$. From NL we may suppose that $\varphi(x,\bar{y})$ is already normalised with respect to p . From 1.3 if $\bar{b}_0$, $\bar{b}_1$ realise p and $|\varphi(M,\bar{b}_0) \cap \varphi(M,\bar{b}_1)|$ exceeds some fixed finite number, then $\varphi(x,\bar{b}_0)$ and $\varphi(x,\bar{b}_1)$ are equivalent. By compactness we can choose n and $\varphi(x,\bar{y})$ such that

$$\models \forall \bar{y}^0 \forall \bar{y}^1 (\exists^{\geq n} x(\varphi(x,\bar{y}^0) \wedge \varphi(x,\bar{y}^1)) \rightarrow \forall x(\varphi(x,\bar{y}^0) \leftrightarrow \varphi(x,\bar{y}^1))) .$$

Since we are assuming there is no uniform decomposition there are only a finite number of distinct instances $\varphi(x,\bar{b})$ of $\varphi(x,\bar{y})$ with $\bar{b}$ realising p . By compactness there is $\psi(x)$ equivalent to the disjunction of these instances. Clearly if $\bar{b}$ realises p then $\psi(x)$ almost contains $\varphi(x,\bar{b})$ in the sense that $\varphi(x,\bar{b}) \wedge \neg \psi(x)$ has at most a finite number of solutions. Also $r(\psi(x)) = 1$. Conversely if there exists a uniform decomposition then we can find $\varphi(x,\bar{y})$ and $\bar{b}_i$, $i < \omega$, all realising the same type such that $\varphi(x,\bar{b}_i)$ is infinite and $\varphi(x,\bar{b}_i)$ $\varphi(x,\bar{b}_j)$ is finite whenever $i \neq j$. Clearly $\varphi(x,\bar{b})$ is not almost contained in any formula $\psi(x)$ of rank 1 .

Until further notice assume that there is no uniform decomposition and that $\langle \psi_i(x) : i < \omega \rangle$ is an enumeration of all formulas of rank 1 containing no parameters. It is clear that $\bigwedge_{i \in \omega} \neg \psi_i(x)$ is s.m.

Let $A \cup \{a\} \subset \bigwedge_{i<\omega} \neg \psi_i(\bar{M})$, $B \subset \bigvee_{i<\omega} \psi_i(\bar{M})$, and $a \in cl(A \cup B)$. We shall prove that $a \in cl(A)$. If not, choose $\bar{a} \in A$, b , $\bar{b} \in B$, $n < \omega$ and $\varphi(x,y,\bar{x},\bar{y})$ such that

$$\models \varphi(a,b,\bar{a},\bar{b}) \wedge \forall y \forall \bar{x} \forall \bar{y} \exists^{<n} x\, \varphi(x,y,\bar{x},\bar{y})$$

and $a \notin cl(A \cup \text{Rng}\, \bar{b})$. Choose i such that $\models \psi_i(b)$ and consider the formula

$$\exists y\, [\varphi(x,y,\bar{a},\bar{b}) \wedge \psi_i(y)] \, .$$

This formula has rank 1 since $\psi_i(x)$ has rank 1 and thus it is almost contained in some $\psi_j(x)$ which gives $a \in cl(A \cup \text{Rng}\, \bar{b})$, contradiction. Intuitively, this means that elements realising $\bigwedge_{i<\omega} \neg \psi_i(x)$ interact only very weakly with elements outside this type. Continuing with the same hypotheses we shall show that in fact $a \in cl(\{a'\})$ for some $a' \in A$. If not, we can choose a', a'', $\bar{a} \in A$, $n < \omega$, and $\varphi(x,x',x'',\bar{x})$ such that $a \notin cl(\{a''\} \cup \text{Rng}\, \bar{a}) \cup cl(\{a'\} \cup \text{Rng}\, \bar{a})$, $\{a',a''\} \cup \text{Rng}\, \bar{a}$ is independent in $\bigwedge_{i<\omega} \neg \psi_i(x)$, and

$$\models \varphi(a,a',a'',\bar{a}) \wedge \forall x \forall x' \forall x'' \forall \bar{x} (\exists^{<n} x\, \varphi(x,x',x'',\bar{x}) \wedge$$

$$\exists^{<n} x'\, \varphi(x,x',x'',\bar{x})) \, .$$

Consider the formula $\exists y\, \varphi(a,x,y,\bar{a})$. If it is finite then a' is algebraic over $\{a\} \cup \text{Rng}\, a$, whence by the exchange lemma $a \in cl(\{a'\} \cup \text{Rng}\, \bar{a})$, contradiction. Hence we can fix i such that $\exists y\, \varphi(a,x,y,\bar{a}) \wedge \psi_i(x)$ is infinite. By choice of φ we see that $\exists y\, (\varphi(a,y,x,\bar{a}) \wedge \psi_i(y))$ is also infinite and thus is satisfied by some b'' , $\models \bigvee_{j<\omega} \psi_j(b'')$. Now choose b' such that $\models \varphi(a,b',b'',\bar{a}) \wedge \psi_i(b')$. From the result above, since $a \in cl(\{b',b''\} \cup \text{Rng}\, \bar{a})$ we get $a \in cl(\text{Rng}\, \bar{a})$, contradiction.

4.1. DEFINITION. An s.m. 1-type p is called <u>disconnected</u> if for any set of elements $A \cup \{a\}$ all realising p such that $a \in cl(A)$ there exists $a' \in A$ such that $a \in cl(\{a'\})$.

What we showed above was that the s,m, formula $\bigwedge_{i<\omega} \neg \psi_i(x)$ is disconnected. Now drop the assumption that there is no uniform decomposition. However suppose that $\langle \psi_i(x) \colon i < \omega \rangle$ is as before. Suppose that $\bigwedge_{i<\omega} \neg \psi_i(x)$ is s.m. then we can easily show that every rank 1 formula $\varphi(x,\bar{a})$ is almost contained in some $\psi_i(x)$. We can summarise our findings:

4.2. THEOREM. Let $r(x = x) = 2$, $d(x = x) = 1$. Let $\langle \psi_i(x): i < \omega \rangle$ be an enumeration of all formulas of rank 1 containing no parameters. The following conditions are equivalent:

(i) There exists no uniform decomposition of $x = x$

(ii) $\bigwedge_{i<\omega} \neg \psi_i(x)$ is s.m.

Further when (i) and (ii) hold $\bigwedge_{i<\omega} \neg \psi_i(x)$ is disconnected.

4.3. The spectrum when there is no uniformdecomposition. Assume there is no uniform decomposition of $x = x$. Choose a sequence $\langle \chi_i(x): i < \omega \rangle$ of rank 1 formulas such that $\chi_i(x)$ is minimal in the sense that there is no rank 1 formula $\chi(x)$ implying $\chi_i(x)$ such that $d(\chi) < d(\chi_i)$, χ_i and χ_j are disjoint for $i \neq j$, and for every rank 1 formula $\chi(x)$ there exists j such that $\chi(x)$ is almost contained in $\bigvee_{i<j} \chi_i(x)$. To simplify the remainder of the discussion suppose that $d(\chi_i) = 1$ for all i; the general case requires no new ideas and is tiresome from the point of view of notation. Now define a sequence $\{i_j: j < \alpha\}$ as follows: $i_0 = 0$ and i_{j+1} = the least $i > i_j$ such that χ_i is not linked to any of $\chi_{i_0}, \ldots, \chi_{i_j}$. Any model M is characterized upto isomorphism by the dimensions of the formulas $\chi_{i_j}(x)$ in M, $j < \alpha$, together with the dimension of $\bigwedge_{i<\omega} \neg \psi_i(x)$. (The dimension of an s.m. type is defined in exactly the same way as that of an s.m. formula.) Further, the values of these dimensions have no interrelationship. Recall that $I(\kappa)$ denotes the number of nonisomorphic models of power κ. If $\alpha = \omega$ we have $I(\aleph_\beta) = |\beta + 1|^{\aleph_0}$ for all $\beta \geq 1$ because $\aleph_\gamma$ is a possible dimension of χ_i for each $\gamma \leq \beta$. Allowing finite dimensions where possible there can be at most $|\beta + \omega|^{\aleph_0}$ models of power $\aleph_\beta$ and $|\beta + \omega|^{\aleph_0} = |\beta + 1|^{\aleph_0}$ for all $\beta \geq 1$. Further, $I(\aleph_0)$ is either $\aleph_0$ or $2^{\aleph_0}$ according as all but a finite number of the χ_{i_j}'s are $\aleph_0$-categorical or not. Notice that $\bigwedge_{i<\omega} \neg \psi_i(x)$ can have any dimension whatever. If $\alpha < \omega$ then the number of independent dimensions is $|\alpha| + 1$. Thus $I(\aleph_\beta) = |\beta| + \aleph_0$ for all β.

4.4. The case in which there is an $\aleph_1$-categorical formula of rank 2. From 1.9 there is an $\aleph_1$-categorial formula $\vartheta(x)$ of rank 2 with no parameters. From 1.7 $\neg\,\vartheta(x)$ breaks up into a finite number of disjoint s.m. components which are almost definable without parameters. Any model is characterised upto isomorphism by the dimension of $\vartheta(x)$ together with the dimensions of the s.m. components of $\neg\,\vartheta(x)$. Suppose the theory is not $\aleph_1$-categorical then at least two of these dimensions are independent. Thus for $\beta \geqslant \omega$ we have $I(\aleph_\beta) = |\beta|$. If the theory is also not $\aleph_0$-categorical we have $I(\aleph_\beta) = \aleph_0$ for $\beta < \omega$. When the theory is $\aleph_0$-categorical each dimension is at least $\aleph_0$, thus we have $I(\aleph_\beta) < \aleph_0$ for $\beta < \omega$. In this case the exact value of $I(\aleph_\beta)$ depends on how many independent dimensions there are and on the group of automorphisms of the s.m. components of $\neg\,\vartheta(x)$. The group in question is the group of permutations on the set of s.m. components of $\neg\,\vartheta(x)$ induced by automorphisms of the prime model.

Until the end of the section we assume that there is a uniform decomposition of $x = x$ and that no rank 2 formula is $\aleph_1$-categorical.

4.5. The case of two infinite orthogonal equivalence relations. From § 3 there exists a formula $\varphi(x,y)$ such that $r(\exists^{\geqslant\omega} y\,\varphi(y,x) = 1)$ and (Φ_1) holds. From 1.7 there is a formula $\varepsilon(x,y)$ splitting $\exists^{\geqslant\omega} y\varphi(y,x)$ into components, i.e. ε defines an equivalence relation with $d(\exists^{\geqslant\omega} y\,\varphi(y,x))$ equivalence classes and if $\models \exists y\,\varepsilon(b,y)$ then $\exists^{\geqslant\omega} y\,\varphi(y,x) \wedge \varepsilon(b,x)$ is s.m. From the remark after Case 1.1 we can choose $\varphi(x,y)$ such that (Φ_3) :

$$\models \forall y \forall y'(\varepsilon(y,y') \wedge y \neq y'. \rightarrow \neg\exists x(\varphi(x,y) \wedge \varphi(x,y'))\ .$$

From Case 2 of § 3 we have $d(\exists^{\geqslant\omega} y\varphi(y,x)) \leqslant 2$. Suppose until further notice that $d(\exists^{\geqslant\omega} y\,\varphi(y,x)) = 2$. Using the strong minimality of the components $\varepsilon(x,b) \wedge \exists^{\geqslant\omega} y\varphi(y,x)$ it is easy to modify $\varphi(x,y)$ such that

$$\models \forall y(\exists x\,\varphi(x,y) \rightarrow. \varepsilon(y,y) \wedge \exists^{\geqslant\omega} y'(\neg\,\varepsilon(y,y') \wedge \exists x(\varphi(x,y) \wedge \varphi(x,y'))).$$

Thus we may suppose that $\exists^{\geqslant\omega} x\,\varphi(x,y)$, $\varepsilon(y,y)$ and $\exists x\,\varphi(x,y)$ are all equivalent. Let $\vartheta(x,y)$ denote $\neg\,\varepsilon(x,y) \wedge \varepsilon(x,x) \wedge$ $\wedge \neg\exists z(\varphi(z,x) \wedge \varphi(z,y))$. Consider nonalgebraic b in $\varepsilon(x,x)$, we know that $\models \exists^{<\omega} x\,\vartheta(x,b)$. If the solution set of $\vartheta(x,b)$ depends on b then it is easy to see that the two s.m. components of $\varepsilon(x,x)$ are linked, whence there is an $\aleph_1$-categorical formula of rank 2, namely $\exists y\, \varphi(x,y)$. It follows that $\vartheta(x,b)$ is independent of b and thus we can choose $\varphi(x,y)$ such that

$$\models \forall y \forall y'(\varepsilon(y,y) \wedge \varepsilon(y',y') \wedge \neg \varepsilon(y,y') \,.\!\rightarrow.\, \exists x(\varphi(x,y) \wedge \varphi(x,y'))) \,.$$

Finally, we may further modify φ such that

$$\models \forall x \forall y(\varphi(x,y) \rightarrow. \exists y'(\neg \varepsilon(y,y') \wedge \varphi(x,y'))) \,.$$

Now $\exists z(\varphi(x,z) \wedge \varphi(y,z))$ defines a relation which is the disjunction of two equivalence relations which are infinite in the sense that each class of one equivalence relation meets each class of the other in a finite nonempty set. Further, each equivalence class is s.m. if it is not algebraic.

Each model is characterized upto isomorphism by the dimensions of the s.m. components of $\neg \exists y\, \varphi(x,y)$ together with the dimensions of the two s.m. components of $\exists y\, \varphi(y,x)$. Thus the possibilities for the spectrum are exactly the same as under 4.4.

Now suppose that $d(\exists^{\geq\omega} y\, \varphi(y,x)) = 1$ and that $\varphi(x,b)$ and $\varphi(x,b')$ are linked when $\{b,b'\}$ is an independent set in $\exists^{\geq\omega} y\, \varphi(y,x)$. By exploiting the linkage we can find another formula $\psi(x,y)$ with the same properties as $\varphi(x,y)$ and such that

$$\models \forall y(\exists^{\geq\omega} y\, \psi(x,y) \rightarrow. \exists^{\geq\omega} y'(\exists^{\geq\omega} x\, \varphi(x,y') \wedge \exists x(\psi(x,y) \wedge \varphi(x,y')))) \,.$$

Really we should write $\psi(x,\bar{y})$ just as $\varphi(x,y)$ is really $\varphi(x,\bar{y})$. If $d(\exists^{\geq\omega} y\, \psi(y,x)) = 2$ we have the case considered above. If $d(\exists^{\geq\omega} y\psi(y,x)) = 1$ we can repeat the argument above with $\exists^{\geq\omega} y\, \varphi(y,x)$ and $\exists^{\geq\omega} y\, \psi(y,x)$ playing the roles of the two components of $\exists^{\geq\omega} y\, \varphi(y,x)$. Everything is as before except that now the two infinite equivalence relations can be distinguished.

4.6. The residual case. As in the last case we have $\varphi(x,y)$ satisfying (Φ_1) and (Φ_3) . However, now $d(\exists^{\geq\omega} y\, \varphi(y,x)) = 1$ and $\varphi(x,b)$, $\varphi(x,b')$ are unlinked s.m. formulas whenever $\{b,b'\}$ is an independent set in the s.m. type $\exists^{\geq\omega} y\, \varphi(y,x)$. Every rank 1 formula is almost contained in some formula $\psi(x) \vee \bigvee_{i<k} \varphi(x,b_i)$ where $r(\psi(x)) = 1$ and $\models \exists^{\geq\omega} y\, \varphi(y,b_i)$ for $i < k$.

With respect to the spectrum this is the most interesting case. Suppose $\exists^{\geq\omega} y\, \varphi(y,x)$ is connected, i.e. not disconnected in the sense of 4.1. We shall show that for $\varkappa \geq \aleph_1$ an arbitrary symmetric reflexive binary relation on $\varkappa$ can be coded into a model T of power $\varkappa$ and thus $I(\varkappa) = 2^{\varkappa}$ for all $\varkappa \geq \aleph_1$. There exist $A = \{b_0, b_1, c_0, \ldots, c_{k-1}\}$, independent in $\exists^{\geq\omega} y\, \varphi(y,x)$, and a such that $\models \exists^{\geq\omega} y\, \varphi(y,a)$ and

$$a \in \mathrm{pcl}\,(A) =_{\mathrm{dfn}} \mathrm{cl}(A) - \{\mathrm{cl}(X) : X \subsetneq A\} .$$

Let R be an arbitrary symmetric reflexive binary relation on $\varkappa \geq \aleph_1$. We form a model M(R) of power $\varkappa$ containing a set $\{b_0, b_1, c_0, \ldots, c_{k-1}\} \cup \{a^i : i < \ \}$ independent in $\exists^{\geq\omega} y\, \varphi(y,x)$. For $n < 2$, $i < \varkappa$ we choose $a^i_n \in \mathrm{pcl}\,(\{a^i, b_n\} \cup \mathrm{Rng}\, \bar{c})$. For $i,j < \varkappa$ we choose $a^{\{i,j\}} \in \mathrm{pcl}\,(\{a^i, a^j\} \cup \mathrm{Rng}\, \bar{c})$. We construct M(R) such that for arbitrary nonalgebraic b in $\exists^{\geq\omega} y\varphi(y,x)$, $\varphi(x,b)$ has dimension $\varkappa$ in M(R) if and only if $\varphi(x,b)$ is linked to $\varphi(x,a)$ where

$$a \in C =_{\mathrm{dfn}} \{ a^i_0, a^i_1 : i < \varkappa\} \cup \{a^{\{i,j\}} : i,j < \varkappa \text{ and } i\, R\, j\} .$$

Given the elements b_0, b_1, $\bar{c}$ we can recover from M(R) the isomorphism type of R. Since there are at most $\varkappa$ choices for $\langle b_0, b_1\rangle ^\frown \bar{c}$ in M(R), $\{M(R) : R \subset \varkappa \times \varkappa\}$ contains $2^{\varkappa}$ nonisomorphic models. In this case $I(\aleph_0)$ can take any of the values 1, $\aleph_0$, $2^{\aleph_0}$.

Now suppose that $\exists^{\geq\omega} y\, \varphi(y,x)$ is disconnected and let $\beta \geq \omega$. We shall show that $I(\aleph_\beta) = 2^{|\beta|}$. Let $\chi(x)$ be an infinite conjunction characterizing the unique nonalgebraic 1-type containing $\exists^{\geq\omega} y\varphi(y,x)$. There is a sequence $\{\psi_i(x) : i < \alpha\}$, $\alpha \leq \omega$, of disjoint formulas of rank ≤ 1, disjoint from $\vartheta(x) = \exists y(\varphi(x,y) \wedge \chi(y))$ such that

$$\models \forall x(\vartheta(x) \vee \bigvee_{i<\alpha} \psi_i(x)) \quad .$$

In any model M we can define an equivalence relation E on $\vartheta(M)$: $a_0\, E\, a_1$ if and only if there exist b_0, b_1 in M such that

$$\models \chi(b_0) \wedge \chi(b_1) \wedge \varphi(a_0,b_0) \wedge \varphi(a_1,b_1)$$

and b_0, b_1 are mutually algebraic. The equivalence classes under E are called blocks. By restricting M to a block we can speak of the isomorphism type of the block. If $\|M\| = \aleph_\beta$ the number of possible isomorphism types of blocks is $\geq |\beta|$ and $\leq 2^{|\beta|}$ because each block may be expressed as the disjunction of $\leq \aleph_0$ disjoint s.m. formulas $\varphi(x,b)$. Because $\exists^{\geq\omega} y\, \varphi(y,x)$ is disconnected any permutation of blocks which respects their isomorphism types is induced by some automorphism of M . Thus to characterize M it suffices to specify the isomorphism type of each structure $M \upharpoonright \psi_i$ obtained by restricting M to ψ_i , $i < \alpha$, and to specify the number of blocks of each kind. It follows easily that $I(\aleph_\beta) = 2^{|\beta|}$. One can also compute the possibilities for $I(\aleph_\beta)$ when $\beta < \omega$. Any values are possible which satisfy: $I(\aleph_0) \leq I(\aleph_1) = \ldots = I(\aleph_n) = \ldots$, $I(\aleph_0) \in \{1, \aleph_0, 2^{\aleph_0}\}$, and $I(\aleph_1) \in \{\aleph_0, 2^{\aleph_0}\}$.

References

[1] J.T. Baldwin , Countable theories categorical in uncountable power, Ph. D. Thesis, Simon Fraser Univesity, 1970.

[2] J.T. Baldwin , α_T is finite for $\aleph_1$-categorial T , Trans. Amer.Math.Soc. 181(1973), 37-52 .

[3] J.T. Baldwin and A.H. Lachlan, On strongly minimal sets, J.Symb.Logic 36(1971), 79-96.

[4] A.H. Lachlan, A property of stable theories, Fund.Math., 77(1972), 9-20.

[5] A.H. Lachlan, On the number of countable models of a countable superstable theory, Logic Methodology and the Philosophy of Science IV, North-Holland, Amsterdam 1973, 45-56.

[6] A.H. Lachlan, Two conjectures regarding the stability of ω-categorical theories, Fund.Math., 81(1974), 133-145.

[7] A.H. Lachlan, Theories with a finite number of models in an uncountable power are categorical, Pacific J.Math., to appear.

[8] D. Lascar, Ranks and definability in superstable theories, preprint.

[9] W.E. Marsh, On ω_1-categorical but not ω-categorical theories, Ph. D. Thesis, Dortmouth College 1966.

[10] M.D. Morley, Categoricity in power, Trans.Amer.Math.Soc. 114(1965), 514-538.

[11] S.Shelah, Stable theories, Israel J. Math. 7(1969),187-202.

[12] S.Shelah, Stability, the f.c.p. and superstability; model--theoretic properties of formulas in first-order theory, Ann.Math. Logic 3(1971), 271-362.

[13] S.Shelah, Uniqueness and characterization of prime models over sets for totally transcendental first-order theories, J.Symbolic Logic 37(1972), 107-113.

[14] S. Shelah, Categoricity of uncountable theories, Proceedings of Symposia in Pure Mathematics 25, Amer. Math. Soc., Providence 1974, 187-203.

[15] S. Shelah, forthcoming book on stability theory.

[16] A. Wierzejewski, On stability and products, to appear.

MEASURE AND CATEGORICITY IN α - RECURSION

by F. Lowenthal

The idea of this work grew out papers by Myhill [5] and Sacks [7 and 8] . It is only later that we heard of Yates' work [15 and 16] and of Martin's [4] .

Our first aim in this paper was to lift well known results in ordinary recursion theory (O.R.T.) to α-recursion : the idea was to show that α-recursion is easy when we use measure and categoricity. Unluckily (or luckily ?) the obvious measure is connected with an obvious topology, and this is not the topology which Kreissel would have mentioned as "The good one" (in [2]) if he had mentioned any topology. This difference will give us the oppertunity to show where and how α-recursion differs fundamentally from ORT for all countable admissible $\alpha > \omega$, and as a corollary why this is not (exactly) the case for $\alpha = \aleph_1$.

It has been claimed that measure and category are not useful tools to solve problems in recursion theory. It is true that the "topological" method and the use of Baire's theorem to prove existence theorems is not always as strong as purely recursion theoretic proofs:

We shall show that there are 2 incomparable α-degrees, but not that they are α-r.e. ; Shoenfield has shown that there are $\aleph_1$ pairwise incomparable degrees (in O.R.T.) [11] and his result is slightly stronger than Myhill's [5].
Nevertheless we think that this "topological" method is easier to understand and gives a better picture of the situation. Moreover, in an unpublished paper, Lacombe has constructed (in O.R.T.) a continuum of functions of minimal degree : his argument combines recursion theoretic and descriptive set-theoretic methods. More recently D.A.Martin

[4] proved, in O.R.T., that there is a degree less than $\underline{0''}$ with no minimal predecessor, using the fact that the minimal degrees form a meager set. (Yates used a delicate priority argument to show that there is such a degree, less than $\underline{0'}$).

It has also been claimed that "α-recursion theory is so nice that every result valid for O.R.T. can be lifted to α-recursion". We think that this is not completely true and that our result concerning the "strange case of the regular sets" might shed little bit more light on the problems with which a student of α-recursion has to struggle.

Our main results in recursion concern the existence of two incomparable α-degrees, the set of all α-degrees $\underline{d}$ such that $\underline{d}$ is the union of 2 incomparable α-degrees less than $\underline{d}$, the collection of all regular subsets of α and of the subgeneric sets. Our proofs will use measure-theoretic and topological arguments combined with pure recursion-theoretic arguments.

The author wishes to express here his thanks to the M.I.T. team who introduced him to the last ones and to his collegues of the University of Brussels, who helped him to find his way through the first ones.

§ 1. INTRODUCTION

First definitions and elementary facts.

DEFINITION 1. $M = \langle M, \in \upharpoonright M, R_1, \ldots, R_k \rangle$ is an admissible structure (or : admissible relative to the relations $R_1, \ldots, R_k$) iff

(0) M is transitive and non empty

(1) M is closed under pairing and union

(2) M satisfies $\Delta_0(R_1, \ldots, R_k)$-Separation
i.e. $M \models (\exists x)(\forall y)(y \in x \rightarrow y \in a \wedge \varphi(y))$
for each $a \in M$ and for each $\Delta_0(R_1, \ldots, R_k)$-formula $\varphi(y)$ (including parameters from M).

(3) M satisfies $\Delta_0(R_1, \ldots, R_k)$-Collection
i.e. $M \models (\forall x \in a)(\exists y)\ \varphi(x,y) \rightarrow (\exists w)(\forall x \in a)(\exists y \in w)\ \varphi(x,y)$
for each $a \in M$ and for each $\Delta_0(R_1, \ldots, R_k)$-formula $\varphi(x,y)$ (with parameters from M).

DEFINITION 2 .

(i) α is an <u>admissible</u> <u>ordinal</u> iff $M_\alpha = \langle L_\alpha , \in \upharpoonright L_\alpha \rangle$ is admissible.

(ii) <u>Let</u> f be a partial function from α to α <u>Then</u> f is <u>partial</u> α-<u>recursive</u> iff f is Σ_1 over M_α

(iii) $A \subseteq \alpha$ is α-<u>recursively</u> <u>enumerable</u> (α-r.e.) iff A is the domain of a partial α-recursive function

(iv) $A \subseteq \alpha$ is α-<u>recursive</u> iff A and $\alpha - A$ are α-r.e.

(v) $A \subseteq \alpha$ is α-<u>finite</u> iff $A \in L_\alpha \cap 2^\alpha$ where 2^α denotes the power set of α .

From now on we will identify any subset of α with its characteristic function. ${}^\alpha 2$ denotes the set of charcteristic functions.

<u>Fact</u> (<u>Folklore</u>)
Let α be admissible, then
i) we can Gödel-number the α-(partial) recursive functions using ordinals less than α
ii) every α-finite set has an index less than α .

DEFINITION 3. $A \subseteq \alpha$ is α-<u>regular</u> iff $A \cap K$ is α-finite for every α-finite K iff $A \cap \gamma$ is α-recursive for all ordinals $\gamma < \alpha$. Note that all subsets of ω are regular.

DEFINITION 4. Let α be an admissible ordinal. $A \subseteq \alpha$ is <u>subgeneric</u> iff $\langle L_\alpha[A], \in \upharpoonright L_\alpha[A], A \rangle$ is admissible.

DEFINITION 5. Let α be admissible and R_ε the εth α-r.e. set. Let $A \subseteq \alpha$ and $\varphi \subseteq \alpha \times \alpha$ a partial function. We have (weakly reducible to) :

$\varphi \leq_w A$ iff there exists an ε such that :

$$\varphi(\gamma) \simeq \delta \iff (\exists K)(\exists H)[\langle K,H,\gamma,\delta\rangle \in R_\varepsilon \wedge K \subseteq A \wedge H \cap A = \Phi]$$

and conversely we define :

$$\{\varepsilon\}^A(\gamma) \simeq \delta \iff (\exists K)(\exists H)[\langle K,H,\gamma,\delta\rangle \in R_\varepsilon \wedge K \subseteq A \wedge H \cap A = \Phi]$$

where K and H range over the α-finite sets.

Unfortunately $\{\varepsilon\}^A$ is **not** necessarily single valued. Machtey proved that for all $\varepsilon < \alpha$ there is an $\varepsilon^* < \alpha$, obtained effectively from ε, such that $\{\varepsilon^*\}^A$ is single valued and such that for all regular $A \subseteq \alpha$, _if_ $\varphi = \{\varepsilon\}^A$, _then_ $\varphi = \{\varepsilon^*\}^A$.

DEFINITION 6. If $\{e\}^A(\gamma) \simeq \delta$ because of the α-finite sets K and H, as in definition 5.
Then we say that (K,H) is an α-finite _neighbourhood_ condition. (In fact, for every set B such that $K \subseteq B$ and $H \cap B = \phi$, we have $\{e\}^B(\gamma) \simeq \delta$ because of (K,H)).

From now on α will denote a countable admissible ordinal, $\alpha > \omega$ and $F = {}^{\alpha}2$ = the set of characteristic functions of subsets of α.

Define a measure μ on F as follows : let ν be the probability measure on $\{0,1\}$ and let μ be the product measure.

By analogy with the usual product topology and the compact-open topology on function spaces, we will define two topologies on F, T_1 and T_0, as follows : a basis for T_1 is the collection of all basic open sets (b.o.s) $N(K,L) = \{f \in F \mid f[K] = \{1\} \wedge f[L] = \{0\}\}$ where K and L are disjoint α-finite sets.
A basis for T_0 is the collection of all b.o.s.
$N(k,l) = \{f \in F \mid f[k] = \{1\} \wedge f[l] = \{0\}\}$, where k and l are disjoint _finite_ subsets of α.

Clearly T_1 is the topology which should be used to study α-recursion but T_0 is the topology which is the most adequate to study the measure μ.

It is clear that $N(K,L)$ (for T_1) and $N(k,l)$ (for T_0) are clopen.

__LEMMA__ (_Folklore_). Two topologies π and π' for a same space X are equal iff there are bases B (for π) and B' (for π') such that the following conditions hold :

1) for each $U \in B$ and each $x \in U$, there is a $U' \in B'$ with $x \in U' \subseteq U$.
2) for each $U' \in B'$ and each $x \in U'$, there is a $U \in B$ with $x \in U \subseteq U'$.

PROPOSITION 1. T_0 is a proper subset of T_1.

Proof. As each finite set is also α-finite, it is clear that condition 1) of the lemma holds. As $\alpha > \omega$, we can choose $U' = N(\omega,\{\omega+1\})$; clearly there is no $N(k,l) \subseteq N(\omega,\{\omega+1\})$ and hence 2) of the lemma does not hold.

Note that F is regular for both topologies.

PROPOSITION 2. i) T_0 is compact ,
ii) T_1 is not compact but it is paracompact.

Proof. i) T_0 is the product topology.

ii,1^0) Assume that T_1 is compact.
Let $i : (F,T_1) \to (F,T_0)$ be the identity. Hence i is a 1-1 continupus map from a compact space onto a Hausdorff space. Hence i is an homeomorphism.

ii,2^0) As there are only countably many α-finite sets, T_1 has a countable basis, hence it is a Lindelöf space. T_1 is a regular space (let $U \subseteq F$ be open for T_1, $x \in U$; then there is a b.o.s. $B \subseteq U$ such that $x \in B$ and B is clopen). As regularity and paracompactness are equivalent for such spaces, (F,T_1) is paracompact.

PROPOSITION 3 (Baire's theorem). For both topologies, F is of second category.

Proof. (i) (F,T_0) is a compact Haussdorf space. Assume (towards a contradiction) that F is of 1st category : F is the union of a denumerable sequence of nondense sets. Let $F_0, F_1, \ldots$ be the closures of the members of this sequences. Then $F = \bigcup_{i=0}^{\infty} F_i$ and $(\forall i)$ [F_i contains no b.o.s.] .

F_0 is closed, thus $F - F_0$ is open and $F - F_0 \supset S_0$ where S_0 is a b.o.s. such that $F_0 \cap S_0 = \Phi$.

F_1 is also closed, so $F_0 \cup F_1$ is closed and $U_1 = F - (F_0 \cup F_1)$ is open.

$U_1 \cap S_0$ is open. $U_1 \cap S_0 \neq \Phi$ (otherwise $S_0 \subset F_1$)

$U_1 \cap S_0$ contains a b.o.s. S_1 and $S_1 \cap (F_0 \cup F_1) = \Phi$

. . .

thus we obtain a sequence of b.o.s. $S_0 \supset S_1 \supset S_2 \supset \ldots$ such that $(\forall n)\ [S_n \cap (\bigcup_{i=0}^{n} F_i) = \Phi]$. Hence any member f of $\bigcap_{i\in\omega} S_i$ belongs to $F - (\bigcup_{i=0}^{\infty} F_i)$.

As each S_n is a set of functions which extend some condition $(f[k] = \{1\}, f[l] = \{0\})$ and as this condition (k,l) is itself an extension of the conditions used for all S_m $(m < n)$ the intersection of all S_n's is not empty : $\bigcap_i S_i$ could be at worst a singleton $\{f\}$, but then $f \in F - (\bigcup F_i)$.
(ii) Use exactly the same proof for (F,T_1) but with b.o.s. for T_1 and α-finite neighbourhood conditions.

Remark. For both topologies F is Haussdorf, Londelöf, paracompact and normal. The $N(k,l)$ $(N(K,L))$ form a countable basis of clopen sets.•

§ 2. GENERALISATION OF CLASSICAL RESULTS

THEOREM 1. If $g \in F$ is not α-recursive and $\mathfrak{A} = \{f \in F | g \leq_w f\}$ ($\mathfrak{A}$ is the cone above g)
Then (i) $\mu(\mathfrak{A}) = 0$
(ii) $\mathfrak{A}$ is of 1st category for both topologies (part (ii) for the topology T_1 was anounced independently by R.JHU) .

Proof.
(i) $\mathfrak{A} = \{f \in F | g \leq_w f\}$
1°) by σ-additivity it is enough to prove that $\mu(S) = 0$ where $S = \{f \in F \mid g \leq_w f$ via $e\}$ (where $e \in \alpha$ is the Gödel Number of same reduction procedure).
2°) S is measurable
$f \in S \iff (\forall \beta < \alpha)(g(\beta) = \{e\}^f(\beta))$
$\iff (\forall \beta < \alpha)(\exists$ an α-finite neighbourhood condition (I,J) such that $g(\beta) = \{e\}^{(I,J)}(\beta))$
$\iff f \in \bigcap_{\beta<\alpha} \bigcup_{\alpha\text{-finite neighb.cond.}} [\Psi]$
* **

(*) : α is countable, this is a countable intersection
(**) : there are only countably many neighbourhood conditions, this is thus a countable union

Ψ is an α-recursive expression, thus it is measurable.
3°) Assume $\mu(S) = 4.m > 0$. We show then that g must be α-recursive.
S is measurable, thus $S \subseteq G$, G open with $\mu(G - S) \leq m$.
G is open, thus there exists a finite union B of b.o.s. such

that $B \subseteq G$ and $\mu(G-B) \leq m$.
Hence $\mu(B-S) \leq m$, $\mu(B \cap S) \geq 3m$.

We try now to compute $g(n)$, for some fixed $n < \alpha$. Remember that, by definition, g is a total function.

4°) Define $T_i = \{f \in B \mid \{e\}^f(n)\downarrow$ and $\{e\}^f(n) = i\}$ for each $i < \alpha$.

We have $\begin{cases} \mu\, T_i \leq m & \text{if } i \neq g(n) \\ \mu\, T_{g(n)} \geq 3m \end{cases}$

[Remark that if $i \neq g(n)$, them $T_i \subseteq B \setminus S$ and if $i = g(n)$, then $T_i \supseteq B \cap S$] .

Let $\{I_\delta \mid \delta < \alpha\}$ be an enumeration of all finite sets of α-finite neighbourhood conditions.

Stage σ

There are two cases :

(σ.1) I_σ is not consistent : i.e. it is not the case that all members of I_σ yield to a convergent computation such that all these computations have the same result.
Then throw I_σ away.

(σ.2) I_σ is comsistent : all conditions yield to the same $k \in \alpha$.
Consider $\{f \mid \{e\}^f(n) = k$ because of one of the conditions enumerated in $I_\sigma\}$.(i.e. consider the set of functions belonging to the b.o.s. $N(K,L)$ defined by some neighbourhood conditions $(K,L) \in I_\sigma$) . This set is measurable and has measure $\neq 0$ iff the α-finite sets used in the condition are really finite.

In fact we want $\{f \mid (\exists (K,J) \in I_\sigma)(\{e\}^f(n) = \{e\}^{(K,J)}(n) = k)\}$

$$= \bigcup_{j=1}^{j=p<\omega} \{f \mid \{e\}^f(n) = \{e\}^{(K_j,J_j)}(n) = k\}$$

(Remember : I_σ is a finite set of conditions).

But if one of the sets K_j, J_j is infinite, then the set of functions extending the condition (K_j,J_j) has measure 0 .

We know that $\mu(T_{g(n)}) \geq 3m$. Hence, for some $\lambda < \alpha$, we will enumerate I_λ : a consistent set of neighbourhood conditions giving rise to a set of functions of size $> m$. We may then conclude that the common value k given by all the pairs (K_j,J_j) is the real value of $g(n)$ [since $\mu(T_i) \leq m$ if $i \neq g(n)$] .

Hence g is α-recursive.

(ii) $\mathcal{A} = \{f \mid g \leq_w f\}$

1°) It is enough to prove that $\mathcal{A}$ is of first category for T_1. It is even enough to show that $B = \{f \mid g \leq_w f \text{ via } e\}$ is nondense.

Let $\bar{B}$ be the closure of B.

Let S be a non empty b.o.s. such that $S \subseteq \bar{B}$. Then $S \cap B \neq \phi$.

there is a function $\tilde{h} \in F$ s.t. $g = \{e\}^{\tilde{h}}$.

2°) Say $S = N(K,J)$.

If (K',J') is a neighbourhood condition extending (K,J), then $\{e\}^{[(K',J')]} \subset g$, where

$$\{e\}^{[(K',J')]}(x) = \begin{cases} \{e\}^{(K',J')}(x) & \text{if } x \in K' \cup J' \\ \uparrow & \text{otherwise} \end{cases}$$

[<u>Subproof</u> : assume $\{e\}^{[(K',J')]} \not\subset g$: then

(1) $(\exists \beta)(\{e\}^{[(K',J')]}(\beta) \neq g(\beta)))$ but in this case

$N(K',J') \subseteq N(K,J) \subseteq \bar{B}$

$N(K',J') \cap A \neq \phi$

but by (1) $N(K',J') \cap A = \phi$]

3°) Hence $g = \{(x,y) \mid (\exists (K,J))(\{e\}^{[(K,J)]}(x) = y)\}$

[Subproof : <u>If</u> $(\exists (K,J))(\{e\}^{[(K,J)]}(x) = y)$ <u>Then</u> $g(x) = y$ by 2°) and <u>if</u> $g(x) = y$ <u>then</u> $g(x) = \{e\}^h(x) = y$ for some h (by 1°), i.e. there is an α-finite computation from h which gives the correct value for g.

Thus, there is an α-finite condition (K,J) such that $\{e\}^{[(K,J)]}(x) = y$.]

But in this case g is α-recursive.

DEFINITION. Let A be a set of α-degrees. Let A^* be the set of function $f \in F$, whose degrees are member of A. We say that A is measurable iff A^* is μ-measurable and we define the measure of A to be μA^*.

(Notation : This measure on sets of α-degrees will also be named μ). We define 2 topologies for sets of α-degrees as follows : A (set of α-degrees) is open iff A^* $(\subseteq F)$ is open.

COROLLARY 1. Let $\underline{d}$ be an α-degree greater than $\underline{0}$. Then the set of all degrees incomparable with $\underline{d}$ has measure 1 and is of 2nd category (for T_0 and T_1).

Using the σ-additivity of the measure or the definition of 1st and 2nd category, it is easy to prove the following result (using ideas similar to that of [5] for O.R.T.).

COROLLARY 2. There exists an uncountable family of pairwise incomparable α-degrees (α countable).

Proof.

1^0) If f is not α-recursive, then the family of all functions of degree $\geq$ the degree of f is of first category (by theorem 1).

2^0) If f is not α-recursive, then the family of all functions of α-degree comparable with the α-degree of f is of first category (by : α is countable and 1^0) and of measure 0.

3^0) If $\mathfrak{A}$ is a finite or countable family of non α-recursive functions, then the family of all functions of α-degree comparable with that of some function belonging to $\mathfrak{A}$ is of measure 0 and of first category (by definition of first category).

4^0) If $\mathfrak{A}$ is a finite or countable family of non α-recursive functions, then there exists a non α-recursive function of α-degree incomparable with the α-degree of every function in $\mathfrak{A}$ (by : F is of 2nd category and $\mu F = 1$).

5^0) There exists an uncountable family of pairwise incomparable degrees.

THEOREM. The set of all α-degrees $\underline{d}$ such that $\underline{d}$ is the union of 2 incomparable degrees less than $\underline{d}$ has measure 1.

Proof. Let $\mathfrak{A} = \{\underline{d}: (\exists \underline{a},\underline{b})(\underline{d} = \underline{a} \cup \underline{b} \wedge \underline{a} \not\leq \underline{b} \wedge \underline{b} \not\leq \underline{a} \wedge \underline{a} < \underline{d} \wedge \underline{b} < \underline{d})\}$
Let $F^2 = F \times F$ and consider the space (F, T_1) with the measure μ.
Put on F^2 the product measure and product topology.
Let $S_e = \{(f,g) \in F^2 \mid f \leq_w g$ via G.N. $e\}$

$1^0.1$ Claim S_e is measurable.

Subproof. let $n, e < \alpha$ be fixed, let $S_{n,e} = \{(f,g) \mid \{e\}^g(n)$ converges and equals $f(n)\}$ consider $\{g \mid \{e\}^g(n)\}$.

$\{e\}^g(n) \iff (\exists\, m < \alpha)(\exists (K,J)$ α-finite nbd condition)
$(g \in N(K,J) \wedge \{e\}^{(K,J)}(n) = m)$

$\{g \mid (\exists\, m < \alpha)(\{e\}^g(n) = m)\}$ is a countable union of b.o.s. now let $m < \alpha$ be fixed.

$T_m = \{g \mid \{e\}^g(n) = m\}$ is a countable union of b.o.s. thus T_m is measurable.

$B_m = \{f \mid f(n) = m\}$ is a b.o.s., hence measurable.

But $(f,g) \in S_{n,e} \iff (\exists m < \alpha)(f(n) = m \wedge \{e\}^g(n) = m)$

$\iff (\exists m < \alpha)[(f,g) \in B_m \times T_m]$

Hence $S_{n,e} = \bigcup_{m<\alpha} (B_m \times T_m) \subseteq F^2$

$S_{n,e}$ is a countable union of products of b.o.s. and countable unions of b.o.s.

$S_{n,e}$ is measurable.

But $S_e = \bigcap_{n<\alpha} S_{n,e}$ is a countable intersection of measurable sets, hence it is measurable.

$1^0.2$ $S = \{(f,g) \in F^2 \mid f \leq_w g \text{ or } g \leq_w f\}$

$S = \bigcup_{e<\alpha} S_e \cup \bigcup_{e<\alpha} S'_{e'}$ with $S'_{e'} = \{(f,g) \in F^2 \mid (g,f) \in S_{e'}\}$

S is measurable.

2^0) $S_e = \{(f,g) \in F^2 \mid f \leq_w g$ via G.N. $e\}$, assuming that f is not recursive we have : S has measure 0 by theorem 1 and Fubini [$\mu(\{g \mid f \leq g\}) = 0$ by thm 1 and $\mu(\{g \mid g \leq f\}) = 0$ by countability].

3^0) Define $\psi : F \to F^2$ such that a subset A, measurable, of F, with measure m has image a subset $\psi(A)$, measurable, of F^2 with measure m :

for each $f \in F$ let $u_f(n) = f(2n)$

$v_f(n) = f(2n+1)$

and $\psi(f) = \langle u_f, v_f \rangle$

ψ is $1-1$ and onto.

It is easy to check that ψ is measure preserving on any b.o.s.:

Let $A \subseteq F$ be a b.o.s.

$A = \{f \in F \mid f \supseteq f_0\}$

$\psi''A = \{\psi(f) \mid f \in A\} = \{\langle u_f, v_f \rangle \mid u_f(n) = f(2n) \wedge$

$\wedge v_f(n) = f(2n+1)$ if $n \in \text{dom } f_0\}$

$\psi''A = \{\langle h,g \rangle \mid h(n) = f_0(2n) \wedge g(n) = f_0(2n+1)$ if $n \in \text{dom } f_0\}$

4^0) $T = \{f \in F \mid u_f \leq_w v_f \text{ or } v_f \leq_w u_f\}$

$\psi T = \{\langle u,v \rangle \in F^2 \mid u \leq_w v \text{ or } v \leq_w u\} = S$

ψT has measure 0

$\mu T = 0$ and $\deg(f) = \deg(u_f) \cup \deg(v_f)$

$\mu(\{\underline{d} \mid (\exists \underline{a},\underline{b})(\underline{d} = \underline{a} \cup \underline{b} \wedge (\underline{a} \leq \underline{b} \text{ or } \underline{b} \leq \underline{a})\}) = 0$

5^0) Thus : the set of degrees which are union of two smaller but incomparable degrees has measure 1.

COROLLARY . The set of all minimal degrees has measure 0 .

Proof. $\underline{d}$ is minimal iff $\underline{a} \leq \underline{d} \Rightarrow \underline{a} = \underline{0}$ or $\underline{a} = \underline{d}$
$\underline{d} = \underline{a} \cup \underline{b} \Rightarrow \underline{a} \leq \underline{b}$ or $\underline{b} \leq \underline{a}$

It is possible to generalise other classical results, but we turn now to some results which are specific of α-recursion, for $\alpha > \omega$.

§ 3. THE STRANGE CASE OF THE REGULAR SETS

For certain parts of this paragraph we will prefer to look at F as the power set of α , this has obviously no importance, as far as the concept themselve are concerned. Notation : we will say that $f \in F$ is regular iff it is the characteristic function of a regular set.

As we have 2 topologies on F we might try to consider dense subsets of F as "big". In fact this fails : very small collections (e.g. the recursive sets) are dense. We will need later a similar fact for the regular sets : (By regular we mean α-regular, througout this chapter).

LEMMA. Let $\mathfrak{A}$ be the collection of all regular subsets of α . Then $\mathfrak{A}$ and $F-\mathfrak{A}$ are dense (for both topologies).

Proof.
1°) $\mathfrak{A}$ is dense.
We want to prove that the closure of $\mathfrak{A}$, $\overline{\mathfrak{A}} = F$.
It is enough to show for every b.o.s. U containing f , $U \cap \mathfrak{A} \neq \phi$.
Let $f \in F$. Assume $f \in N(K,L)$ [the proof would be similar for T_0], then define $g \in F$ as follows :

$$g(\beta) = \begin{cases} 0 & \text{if } \beta \in K \\ 1 & \text{o.w.} \end{cases}$$

g is an α-recursive function : it is the characteristic function of K , hence of regular set.
Thus $N(K,L) \cap \mathfrak{A} \neq \phi$.

2°) $F-\mathfrak{A}$ is dense.
Let $f \in F$. Let $N(K,L)$ be a b.o.s. (for T_1) containing f .
Let β be the least limit ordinal $\geq U(K \cup L)$.

Define g as follows :

$$g(\gamma) = \begin{cases} 0 & \text{if } \gamma \in K \\ 1 & \text{if } \gamma \in \beta - K \text{ or if } \gamma \geq \beta + \omega \\ \text{a copy of the characteristic function of a real} & \\ X \notin L \quad (\omega < \alpha < \aleph_1) & \text{if } \beta \leq \gamma^{\alpha} < \beta + \omega \end{cases}$$

Hence g is not α-regular.
Then $g \in N(K,L)$ and $N(K,L) \cap (F \setminus \mathcal{Q}) = \phi$.

COROLLARY. $\mathcal{Q}$ and $F - \mathcal{Q}$ have both empty interiors.

Note that it is easy to prove similarily that the collection of all α-recursive functions is also dense in F , hence the notion of density alone cannot be very powerful (for T_1 or for T_0).

THEOREM. Let $\mathcal{Q} = \{f \in F \mid f$ is regular$\}$.
f is regular iff $(\forall \beta < \alpha)$ $[f \upharpoonright \beta$ is α-finite$]$
iff f is the characteristic function of a set $X \subset \alpha$ such that $(\forall \beta)(X \cap \beta$ is α-recursive).
From now on we will speak in term of sets instead of characteristic functions and we will denote $X \cap \beta$ by X_β .
i) $\mathcal{Q}$ is measurable and $\mu(\mathcal{Q}) = 0$
ii) $\mathcal{Q}$ is of first category for T_0
iii) $\mathcal{Q}$ is of 2nd category for T_1 .

<u>Proof</u>. i) X is regular $\Leftrightarrow (\forall \beta < \alpha)(X_\beta$ is α-recursive)
$\Leftrightarrow (\forall \beta < \alpha)(\exists (I,J))(\Psi)$
where (I,J) is an α-finite neighbourhood condition and Ψ is an α-recursive expression.
Thus $\mathcal{Q} = \bigcap_{\beta<\alpha} \bigcup_{(I,J)} [\alpha\text{-recursive expression}]$
Hence $\mathcal{Q}$ is clearly measurable.
But $\mathcal{Q} \subseteq \mathcal{C} = \{X \mid X_\omega$ is α-recursive$\}$.
As $\mathcal{C}$ is obviously measurable, it is enough to show that $\mu(\mathcal{C}) = 0$.
Assume $X \in \mathcal{C}$, then X_ω is α-recursive (say via procedure e , because of the α-finite neighbourhood condition (K,J)). As ω is infinite, the set $\{Y \mid Y_\omega = X_\omega\}$ has measure 0 (this set is simply $N(K,L)$ where $K = \{n \mid n \in X \cap \omega\}$ and $L = \omega - K$, thus at least one of K or L must be infinite).
As we must only consider countably many such sets (one for each reduction procedure), we have proved that $\mu(\mathcal{C}) = 0$.

ii) By a similar argument it is easy to show that $\mathcal{A}$ is of 1st category for the topology T_0 : it is enough to exhibit a countable covering of $\mathcal{C}$ (hence of $\mathcal{A}$) by closed nondense sets.
Let K_e be the eth α-finite subset of ω

$$\mathcal{C} = \bigcup_e \{X \mid X \supset K_e \wedge X \cap (\omega - K_e) = \Phi\}$$

$$= \bigcup_e \{X \mid X_\omega = K_e\}$$

Consider, for a given e , $\{X \mid X_\omega = K_e\} = V$.
For each $n \in K_e$, consider the b.o.s. (for T_0) $V_n = \{X \mid n \in X\}$ and for each $n \in (\omega - K_e)$ consider the b.o.s. (for T_0) $V_n = \{X \mid n \notin X\}$. For each n , V_n is closed, hence $V = \bigcap_{n<\omega} V_n$ is closed.

<u>Claim</u>. V is nondense : the closure of V is V . Assume $V \supset U \neq \Phi$ with U open for T_0 , then $U \supseteq D \neq \Phi$ where D is a b.o.s. for T_0 . But then $\mu(D) > 0$ and thus $\mu(V) > 0$. As $V \subseteq \mathcal{C}$ this implies that $\mu(\mathcal{C}) > 0$.

Hence, for each $e < \alpha$, $\{X \mid X_\omega = K_e\}$ is nondense and $\mathcal{C}$ is a countable union of such sets. Thus $\mathcal{C}$ is of 1st category.

This "topological" result is in fact based upon measure theoretic notions, i.e. : each b.o.s. for T_0 has positive measure this establishes a link between category and measure.! Note that this argument cannot be used for T_1 , as there are b.o.s. (for T_1) of measure 0 .

LEMMA A. Let $\{K_e \mid e < \alpha\}$ be an enumeration of the α-finite subsets of α .
X is regular iff $(\forall \beta < \alpha)$ (X_β is α-recursive).
Hence to show that X is regular, it is enough to exhibit for every β a set K_e such that $X_\beta = K_e \cap \beta$.

Thus $\mathcal{A} = \bigcap_\beta \bigcup_e \{X \mid X_\beta = K_e \cap \beta\}$

Now, given e and β , $K_e \cap \beta$ is clearly α-finite. Hence $\{X \mid X_\beta = K_e \cap \beta\}$ is a b.o.s. for T_1 and $H_\beta = \bigcup_e \{X \mid X_\beta = K_e \cap \beta\}$ is open.
It is even dense open in F (for T_1) .
(If $Y \subseteq \alpha$ and $Y \in N(U,V)$, then $U \in N(U,V)$ and $U \in H_\beta$) .
Hence $\mathcal{A} = \bigcap_\beta H_\beta$ is a G_δ set, such that for all β , H_β is dense open.
Note that this argument does not work for (F,T_0) as $\{X \mid X_\omega = \omega\}$ is <u>not</u> open in T_0 .

Note also that it is trivial to prove (e.g. by Baire's theorem) that $\mathcal{A}$ is dense in F for T_1 .

LEMMA B. Let (Y,τ) be a topological space.

<u>If</u> $\left[\begin{array}{l} F \text{ is a closed nondense subset of } Y \\ X \text{ is dense in } Y \end{array}\right.$

<u>Then</u> $F \cap X$ is nondense in X .

<u>Proof</u>. Let $U = Y - F$.

U is open and dense in Y .

<u>Claim.</u> $U \cap X$ is open and dense in X .

<u>Proof of the claim</u>.

1^0) U is open in Y

$U \cap X$ is open in X .

2^0) $U \cap X$ is dense in X iff

$(\forall a \in X)(\forall V_X$ open neighbourhood in X of a) we have

$V_X \cap (U \cap X) \neq \Phi$

let $a \in X$, V' be an open neighbourhood in X of a.

$V' = V \cap X$ for some V , open in Y , <u>with $a \in V$</u>

$V \cap U \neq \Phi$

Assume $V \cap (U \cap X) = \Phi$

but $V \cap U \neq \Phi$ (say $b \in V \cap U$)

V and U are open in Y

$V \cap U$ is an open neighbourhood of b in Y

but $(V \cap U) \cap X = V \cap (U \cap X) = \Phi$

X is not dense in Y

thus $V \cap (U \cap X) = (V \cap X) \cap (U \cap X) \neq \Phi$

$V' \cap (U \cap X) \neq \phi$

a is in the closure of $(U \cap X)$ in X

$U \cap X$ is open and dense in X

Assume that $F \cap X$ is not nondense in X

1^0) $F \cap X$ is closed in X

2^0) we assume that there exists a set $V \subseteq F \cap X$, $V \neq \Phi$ with V open in X .

Let $a \in V$: V is an open neighbourhood of a in X , disjoint of $U \cap X$

$U \cap X$ is not dense in X .

LEMMA C. Let (Y,τ) be a topological space.

If $\left[\begin{array}{l} Y \text{ is of 2nd category in itself} \\ X \text{ is dense in } Y \\ X \text{ is a Baire space for the induced topology} \end{array}\right.$

Then X is of 2nd category in Y.

Proof. Assume it is not

$$X \subseteq \bigcup_{n<\omega} F_n \text{ , with } F_n \text{ closed nondense in } Y$$

$$X \subseteq \bigcup_{n<\omega} (F_n \cap X)$$

but by lemma B , $F_n \cap X$ is nondense in X (for each n)

X is not a Baire space for the induced topology.

We turn now to the proof of part (iii) of the theorem. By the previous lemma it is enough to show that $\mathfrak{A} = \{X \subseteq \alpha \mid X \text{ is regular}\}$ is a Baire space for the induced topology.

Let $\mathfrak{A} = \bigcap_{n<\omega} H_n$. where each H_n is dense open in F (possible by lemma A).

We proved previously that $\mathfrak{A}$ is dense in F .

Now, if $G_1, G_2, \ldots$ is a sequence of dense open sets in $\mathfrak{A}$ (for the induced topology), then for each k , $G_k = J_k \cap \mathfrak{A}$, where J_k is dense open in F .

But now $J_1, H_1, J_2, H_2, J_3, H_3, \ldots$ is a sequence of dense open sets in the Baire space F , and hence $\bigcap (J_k \cap H_k) = (\bigcap J_k) \cap (\bigcap H_k) = (\bigcap J_k) \cap \mathfrak{A} = \bigcap G_k$ is dense in F , and hence in $\mathfrak{A}$.

Thus $\mathfrak{A}$ is a Baire space.

One could ask now the following question : ω is "nice" because all subsets of ω are regular ; does there exist another "nice" countable admissible ordinal ?

The preceding theorem has the following consequence:

COROLLARY. For all admissible and countable ordinals α , such that $\alpha > \omega$, the set of all regular subsets of α is not equal to 2^{α} . This is nevertheless the case for $\alpha = \omega$ or (assuming $V = L$) for $\alpha = \aleph_1$.

Proof. The result is trivial for $\alpha = \omega$: every bounded subset of ω is finite, hence recursive.

For $\omega < \alpha < \aleph_1$ the result is a consequence of the fact that $\mu(\mathfrak{A}) = 0 < 1 = \mu(2^{\alpha})$.
For $\alpha = \aleph_1$ it is enough to show that every bounded initial segment of α can be constructed before level $\aleph_1$ (in the L-hierarchy).

§ 4. FURTHER RESULT AND OPEN PROBLEMS

THEOREM. The set of all subgeneric subsets of α has measure 1.

<u>Sketch of the proof</u>. A subgeneric subset of α , X , is a set such that all computations from X are "short" (i.e. take less than α steps, whenever convergent) but not necessarily α-finite (i.e. the computation tree is <u>not</u> necessarily a member of L_α) .
In O.R.T., Sacks proved [8] that the set of reals X such that $\omega_1^X = \omega_1^{cK}$ has measure 1 ; he also proved in [10] that every countable admissible α is of the form ω_1^T , for some real T .
In O.R.T., the reals X such that $\omega_1^X = \omega_1^{cK}$ are clearly the ω-subgeneric sets. Assume $\alpha = \omega_1^T$, then the reals X such that

$$\omega_1^{(T,X)} = \alpha = \omega_1^T$$

are α-subgeneric reals. By a simple modification (relativisation) of the proof given in [8], it is easy to show that these reals form a set of measure 1. Hence, the set of all subgeneric sets has measure 1.

<u>Problem</u> 1. How can we generalise the arguments used by Martin and Yates in [4, 15, 16, 17] ? What kind of topological notion should we introduce to generalise the "Banach-Mazur" games, the "prioric games" and the "minimal degrees below $\underline{0}'$" .

<u>Problem</u> 2. How can we use the result of § 3 (and mostly the corollary to the main theorem) ? Is it possible to generalise other notions (as "effective algebras", "recursively presented structures", "effective isomorphisms", ...) which were defined in O.R.T., to 1°) all countable admissible ordinals or even all admissible ordinals and 2°) all (some) admissible ordinals $\beta > \omega$ such that all the subsets of β are regular ?

References

[1] J.L. Kelley, General topology, Van Nostrand-Reinhold, New York, 1955.

[2] G. Kreisel, Some reasons for generalising recursion theory in "Logic Colloquium '69", North Holland, Amsterdam, 1971, 139-198.

[3] G. Kreisel and G.E. Sacks, Metarecursive sets, Journal of Symbolic Logic, 30, (1965), 318-338.

[4] D.A. Martin, Measure, Category and degrees of unsolvability (mimeograph notes).

[5] J. Myhill, Category methods in recursion theory, Pacific Journal of Mathematics, 11 (1961), 1479-1486.

[6] H. Rogers, Theory of recursive functions and effective computability, Mc Graw-Hill, New York, 1967.

[7] G.E. Sacks, Degrees of Unsolvability, Annals of Mathematics, Study n° 55, Princeton, 1963.

[8] G.E. Sacks, Measure-theoretic uniformity in recursion theory and set theory, Transactions of the American Mathematical Society, 142 (1969), 381-420.

[9] G.E. Sacks, Higher recursion theory (mimeograph notes).

[10] G.E. Sacks, Countable admissible ardinals and hyper-degrees (mimeograph notes).

[11] J. Shoenfield, On degrees of unsolvability, Annals of Mathematics, 39 (1959), 644-653.

[12] H. Tanaka, Some results in the effective set theory, Publ. R.I.M.S. Kyoto University, Ser. A, 3 (1967), 11-52.

[13] H. Tanaka, Notes on measure and category in recursion theory, Annals of the Japan Association of Philisophical Sciences, 3 (1970), 43-53.

[14] S. Willard, General topology, Addison-Wesley Reading, 1970.

[15] C.E.M. Yates, Banach Mazur Games, comeager sets and degrees of unsolvability (mimeograph notes).

[16] C.E.M. Yates, A General framework for Δ_2^0 and Σ_1^0 priority arguments(mimeograph notes).

[17] C.E.M. Yates, Prioric games and minimal degrees below 0′, Fundamenta Mathematicae, LXXXII, (1974), 217-237.

URELEMENTS AND EXTENDABILITY

by W. Marek and M. Srebrny

Section 0. Introduction.

Recent development of the theory of admissible sets with individuals allows to treat problems of nonstandard models for various set theories. Here we use this approach to settle the problem of the expandability of denumerable models of Peano arithmetic to models of second order arithmetic and ZF set theory and of denumerable models of ZFC set theory to models of Kelley-Morse class-set theory.

We find that the results obtained in [11] for the case of transitive models have natural counterparts for the case of arbitrary models. Moreover, an application of the theory of admissible sets with urelements allows us to obtain analogous results for models of Peano arithmetic and their extendability to models of ZF set theory. To do this we construct an interpretation of some weak set theories within second order arithmetic (or KM theory of clases). We present these interpretations in sections 1 and 2. In the remaining sections we apply these results to the problem of extendability.

A few words about notation.

If $\mathfrak{M} = \langle \ldots, M, E\rangle$ is a (possibly nonstandard) model and $p \in M$ then $p_E = \{q : q \mathrel{E} p\}$. For a function f we use Df to denote the domain of f, Rf to denote the range of f and if $x \subseteq Df$ then $f * x$ is the image of x under f. $P(x)$ denotes the power set of x. A_2 is second order arithmetic (including the scheme of choice), cf. [1]. KM is Kelley-Morse impredicative theory of classes (again with choice scheme), cf. [10].

The method of admissible sets with atoms (urelements, individuals) was introduced in [3], to which we refer the reader as well as to [4].

We deal with several systems of set theory with atoms. Given a system S of set theory (e.g. ZF, ZFC^-, KP) we use SA to denote the set theory with the axioms of S but expressed in the language of set theory with atoms and enriched by the axiom $(Ea)(x)(x \in a \leftrightarrow (Ep)(p = x))$ asserting that the collection of all atoms form a set (not a proper class). In this case extensionality is restricted to sets (or sets and classes) only, not atoms. Similarly the axiom scheme of regularity postulates the existence of an $\in$-minimal set, not atom.

We would like to express here our gratitude to the late Professor Mostowski. It is sad that this paper is the last one such that while working on it we had chance to use his experience and help. Our thanks go also to our colleagues from Warsaw for a number of fruitful discussions as well as to John Bell, Alistair Lachlan and George Wilmers.

Section 1. Trees with urelements.

Let $\underline{F} = \langle F,M,E\rangle$ be a model of KM and let $\mathfrak{M} = \langle M, E \upharpoonright M\rangle$ be its set-part. We are going to prove the following theorem:

<u>Theorem</u> 1.1. There is a least admissible set $B_{\mathfrak{M}}$ over $\mathfrak{M}$ such that $P(M) \cap |B_{\mathfrak{M}}| = F \cup G$ (where $G = \{x_E : x \in M\}$) .

The proof needs some definitions and lemmas.
We define first the notion of a tree. (This notion resembles strongly the notion of tree in [10]).
Let

$$K_0 = \{0\} \times V \times (\{\{x\} : x \in V\} \cup \{\emptyset\})$$

$$K_1 = \{1\} \times V$$

<u>Definition</u>. A class X is called a tree iff it satisfies the following conditions:

(1) X is a function, $DX \subseteq K_0 \cup K_1$
$DX - RX \subseteq K_0$, $RX \subseteq K_1$

(2) X is well founded, i.e.

$$(Y)(Y \subseteq DX \wedge Y \neq \emptyset \Rightarrow (Ez)_Y(t)_Y(z \neq X(t))$$

(3) $\overline{\overline{RX - DX}} = 1$ and its unique element, called MAX_X, has the following property

$$(x)_{DX}(Ex)_{\omega}\,(X^{(n)}(x) = MAX_X)$$

where $X^{(n)} \overset{df}{=} X \circ \ldots \circ X$ (n times)

(4) $(x)(X^{-1} * \{x\} \subseteq DX - RX \Rightarrow \overline{\overline{X^{-1} * \{x\}}} = 1)$

(5) X has no nontrivial automorphism of certain type, i.e.

$$(Z)[(Z : \text{Fld } X \xrightarrow[\text{onto}]{1-1} \text{Fld } X) \wedge (y)_{\text{Fld}X}\,(X(Z(y)) = Z(X(y))) \wedge$$
$$\wedge (s)_{DX-RX}\,(pr_2(s) = pr_2(z(s))) \Rightarrow Z = Id \restriction \text{Fld } X$$

To make this obscure definition a bit clearer we give an example

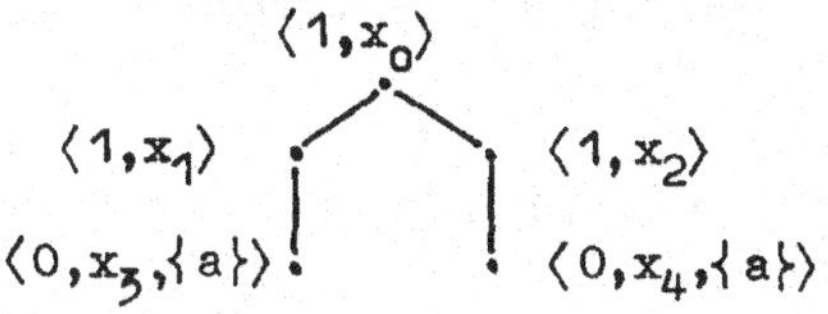

is not a tree since it has an automorphism of the wrong kind

whereas

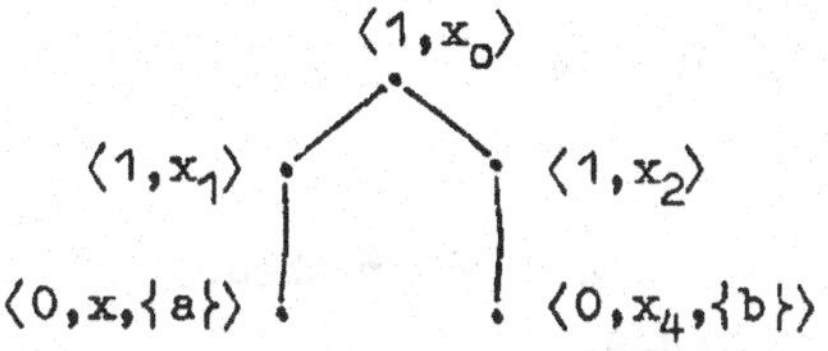

is a tree (providing $a \neq b$)

This tree serves as code for the set $\{a,b\}$ consisting of urelements a and b .

Definition. (a) If X is a tree then $AMAX_X = X^{-1} * \{MAX_X\}$
(b) If X is a tree and $x \in RX$ then X_x is the tree defined as follows: $DX_x = \{y : (En)_{\omega}\, X^{(n)}(y) = x\}$; $X_x = X \restriction (DX_x \cup \{x\})$
(c) If X and Y are trees, we say that X is isomorphic to Y (we denote it X Eq Y) iff there is a Z such that

$$Z : \text{Fld}X \xrightarrow[\text{onto}]{1-1} \text{Fld } Y \wedge (x)_{\text{Fld } X}\,(Z(X(x)) = Y(Z(x)) \wedge$$
$$\wedge (x)_{DX-RX}\,(Pr_2(x) = Pr_2(Z(x))\;.$$

Thus $\begin{array}{c} \langle 1,x\rangle \\ \langle 0,z,\{a\}\rangle \end{array}$ | Eq | $\begin{array}{c} \langle 1,y\rangle \\ \langle 0,t,\{a\}\rangle \end{array}$ but if $a \neq b$

then | $\begin{array}{c} \langle 1,x\rangle \\ \langle 0,z,\{a\}\rangle \end{array}$ non Eq | $\begin{array}{c} \langle 1,y\rangle \\ \langle 0,t,\{b\}\rangle \end{array}$.

<u>Definition</u>. If X and Y are trees then we say that X belongs to Y (we denote if X Eps Y) iff $(Ey)_{AMAX_Y}(X \text{ Eq } Y_y)$.

Now following Barwise idea we define the theory ZFC^-A as follows (we assume the reader is familiar with [3]). It differs from KPU^+ by the following: Instead of collection scheme for Δ_0-formulas we add the full scheme of replacement choice

$$(x)_a(Ey)\Phi(x,y) \rightarrow (Ef)(\mathrm{Func}(f) \wedge Df = a \wedge (x)_a \Phi(x,fx))$$

for all formulae Φ .
We have the following main theorem:

<u>Theorem</u> 1.2. ZFC^-A is interpretable in KM by Tr, Eps and Eq .

<u>Proof</u>. We refer the reader to [10] where a similar theory (ZFC^-) is interpreted in KM by similar means.
Instead, we look more closely at one particular tree:

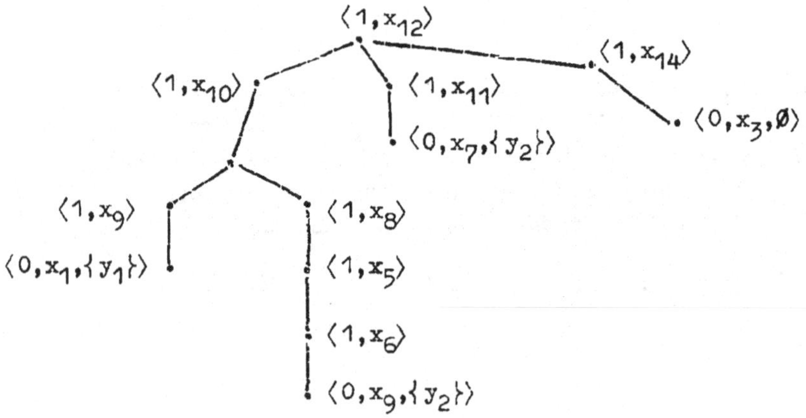

this tree serves as a code for the following set built from urelements y_2 and y_1 :

$$\{\emptyset , y_2 , \{\{y_1\} ,\{\{y_2\}\}\}\} .$$

Now we pass to a model theoretic version if our theorem. Consider $N = \langle Tr^{\underline{F}} , Eps^{\underline{F}} , Eq^{\underline{F}} \rangle$. It is a model for $ZF\dot{C}^{-}A$ (although equality is a non-logical predicate here). Let us see what represents individuals in N .

These are of two sorts:

1) $\langle 1,x\rangle$ — $\langle 0,y,\{z\}\rangle$ representing z ,

2) $\langle 1,x\rangle$ — $\langle 0,y,\emptyset\rangle$ representing $\emptyset$, the empty set.

The sets of individuals (urelements) are, in the structure $\langle Tr^{\underline{F}} , Eps^{\underline{F}} , Eq^{\underline{F}} \rangle$ represented by trees of the form

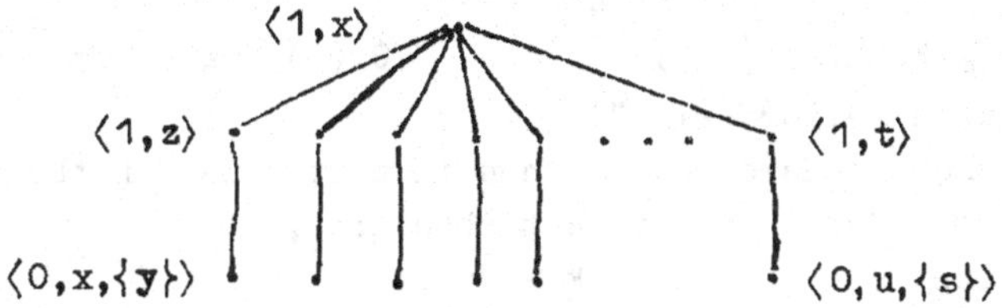

representing $\{y,\ldots,s\}$.

Notice that at the bottom different triples have different third coordinates.

Consider now $\underline{N} \models \langle Tr^{\underline{F}} , Eps^{\underline{F}} \rangle / _{Eq^{\underline{F}}}$. $\underline{N}$ is a model with absolute identity.

Clearly $\underline{N} \models ZF\dot{C}^{-}A$.

We consider now the standard part of the model $\underline{N}$.

The following fact slightly extends lemma 7.4 of [3].

Lemma 1.3. If $\langle M,E\rangle$ is a model of KPU^+ then $\langle SpM, E \restriction SpM\rangle$ is also a model of KPU^+ with the same urelements and sets of urelements.

Now it is necessary to show that the sets of individuals in $\underline{N}$ are represented by $F \cup G$. This however follows directly from our considerations on the form of the sets of individuals in N. Still it seems interesting to discover what trees in $F \cup M$ produce elements of $Sp\ \underline{N}$. To see this we introduce the notion of the rank of a tree.

For X a tree, T a wellordering $x \in RX$, $t \in Fld\ T$ we define a relation $Rank\ (X,T,x,t)$ (read: "the initial segment of T generated by t is a rank of x in X") by the following clause:

$$Rank\ (X,T,x,t) \Leftrightarrow (X^{-1} * \{x\} \subseteq DX - RX \wedge \text{"}t \text{ is the first element of } T\text{"}) \vee (t = Sup_T \{Sc_T(s) : (Ey)_{X^{-1}*x}\ Rank\ (X,T,y,s)\})$$

(where $Sc_T(s)$ denotes "successor of s in T". This definition (an implicit one) may be transformed into an explicit one.
With help of the predicate $Rank\ (.,.,.,.)$ we define the predicate $Rk(X,T)$ (read "T is a rank of X") as follows:
$Rk(X,T) \Leftrightarrow Tr(X) \wedge W.O.(T) \wedge \neg\ (Et)_{Fld\ T}\ Rank(X,T,MAX_X,t) \wedge$

$\wedge\ (x)_{AMAX_X}\ (Et)_{Fld\ T}\ Rank(X,T,x,t)$. [Here $W.O.(\cdot)$ is a formula expressing: "$(\cdot)$ is a wellordering"]

If we consider nonstandard models then in particular in these models, the notion of ordered pair is not absolute ;
For $Z \in F \cup M$ define $Z^o = \{\langle s,t\rangle : \underline{F} \models \text{"}\langle s,t\rangle \in Z\text{"}\}$.
Let $\underline{F}$ be a model of KM, $\underline{F} \models Tr\,[X] \wedge W.O.\,[T] \wedge Rk\,[X,T]$.

Lemma 1.4. Under the above assumptions:

$$Tr\ (X^o) \Leftrightarrow W.O.\ (T^o)$$

After proving this lemma one immediately sees

Proposition. The elements of $sp\ \underline{N}$ are exactly those $[X]$ for which X^o is a tree.

To complete the proof of our theorem it is enough to show that sp $\underline{N}$ is the least admissible set containing $F \cup G$. To see this, for a tree being a set, we define its realisation, $\|X\|$, as follows. (The realisation is defined within the Fraenkel-Mostowski universe $V_{\mathfrak{M}}$) . We define it by steps:
For $x \in RX$ we define $\|x\|_X$ as follows:
$\|x\|_X$ = the unique z such that $(Et)(X(\langle 0,t,\{z\}\rangle) = x))$ if $X^{-1} * \{x\} \subseteq DX - RX$ and $\|x\|_X = \{\|y\|_X : y \in X^{-1} * \{x\}\}$ otherwise.
Then $\|X\| = \|MAX_X\|_X$.
Notice the close connection between the realization function and Mostowski's collapsing function

Theorem 1.5. If $C_{\mathfrak{M}}$ is a transitive model of KPA , Tr(X) , $X \in C_{\mathfrak{M}}$, then $\|X\| \in C_{\mathfrak{M}}$.

Proof. If X is a tree then X has a rank (outside of $C_{\mathfrak{M}}$). This rank is an ordinal. Using Σ-recursion we show that this rank is actually in $C_{\mathfrak{M}}$. Now, by induction (on the rank of x in X) we show that $\|x\|_X \in C_{\mathfrak{M}}$.

This theorem leads to the remaining part of the proof of our theorem 1.1. Indeed, since $B_{\mathfrak{M}} =$ sp $\underline{N}$ arises from $\underline{F}$ as follows: $|B_{\mathfrak{M}}| = \{\|X^0\| : x \in F \cup M \wedge X$ is a tree$\}$, thus $|B_{\mathfrak{M}}| \subseteq |C_{\mathfrak{M}}|$ whenever $F \cup G \subseteq |C_{\mathfrak{M}}|$. Thus theorem 1.1 is proved.

Consider now $h(\underline{F})$ = the supremum of types of wellorderings in $\underline{F}$. Then we have

Proposition 1.6. $B_{\mathfrak{M}} \cap On = h(\underline{F})$.

Proof. $L \geq R$ follows from wellknown facts on admissible sets and Theorem 1.1. To prove $R \geq L$, Let $\alpha \in B_{\mathfrak{M}} \cap On$. Then α is a value of a tree X . Consider in $AMAX_X$ the relation
$xTy \iff X_x$ Eps X_y . As α is on ordinal, T is a wellordering of type α . Thus $R \geq L$.
Thus if $\mathfrak{M} = \langle M, E \upharpoonright M\rangle$ is nonstandard then by comparability of wellorderings in F we have $h(\mathfrak{M}) = h(\underline{F})$. Thus

Proposition. If $\mathfrak{M} = \langle M, E \upharpoonright M\rangle$ is nonstandard then $h(B_{\mathfrak{M}}) = h(\mathfrak{M})$

If however $\mathfrak{M}$ is standard then $h(\underline{F}) \geq (On \cap M)^+$ and so $h(B_{\mathfrak{M}}) > h(\mathfrak{M})$.
We notice that $B_{\mathfrak{M}}$ coincide with $\underline{N}$ iff $\underline{N} = sp\ \underline{N}$ iff $\underline{F}$ is a β-model.
Finally let us remark that the model $\underline{N}$ has another interesting property. It satisfies the following statement called Inacc (A) .

$$(Ep)_A \ (\text{"}p \text{ is a limit ordinal in the sense of } \langle A,E\rangle\text{"}) \wedge$$

$$\wedge (p)_A (Eq)_A (q_E = \{x : x_E \subseteq p_E\}) \wedge$$

$$\wedge (f)(f \subseteq A^2 \wedge \text{Func } (f) \Rightarrow (p)_A (Eq)_A (q_E = f \cdot p_E)) \wedge$$

$$\wedge (x)(x \in A \Leftrightarrow (Ep)(x = p))$$

Lemma 1.7. If $C_{\mathfrak{M}}$ is a (possibly nonstandard) model of ZFC^-A + Inacc (A) then $\mathfrak{M}$ is extendable to a model of KM.

Proof. $F = P(|\mathfrak{M}|) \cap C_{\mathfrak{M}}$ is an extension of $\mathfrak{M}$.

Thus we have the following

Theorem 1.8. Let $\mathfrak{M} = \langle M,E\rangle$ be a model of ZFC . Then $\mathfrak{M}$ is extendable to a model of KM iff there is a (possibly nonstandard) model $C_{\mathfrak{M}}$ of ZFC^-A + Inacc (A) over $\mathfrak{M}$.

Proof. If $\mathfrak{M}$ is extendable then $B_{\mathfrak{M}}$ is a suitable model $C_{\mathfrak{M}}$ (as shown in the proof of the theorem 1.1). The other part of equivalence follows from the above lemma.
Using the same method of trees with urelements we get an analogous result for the case of second order arithmetic A_2 .
Let $\omega \simeq \mathfrak{M}$ denote the following statement:

$$(Ef)(\text{func}(f) \wedge \text{Dom}(f) = \omega \wedge \text{Rg}(f) = A \wedge \text{"}f \text{ is one-to-one"} \wedge$$

$$(n)_\omega \ (m)_\omega \ (n \in m \rightarrow f(n) < f(m))$$

(where $<$ is the symbol for the "less-than" relation of $\mathfrak{M}$)

Theorem 1.9. Let $\mathfrak{M}$ be a model of Peano arithmetic. Then $\mathfrak{M}$ is extendable to a model of A_2 iff there exists a (possibly nonstandard) model $D_{\mathfrak{M}}$ of $ZFC^{-}A + \omega \simeq \mathfrak{M}$ over $\mathfrak{M}$.

<u>Proof</u>. This is straightforward application of the method applied in the proof of theorem 1.8.

We notice that:

(a) The model $C_{\mathfrak{M}}$ of the theorem 1.8. may be assumed standard iff $\mathfrak{M}$ is β-extendable (and so $\mathfrak{M}$ must be standard).

(b) The model $D_{\mathfrak{M}}$ of theorem 1.9. may be standard just in case $\mathfrak{M} \simeq \langle \omega, +, \cdot, <, 0 \rangle$

Section 2. Barwise interpretations.

In this section we study a general method of constructing models of set theory over models of arithmetic. Given a model $\underline{M} = (M,E)$ of ZF denote $\omega^{\underline{M}}$ its arithmetical part, i.e.

$$\omega^{\underline{M}} \overset{\mathrm{df}}{=} (\{m \in M : \underline{M} \models \text{"m is a non-negative integer"}\}, +^{\underline{M}}, \cdot^{\underline{M}}, 0^{\underline{M}}) ,$$

where

$$+^{\underline{M}} \overset{\mathrm{df}}{=} \{(m,n,l) : m,n,l \in M \text{ and } \underline{M} \models \text{"m,n and l are non-negative integers and } m + n = l\text{"}\}$$

denotes the addition of $\underline{M}$ and $\cdot^{\underline{M}}$, defined similarly, denotes the multiplication of $\underline{M}$. Let $\mathfrak{M} = \omega^{\underline{M}}$
We construct a model $\mathfrak{A}_{\mathfrak{M}} \models ZFA$ over $\mathfrak{M}$.

In general, let $\mathfrak{M}$ be a structure for a first order language and suppose $\mathfrak{M} \in M$. Define two auxiliary functions on M :

$$\hat{x} = \langle 0,x \rangle$$

$$\check{x} = \langle 1,x \rangle$$

for each $x \in M$, where $\langle \cdot,\cdot \rangle$ is a fixed pairing in $\underline{M}$. Now, let

$$\text{Point } (z) \overset{\mathrm{df}}{=} (Ex)(x \in |\mathfrak{M}| \quad z = \hat{x}) ,$$

$$\text{Set } (z) \overset{\mathrm{df}}{=} (Ex)[z = \check{x} \wedge (\forall y \in x)(\text{Set } (y) \vee \text{Point } (y))] ,$$

$$z \mathrel{\bar{\varepsilon}} t \stackrel{\text{df}}{=} (Ex)\,[t = \check{x} \wedge z \in x]\ .$$

In this way we define an interpretation I of the language of set theory with atoms in the theory of $\underline{M}$. Denote by Φ^I the interpretation of a formula Φ .

LEMMA 2.1. If Φ is a theorem of ZFA , then $\underline{M} \models \Phi^I$.

Proof is straightforward by checking all the axioms of ZFA using adequate axioms holding in $\underline{M}$ and the idea that the whole structure of $\underline{M}$ is reflected in the $\langle 1,x\rangle$ part of $\underline{M}$; while the $\langle 0,x\rangle$ part is reserved for atoms. We leave the fairly laborious details to the reader. The following picture clarifies the idea

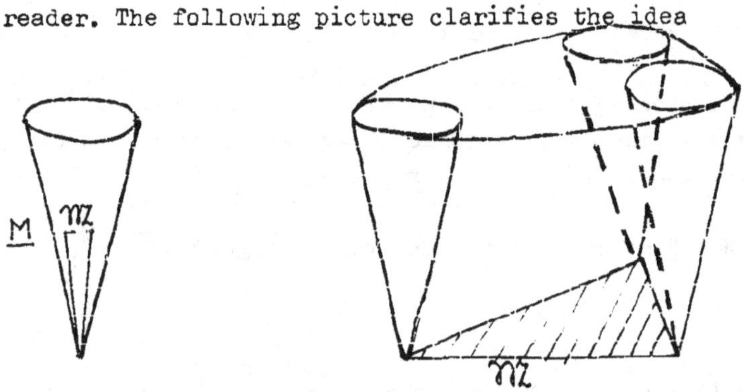

By virtue of our construction the set of atoms in the new model is just equal to $|\mathfrak{M}|$, up to isomorphism. Moreover, the ordinal standard part of the new model is equal to the o.s.p. of $\underline{M}$. If $\underline{M}$ is standard then also the new model is standard.

In the same way can construct models $\mathfrak{A}_{\mathfrak{M}}$ of weaker set theories with atoms whenever $\underline{M}$ models the appropriate set theory. The assumption that $\mathfrak{M} \in M$ is needed only to assure that atoms form a set in $\mathfrak{A}_{\mathfrak{M}}$, although the whole construction works in case of $\mathfrak{M}$ being a definable substructure of $\underline{M}$ as well. Similarly we can deal with $\mathfrak{M} = (|\mathfrak{M}|, F)$ being a model of set theory. Then the covering function is denoted by C_F and defined by

$$C_F(m) = \{n \in |\mathfrak{M}| : n F m\}\ ,$$

whenever $m \in |\mathfrak{M}|$. In the interpretation I one then has to use

$$f(z) = \{\hat{x} : x \in z\}^{\vee}\ ,$$

whenever $z \in |\mathfrak{M}|$.

Section 3. Extendability of models of Peano arithmetic to models of ZF set theory.

We say that a model $\mathfrak{M}$ of P is ZF-extendable iff there exists a model $\underline{M}$ of ZF with $\omega^{\underline{M}} \simeq \mathfrak{M}$.

The question of whether a given $\mathfrak{M} \models P$ is ZF-extendable is a model-theoretic counterpart to the question of how far the ZF axioms affect the structure of the natural numbers. Usually one has in mind Peano's postulates as describing the properties of the natural numbers. On the other hand, the natural number are genuine objects in the genuine universe of all sets. Thus their properties can be deduced in particular from the ZF axioms. Clearly, the ZF axioms are much stronger as far as metamathematical statements are concerned. For example, in ZF one can prove the consistency of P as well as the existence of many models of P. This leads us to repeat an old but still good philosophical question: Is there a "genuine" mathematical statement about natural numbers provable in ZF but not in P ?

So as to have an example consider the standard model $\mathfrak{M} = (\omega, +, \cdot, <, 0)$ in the theory ZF + there exists a standard model of ZF. The assumption of the existence of a standard model of ZF is not surprising here, since we would have assumed existence of some models of ZF in any case. For sake of convenience throughout this section we always work with the assumption that there exist standard models of ZF of large cardinality. Obviously, $\mathfrak{M}$ is extendable. Any standard model of ZF provides an extension in this case. On the other hand, there do exist non ZF-extendable models, by both Gödel's completeness and incompleteness results. Ehrenfeucht and Kreisel [6] proved that every countable model of P has a non ZF - extendable elementary extension. Thus the extendable models cannot be characterized in any elementary way. There is no elementary sentence, not even any set sentences, Φ , such that $\mathfrak{M} \models \Phi$ iff $\mathfrak{M}$ is ZF-extendable.

In this section we show that there are some fairly natural structures associated with models of arithmetic in which this question can be solved in elementary way. These are the admissible sets over models of arithmetic. More precisely, we mean the smallest admissible set over $\mathfrak{M}$ in which the atoms form a set, denoted by $HYP_{\mathfrak{M}}$. The basic properties of these structures were investigated in Barwise [3] and [4] .

LEMMA 3.1. $\mathfrak{M} \models P$ is ZF-extendable iff there exists a model

$$\mathfrak{A}_{\mathfrak{M}} \models \text{ZFA} + \text{"}\omega \simeq \mathfrak{M}\text{"} \quad \text{over } \mathfrak{M} .$$

Proof. $\Longrightarrow$. Immediate by lemma 2.1.
$\Longleftarrow$. Take the universe of pure sets of $\mathfrak{A}_{\mathfrak{M}}$. This is the required extension, up to isomorphism.

In section 4 we give a similar result for the case of extendability to a model of analysis. Though one has to apply a slightly more involved argument there.

LEMMA 3.2. If $\mathfrak{A}_{\mathfrak{M}} \models$ ZFA then $\mathfrak{A}_{\mathfrak{M}}$ is an end extension of $\text{HYP}_{\mathfrak{M}}$.

Proof. The standard part of $\mathfrak{A}_{\mathfrak{M}}$ is a transitive model of KPA over $\mathfrak{M}$. On the other hand, $\text{HYP}_{\mathfrak{M}}$ is the smallest transitive model of KPA over $\mathfrak{M}$.

An interesting fact is that although the ordinal standard part of $\mathfrak{A}_{\mathfrak{M}}$ may be that of $\text{HYP}_{\mathfrak{M}}$, the standard part of $\mathfrak{A}_{\mathfrak{M}}$ is certainly different from $\text{HYP}_{\mathfrak{M}}$.

Let us recall, following Barwise, that $\text{HYP}_{\mathfrak{M}}$ is countable whenever $\mathfrak{M}$ is countable and that all the model theoretic properties of the countable admissible fragments of infinitary logic hold also in the case of admissible sets with atoms.

THEOREM 3.3. There exists a single finitary sentence Φ such that for countable models $\mathfrak{M} \models P$ we have: $\mathfrak{M}$ is ZF-extendable iff $\text{HYP}_{\mathfrak{M}} \models \Phi$.

Proof. Let θ be a Σ_1-formula defining the following theory T . T consists of

(a) $\bigwedge$ ZFA

(b) $\omega \simeq \mathfrak{M}$

(c) $\in$ - diagram of the universe, i.e. infinitary sentences guaranteeing that every model is an end extension of the universe in

which we interprete these sentences.
Few words must be added to the description of (c). It is the analogue for the case of urelements of the formula EE of [8].
Now, set $\Phi \overset{df}{\equiv} \text{Consis}_\theta$, i.e. there is no (infinitary) proof of inconsistency from the axioms defined by θ . Obviously, Φ is a finitary sentence. Moreover Φ is Σ_1 .

$\Longrightarrow$. Suppose $\mathfrak{M}$ is ZF-extendable. Denote by T' the theory defined by θ over $HYP_{\mathfrak{M}}$. T' has a model, by lemma 3.1. and 3.2. Thus T' is consistent. Therefore there is no proof of inconsistency from T' in HYP , since this formula is persistent.
Hence $HYP_{\mathfrak{M}} \models \Phi$.

$\Longleftarrow$. Suppose $HYP_{\mathfrak{M}} \models \text{Consis}_\theta$. Denote by T' the theory defined by θ over $HYP_{\mathfrak{M}}$. It suffices to show that every $HYP_{\mathfrak{M}}$- finite fragment of T' is consistent, by the Barwise Compactness theorem. Suppose on the contrary that there is an inconsistent $HYP_{\mathfrak{M}}$ - finite $T_0 \subseteq T'$. Then there is a proof of inconsistency of T_0 in $HYP_{\mathfrak{M}}$. This contradicts the assumption that $HYP_{\mathfrak{M}} \models \Phi$.

Section 4. Expandability of models of arithmetic to models of analysis.

The results of section 3 can be proved also in the case of extendability of models of Peano arithmetic to models of analysis and its fragments. The key lemma, which is lemma 3.1 might be proved either by the methods of section 1 or by the Barwise interpretations of section 2.

Recall that a model $\mathfrak{M} \models P$ is A_2-expandable iff there exists an $\mathcal{F} \subseteq \mathcal{P}(|\mathfrak{M}|)$ such that $(\mathcal{F}, \mathfrak{M}) \models A_2$. It is well-known that A_2 commutes with ZFC^- . One proves this by applying the method of well founded trees on the integers. Thus $\mathfrak{M} \models P$ is A_2-expandable iff $\mathfrak{M}$ is ZFC^--extendable, i.e. there exists a model $\underline{M} \models ZFC^-$ such that $\omega^{\underline{M}} \simeq \mathfrak{M}$. Thus one might prove theorem 1.9. also applying the Barwise interpretation 2.1.

In both proofs, the use of the choice in A_2 is relevant, although we can deal with subsystems of A_2 as well. The result of Kreisel that Δ^1_2-CA implies Σ^1_2-AC_{01} (see [7]) gives the main step in proving the commutability of subsystems of analysis with appropriate fragments of set theory. The reader is referred to [9] and [2] for details. What we need here is only the fact that these fragments of set theory contain the Kripke-Platek axioms. Denote by S_n the fragment commuting with Δ_n^1 - analysis, i.e. Δ_n^1 - comprehension axioms, for $n \geq 2$. This S_2 = KP + Infinity + Mostowski's collapsing lemma ; for $n \geq 2$ S_{n+1} = KP + Infinity + Δ_n - separation ; and S_∞ = KP + Infinity + full separation , which is $Z^- + \Delta_0$ -collection where by Δ_∞^1 -analysis we mean the full comprehension axioms, CA , denoted by A_2^* in [1] .

We say that $\mathfrak{M} \models P$ is Δ_n^1 -expandable iff there exists an $\mathcal{F} \subseteq \mathcal{P}(|\mathfrak{M}|)$ such that $(\mathcal{F}, \mathfrak{M}) \models \Delta_n^1$ - analysis , $0 \leq n \leq \infty$.

LEMMA 4.1. Let $n \geq 2$. $\mathfrak{M} \models P$ is Δ_n^1 -expandable iff there exists a model $\mathfrak{A}_{\mathfrak{M}} \models S_nA$ + "$\omega \simeq \mathfrak{M}$" over $\mathfrak{M}$.

Barwise and Schlipf [5] showed that every countable model of arithmetic has an elementary extension which is not Δ_1^1 -expandable. The following theorem gives the full characterization.

THEOREM 4.2. Let $0 \leq n \leq \infty$. There is a single finitary sentence Φ_n such that for countable models $\mathfrak{M} \models P$ we have: $\mathfrak{M}$ is Δ_n^1-expandable iff $HYP_{\mathfrak{M}} \models \Phi_n$.

Proof. Case $n = 0$ is trivial, since every model $\mathfrak{M} \models P$ is predicatively expandable.
Case $n = 1$ was done by Barwise and Schlipf [5]. Their sentence Φ_1 is particularly simple; namely $\Phi_1 = \neg$ Inf , i.e. the negation of axiom of infinity. Since $HYP_{\mathfrak{M}}$ is a transitive model, $HYP_{\mathfrak{M}} \models \Phi_1$ iff the ordinal height of $HYP_{\mathfrak{M}}$ is ω .
Case $n = 2$. Recall that S_n denotes the fragment of set theory commuting with Δ_n^1- analysis. By S_nA we denote the S_n set theory with a set of atoms. Then the proof follows the same lines as the proof of theorem 3.3 with the exception that part (a) of T is replaced by

(a') $\bigwedge\!\!\!\bigwedge S_nA$

One can reformulate the above result also for the case of Σ_n^1 - analysis, i.e. Σ_n^1 -comprehension axioms. For $n \geq 1$, Σ_{n+1}^1 - analysis commutes with KP + Infinity + Σ_n -separation.

Let us also announce two recent results of G. Wilmers (private communication).

THEOREM 4.3. There exists a recursive infinitary sentence $\Phi \in L_{\omega_1,\omega}$ such that countable $\mathfrak{M} \models P^{ZF}$ is ZF-extendable iff $\mathfrak{M} \models \Phi$ where P^{ZF} is the set of arithmetical consequences of ZF axioms.

THEOREM 4.4. If $\mathfrak{M} \models P^{ZF}$ is countable, then the following are equivalent:

(a) $\mathfrak{M}$ is ZF - extendable

(b) $\mathfrak{M}$ is A_2 - expandable

(c) $\mathfrak{M}$ is Δ^1_n - expandable, for some $n \geq 2$

(d) $\mathfrak{M}$ is Δ^1_n - expandable, for every $n \geq 2$.

H. Friedman announced that for countable models of true arithmetic the A_2 - expandability is equivalent to the expandability to models of type theory.

Section 5. Expandability of models of ZF set theory to models of KM class theory

A model $\mathfrak{M}$ is KM-expandable iff there exists an $\mathcal{F} \subseteq \mathcal{P}(|\mathfrak{M}|)$ such that $(\mathcal{F}, \mathfrak{M}) \models$ KM . The reader might want to consult Marek and Mostowski [11] for an exposition of the subject. Let us recall only that as proved by S. Krajewski every model of ZF contains an elementary submodel which is not KM-expandable. Applying analogous reasoning as in previous sections one proves the following result.

THEOREM 5.1. There exists a single finitary sentence Φ such that for countable models $\mathfrak{M} \models$ ZFC , $\mathfrak{M}$ is KM-expandable iff $HYP_{\mathfrak{M}} \models \Phi$

The key lemma might be proved either as delivered in section 1 or by the methods of section 2 and the easy classical application of well founded trees that $(\mathcal{F}, \mathfrak{M}) \models$ KM iff there exists a model $\underline{M} \models ZFC^-$ such that $\mathfrak{M} \in |\underline{M}|$, $\mathcal{F} = |\underline{M}| \cap \mathcal{P}(|\mathfrak{M}|)$ and $\underline{M} \models$ "$\mathfrak{M} \simeq (R_\varkappa, \in)$ for an inaccessible cardinal $\varkappa$"

We close this paper with some open questions.

(1) Do these results hold in the case of expandability to models of Kelley-Morse theory of classes without the scheme of choice? Unfortunately, the Σ_2^1-scheme of choice is not provable even from the full impredicative, i.e. Δ_∞^1, comprehension of Kelley-Morse theory, in contrast to the Kreisel's result for analysis. We can only obtain the following partial result applying the methods of Barwise and Schlipf [5].

THEOREM 5.2. If $\mathfrak{M}$ is a non-ω-model of ZF, then $\mathfrak{M}$ is Δ_1^1-expandable iff $o(HYP_{\mathfrak{M}}) = \omega$.

Recently, K. Bieliński obtained the following.

THEOREM 5.3. Suppose $\mathfrak{M} \models ZF$ is nonstandard $\alpha = o.s.p.(\mathfrak{M})$. Then $\mathfrak{M}$ is Δ_1^1-expandable iff $o(HYP_{\mathfrak{M}}) = \alpha$ and $\mathfrak{M} \models ZF_\alpha$ where ZF_α is the ZF set of axioms expressed in the $\mathcal{L}_{L_\alpha}$ fragment of infinitary logic.

(2) (J. Bell). Is there an infinitary sentence characterizing KM-expandability ?

(3) Do the results of this paper hold for uncountable models $\mathfrak{M}$?

References

[1] K. R. Apt, W. Marek, Second order arithmetic and related topics. Ann of Math. Logic 6 (1974), pp. 177-229.

[2] M. Artique, E. Isambert, M. Perrin, A. Zalc, Some remarks on bicommutability, to appear in Fund. Math.

[3] K. J. Barwise, Admissible sets over models of set theory, in Generalized Recursion Theory, North-Holland, 1974, pp. 97-122.

[4] K. J. Barwise, Admissible sets and structures, Springer Verlag, to appear.

[5] K. J. Barwise and J. Schlipf, On recursively saturated models of arithmetic, In Springer L. N. 448 .

[6] A. Ehrenfeucht, G. Kreisel, Strong models of arithmetic, Bull. Acad. Pol. Sci., Sér. Sci. Math. Phys. Astron., XV (1966), pp. 107-110.

[7] G. Kreisel, Survey of proof theory. J. of Symb. Logic 33 (1968).

[8] J. L. Krivine, K. Mc Aloon, Some true unprovable formulas of set theory, The Proceedings of the Bertrand Russel Memorial Logic Conference, pp. 332-341.

[9] W. Marek, ω-models of second order arithmetic and admissible sets, to appear in Fund. Math.

[10] W. Marek , On the metamathematics of impredicative theory of classes, Diss. Math. XCVIII .

[11] W. Marek, A. Mostowski, On extendability of models of ZF set theory to models of KM theory of classes. In Springer L.N. 449 .

INSTITUTE OF MATHEMATICS , UNIVERSITY OF WARSAW , WARSZAWA ,
MATHEMATICAL INSTITUTE , POLISH ACADEMY OF SCIENCES , WARSZAWA .

ON LIMIT REDUCED POWERS, SATURATEDNESS AND UNIVERSALITY.

by L. Pacholski (Wrocław)

Dedicated to the memory of Professor A.Mostowski

This paper is a continuation of [9] and contains a characterisation of those pairs (F,G) which have the property that every limit reduced power $A^I_F|G$ is $\varkappa$-saturated provided the language of A has cardinality less than $\varkappa$. The characterisation is based on a notion of a pair (F,G) being $\varkappa$-good, which is a generalisation of Keisler's notion of a $\varkappa$-good filter. A similar characterisation is given for universal limit reduced powers (Th. 2.8). Here the result is not so nice as Keisler's original characterisation of ultrapowers which are universal. An example is provided to show that a straightforward generalisation of Keisler's theorem is not true.

Because of space and time limits chapters 3 and 4 of the first draft of this paper will be published later in [10]. In chapter 3 it is proven that if for every relational structure A, the limit reduced power $A^I_F|G$ is $\varkappa$-homogeneous, then it is $\varkappa$-saturated. Chapter 4 contains theorems on the existence of $\varkappa$-good pairs (F,G) and deals with limit reduced powers which are not reduced powers.

Since one should expect that [9] will be published with a delay we sketch here the proof of Theorem 1.2 which is proved in detail in [9].

0. Terminology.

We assume notation and terminology of [1]. Below we recall some less common notion **and** we present briefly some concepts introduced in [9]. Let I be a non-empty set. By $E(I)$ we denote the set of all equivalence relations in I. (F,G) is a filter pair if F is a filter over a set I and G is a filter in $E(I)$ (i.e. if $\rho_1,\rho_2 \in G$, then $\rho_1 \cap \rho_2 \in G$ and if $\rho_1 \in G$, $\rho_2 \in E(I), P_1 \subseteq P_2$, then $\rho_2 \in G$). Let A, I be non-empty sets. If $f \in A^I$, then $eq(f) = \{(i,j) \in I^2 : f(i) = f(j)\}$. Clearly $eq(f)$ is an equivalence relation in I. Let G be a filter in $E(I)$. The limit power $A^I|G$ of a relational structure A is the substructure of A^I with the universe $\{f \in A^I : eq(f) \in G\}$. Now let (F,G) be a filter pair. The limit reduced power $A^I_F|G$ is the substructure of the reduced power A^I_F with the universe $\{f/F : f \in A^I|G\}$, where $f/F = \{g \in A^I : \{i \in I : f(i) = g(i)\} \in F\}$.

By 2 we denote the two element Boolean algebra, 0 and 1 denote respectively the minimal and the maximal element of a Boolean algebra.

Let $\rho \in E(I)$. If $i \in I$ then by i/ρ we denote the set $\{j : (i,j) \in \rho\}$ and if $X \subseteq I$, then X/ρ denotes the set $\{i/\rho : i \in X\}$. We say that X is composed of equivalence classes of ρ if $X = \bigcup(X/\rho)$. By $2^I|\rho$ we denote the Boolean algebra of subsets of I which can be composed of equivalence classes of ρ. Thus $2^I|G$, the limit power of the two element Boolean algebra can be identified with the algebra of subsets if I which can be composed from equivalence classes of relations in G. More exactly $2^I|G = \bigcup_{\rho\in G} 2^I|\rho$.

We say that a filter pair (F,G) is $\varkappa$-regular if there is $\rho \in G$ and a family $E \subseteq F \cap 2^I|\rho$ such that E has power $\varkappa$ and every infinite subfamily of E has the empty intersection. Recall that $S_\omega(\varkappa)$ is the set of all finite subsets of $\varkappa$. If $f: S_\omega(\varkappa) \to 2^I$, then we say that f is monotonic if $f(s) \subseteq f(t)$ provided $s \supseteq t$ and we say that f is additive if $f(s \cup t) = f(s) \cap f(t)$. If $g: S_\omega(\varkappa) \to 2^I$ and $f(s) \subseteq g(s)$ for every $s \in S_\omega(\varkappa)$, then we write $f \leq g$. If X is a set and f is a function with $\mathrm{dom}\, f \supseteq X$, then f^*X is the image of X by f.

DEFINITION. A filter pair (F,G) is $\varkappa$-good if and only if it is ω-regular and for every $\lambda < \varkappa$, every monotonic function

$f\colon S_\omega(\lambda) \to F \cap 2^I|G$ and every additive function $H\colon S_\omega(\lambda) \to G$ there is $\rho \in G$ and additive $g\colon S_\omega(\lambda) \to 2^I|\rho$ such that

(g.1) $g \leq f$

(g.2) $g^* S_\omega(\lambda) \subseteq F \cap 2^I|\rho$

and

(g.3) for every $x \in I/\rho$ and every $i,j \in x$ if $x \subseteq g(s)$, then $(i,j) \in H(s)$ (i.e. on $g(s)$ the relation ρ is finer then $H(s)$).

If A is a relational structure, then $L(A)$ denote the first language of A, L_B is the language of the elementary theory of Boolean algebras. If L is a first order language and $\varkappa$ is a cardinal number, then $L(\varkappa)$ is the language obtained by adjoinig to L a set $\{c_\alpha : \alpha < \varkappa\}$ of new distinct individual constants.

Finally we introduce some abbreviations. By I_F we denote the ideal dual to F (i.e. $I_F = \{X \in I\colon I - X \in F\}$). If X is a subset of a set I, then $X^1 = X$ and $X^0 = I - X$. A similar convention is assumed for elements of a Boolean algebra. Let $C^+\colon \varkappa \to 2^I$ and let $C^-\colon \varkappa \to 2^I$. Then we put

$$\overline{C}(s,t) = \bigcap\left\{C^+(\alpha) : \alpha \in s\right\} \cap \bigcap\left\{C^-(\alpha) : \alpha \in t\right\}.$$

Given $C\colon \varkappa \to 2^I$ we put $C^+(\alpha) = C(\alpha)$ and $C^-(\alpha) = I - C(\alpha)$, but we do not assume that C^+ and C^- are always of this form.

If n is a natural number then $n = \{0,1,2,\ldots,n-1\}$. If z is a finite subset of 2^I then by $\rho(z)$ we denote the greatest equivalence relation on I such that if $x \in z$ then $x \in 2^I|\rho$. Of course if $z \subseteq 2^I|G$, then $\rho(z) \in G$.

1. Saturatedness.

In this part we characterize those filter pairs (F,G) which have the property that for every relational structure A the limit reduced power $A^I_F|G$ is $\varkappa$-saturated.

DEFINITION 1.1. Let (F,G) be a filter pair. We say that (F,G) is $\varkappa$-saturative if $A^I_F|G$ is $\varkappa$-saturated for every relational structure satisfying $|L(A)| < \varkappa$.

In the proof of the main theorem we use the following fact (see [9]).

THEOREM 1.2. If a filter pair is $\varkappa$-saturative, then it is $\varkappa$-good.

<u>Sketch of the proof</u>. The proof is divided into four lemmas. We assume that for every relational structure A with $|L(A)| < \varkappa$ the limit reduced power $A^I_F \mid G$ is $\varkappa$-saturated. Let $\lambda < \varkappa$. Let f and H be arbitrary (but fixed) functions which satisfy the hypotheses of the definition of $\varkappa$-goodness.

LEMMA 1.2.a. There are $\rho_1 \in G$ and $d\colon \lambda \to F$ such that

(a1) $\quad d(\alpha) \in 2^I \mid G \qquad$ for $\alpha \in \lambda$

and

(a2) $\quad$ if $y \in I/\rho_1$ and $i,j \in d(\alpha) \cap y$ then $(i,j) \in H(\{\alpha\})$.

<u>Hint for the proof</u>. Consider $B = \langle S(I), \subseteq, \neq \rangle$ and $\Sigma = \{v \subseteq b_\alpha/F\}_{\alpha<\lambda} \cup \{v \neq 0\}$ where $b_\alpha(i) = \{j \in I\colon (i,j) \in H(\{\alpha\})\}$

LEMMA 1.2.b. Let $f_0\colon S_\omega(\lambda) \to F \cap 2^I \mid G$ be monotonic. Then there are $\rho_2 \in G$ and $f_1\colon S_\omega(\lambda) \to S(I)$ such that for $s \in S_\omega(\lambda)$

(b1) $\quad f_1(s) \in F \cap 2^I|\rho_2$, $\quad f_1(s) \subseteq f_0(s)$,

and

(b2) $\quad |\{s \in S_\omega(\lambda)\colon i \in f_1(s)\}| < \omega \qquad$ for every $i \in I$.

<u>Hint for the proof</u>. Let $S = \langle S_\omega(S_\omega(\lambda)) \cup \{S_\omega(\lambda)\}, \subseteq, \neq \rangle$, for $s \in S_\omega(\lambda)$, $i \in I$ let

$$c_s(i) = \begin{cases} \{s\} & \text{if } i \in f_0(s) \\ S_\omega(\lambda) & \text{if } i \notin f_0(s) \end{cases}$$

and let $c_\lambda(i) = S_\omega(\lambda)$.

Consider the set $\Sigma = \{v \supseteq c_s/F\} \cup \{v \neq c_\lambda/F\}$.

LEMMA 1.2.c. There exist monotonic $f_2\colon S_\omega(\lambda) \to F$ and $\rho_1, \rho_2 \in G$ such that

(c1) $f_2 \leq f$

(c2) $f_2(s) \in 2^I|\rho_2$ for $s \in S_\omega(\lambda)$

(c3) if $y \in I/\rho_1$ and $i,j \in y \cap f_2(s)$ then $(i,j) \in H(s)$.

<u>Proof</u>. Put $f_0(s) = \bigcap_{\alpha \in s} d(\alpha) \cap f(s)$ and apply Lemma 1.2.b to get $\rho_2 \in G$ and $f_1\colon S_\omega(\lambda) \to F$ satisfying (b1) and (b2). Then put $f_2(s) = \bigcap_{t \subseteq s} f_1(t)$.

LEMMA 1.2.d. There are additive $g\colon S_\omega(\lambda) \to F$ and $\rho_3 \in G$ such that $g \leq f_2$ and $g(s) \in 2^I|\rho_3$ for $s \in S_\omega(\lambda)$.

<u>Hint for the proof</u>. Let $A = \langle S_\omega(S_\omega(\lambda)), \subseteq, \neq \rangle$ and for $\alpha < \lambda$, $i \in I$ let

$$a_\alpha(i) = \{t \in S_\omega(\lambda) : i \in f_2(t) , \alpha \in t\} .$$

By Lemma 1.2.c a_α a is well defined element of $A^I|G$. Consider the set $\Sigma = \{v \subseteq a_\alpha/F\}_{\alpha<\lambda} \cup \{v \neq 0\}$.

To complete the proof of Theorem 1.2 put $\rho = \rho_1 \cap \rho_3$.

THEOREM 1.3. Let (F,G) be a filter pair. Then (F,G) is $\varkappa$-saturative if and only if

(1) $2^I_F|G$ is $\varkappa$-saturated

and

(2) (F,G) is $\varkappa$-good.

<u>Proof of necessity</u>. Assume that (F,G) is $\varkappa$-saturative. Of course $2^I_F|G$ is $\varkappa$-saturated. Moreower, by Theorem 1.2. (F,G) is $\varkappa$-good.

To prove that (1) and (2) imply that (F,G) is $\varkappa$-saturative we need the following fact.

LEMMA 1.4. Assume that (F,G) is $\varkappa$-good. Let $\lambda < \varkappa$ and let p and C be functions such that

(3) $p: S_\omega(\lambda) \times S_\omega(\lambda) \to 2^I|G$

(4) for every $(s,t) \in S_\omega(\lambda) \times S_\omega(\lambda)$ and every $u \supseteq s \cup t$, $u \in S_\omega(\lambda)$,

$$p(s,t) = \bigcup\{p(s_1,t_1): s \subseteq s_1,\ t \subseteq t_1,\ t_1 \cup s_1 = u\}$$

(5) $p(0,0) = I$, $p(\{\alpha\}, \{\alpha\}) = 0$

(6) $C : \lambda \to 2^I|G$

and

(7) for every $(s,t) \in S_\omega(\lambda) \times S_\omega(\lambda)$, $(\overline{C}(s,t) - p(s,t)) \in I_F$,

Then there is an equivalence relation $\rho \in G$ and functions $D^+ : \lambda \to 2^I|\rho$ and $D^- : \lambda \to 2^I|\rho$ such that $D^+(\alpha) \cap D^-(\alpha) = 0$ for $\alpha < \lambda$

(8) $D^+(\alpha) \triangle C^+(\alpha) \in I_F$, $D^-(\alpha) \triangle C^-(\alpha) \in I_F$ for every $\alpha < \lambda$

and

(9) if $s,t \in S_\omega(\lambda)$, then $\overline{D}(s,t) \subseteq p(s,t)$

Moreover

(10) for every infinite $E \subseteq \lambda$,

$$\bigcap\{D^+(\alpha) : \alpha \in E\} = \bigcap\{D^-(\alpha) : \alpha \in E\} = 0 .$$

First of all we shall prove the following statement.

PROPOSITION 1.4.1. There are $\rho_0 \in G$ and $p_0: S_\omega(\lambda) \times S_\omega(\lambda) \to 2^I$ such that

(11) for $s,t \in S_\omega(\lambda)$ $p_0(s,t) \subseteq p(s,t)$, $p(s.t) - p_0(s,t) \in I_F$

and $p_0(s,t) \in 2^I|\rho_0$

and

(12) for every infinite $E \subseteq \lambda$

$$\bigcap\{p_0(\{\alpha\}, 0) \cup p_0(0, \{\alpha\}) : \alpha \in E\} = 0$$

Proof. The filter pair (F,G) is $\varkappa$-good whence ω-regular. Consequently there is $\rho \in G$ and a subset $\{E_i : i < \omega\}$ of F such that

(13) $\bigcap\{E_i : i < \omega\} = 0$ and $\{E_i : i < \omega\} \subseteq 2^I|\rho$.

Let e be a $1-1$ function on $S_\omega(\lambda) \times S_\omega(\lambda)$ onto λ . For $u \in S_\omega(\lambda)$ we put $H(u) = \rho(\{p(s,t) : e(s,t) \in u\})$ and $f(u) = E_{|u|}$. Clearly H is additive. By (3) we get $H^*(S_\omega(\lambda)) \subseteq G$. By (13) and the definition of f it easily follows that $f^* S_\omega(\lambda) \subseteq F \cap 2^I|G$ and that f is monotonic. Now from the $\varkappa$-goodness of (F,G) it follows that there are $\rho_0 \in G$ and an additive function $g: S_\omega(\lambda) \to F \cap 2^I|\rho_0$ such that $g \leqslant f$ and (g.3) holds. We put $p_0(s,t) = p(s,t) \cap g(\{e(s,t)\})$ It easily follows from (13) the definiton of f and the definition of e that (12) holds. Also it is clear that $p_0(s,t) \subseteq p(s,t)$ and $p(s,t) - p_0(s,t) \in I_F$. To prove that $p_0(s,t) \in 2^I|\rho_0$ notice that $p_0(s,t) = p_0(s,t) \cap g(\{e(s,t)\}) = p(s,t) \cap g(\{e(s,t)\})$ and that on $g(\{e(s,t)\})$ the equivalence relation ρ_0 is finer then $H(\{e(s,t)\}) = \rho(\{p(s,t)\})$.

Using similar arguments we can prove the following.

PROPOSITION 1.4.2. There is $\rho_1 \in G$ and $C_0 : \lambda \to 2^I|\rho_1$ such that

(14) $C(\alpha) \triangle C_0(\alpha) \in I_F$ for every $\alpha < \lambda$.

Proof. Let $H(s) = \rho(\{C(\alpha) : \alpha \in s\})$ and $f(s) = I$ for every $s \in S_\omega(\lambda)$. Let $\rho_1 \in G$ and $g: S_\omega(\lambda) \to F$ be such that (g.1)-(g.3) holds. If for $\alpha < \lambda$ we put $C_0(\alpha) = g(\{\alpha\}) \cap C(\alpha)$, then ρ_1 and C_0 are as required.

<u>Proof of Lemma 1.4.</u> Let $f^-(u) = \bigcup\{\bar{C}_0(s,t) - p_0(s,t):\ s,t \subseteq u\}$. By (7), (14), the definition of p_0 and the fact that $g^*S_\omega(\lambda) \subseteq F$ it follows that $f^-(u) \in I_F$ for every $u \in S_\omega(\lambda)$. Whence, if we put $f(u) = I - f^-(u)$, then $f^*S_\omega(\lambda) \subseteq F$. Of course f is a monotonic function. It is easy to check that $f^*S_\omega(\lambda) \subseteq 2^I|G$. Now let us put $H(s) = I \times I$ for every $s \in S_\omega(\lambda)$. Then, since (F,G) is $\varkappa$-good, there are $\rho_3 \in G$ and an additive function $g: S_\omega(\lambda) \to 2^I|\rho_3$ such that $g^*S_\omega(\lambda) \subseteq F$ and $g \leq f$. We put $g^-(s) = I - g(s)$. Of course $(g^-)^*S_\omega(\lambda) \subseteq I_F$. Let $\rho = \rho_1 \cap \rho_2 \cap \rho_3$. We shall define functions D^+ and D^- such that

(15) $\quad D^+ : \lambda \to 2^I|\rho\,, \qquad D^- : \lambda \to 2^I|\rho$

(16) $\quad \bar{D}(s,t) \subseteq p_0(s,t) \qquad$ for every $s,t \in S_\omega(\lambda)$

(17) $\quad D^+(\alpha) \,\Delta\, C_0^+(\alpha) \subseteq g^-(\{\alpha\})$,

$\qquad D^-(\alpha) \,\Delta\, C_0^-(\alpha) \subseteq g^-(\{\alpha\}) \qquad$ for every $\alpha < \lambda$,

(18) $\quad \bigcup\{D^+(\alpha) \cap D^-(\alpha) : \alpha < \lambda\} = 0$.

We claim that if D^+ and D^- satisfy (15), (16), (17) and (18), then D^+, D^- and ρ satisfy the conclusion of Lemma 1.4. In fact (8) is a consequence of (14) and (17). Since $p_0 \leq p$ (9) easily follows from (16). Now to get (10) let us notice that by (16) $D^+(\alpha) \cup D^-(\alpha) \subseteq$ $\subseteq p_0(\{\alpha\}, 0) \cup p_0(0, \{\alpha\})$. This by (12) implies (10) and proves that to complete Lemma 1.4 it suffices to find D^+ and D^- which satisfy (15) - (18).

For $\alpha < \lambda$ we put

(19) $\quad D^+(\alpha) = C_0^+(\alpha) \cap g_0(\{\alpha\})$

and

(20) $\quad D^-(\alpha) = C_0^-(\alpha) \cap g_0(\{\alpha\})$.

It is obvious that (17) and (18) are satisfied. To check that (15) holds let us take $\alpha < \lambda$. Since $g_0: S_\omega(\lambda) \to 2^I|\rho_3$ and $\rho = \rho_1 \cap \rho_2 \cap \rho_3$ we have $g_0(\{\alpha\}) \in 2^I|\rho$. Moreover $C_0(\alpha) \in 2^I|\rho_1$, whence $C_0(\alpha) \in 2^I|\rho$. Now it remains to prove (16). Let $i \in I$ and assume that $i \in \bar{D}(s,t)$ for some $s,t \in S_\omega(\lambda)$. Then $i \in D^+(\alpha)$ for $\alpha \in s$ and $i \in D^-(\alpha)$ for $\alpha \in t$. Consequently by (19) and (20) $i \in \bigcap\{g_0(\{\alpha\}) : \alpha \in s \cup t\}$. But g_0 is additive, whence $i \in g_0(s \cup t)$. Thus, since $g_0 \leq f$, we get $i \in f(s \cup t)$ and

(21) $\quad i \notin f^-(s \cup t)$.

On the other hand, since $i \in \bar{D}(s,t)$, we also have $i \in \bar{C}_0(s,t)$. From the last fact, the definition of f^- and from (21) we get $i \in p_0(s,t)$ which finishes the proof of (16) and also of Lemma 1.4.

Now we go back to the proof of Theorem 1.

<u>Proof of sufficiency</u>. Assume that F and G satisfy (1) and (2). Let A be a relational structure such that $|L(A)| = \mu < \varkappa$. By B we denote the limit reduced power $A^I_F|G$. Let $\mu \leq \lambda < \varkappa$ and let $\{b_\xi : \xi < \lambda\}$ be a subset of $A^I|G$. Let us put $B_0 = (B, b_\xi/F)_{\xi<\lambda}$ and $A_i = (A, b_\xi(i))_{\xi<\lambda}$ for $i \in I$. Finally let $\Sigma = \{\sigma_\alpha : \alpha < \lambda\}$ be a set of formulas of $L(B_0)$ with one free variable v. We assume that Σ is consistent with $Th(B_0)$ i.e.

(22) $\quad B_0 \models \exists v \bigwedge \Sigma_0$ for every finite $\Sigma_0 \subseteq \Sigma$.

We shall prove that Σ can be satisfied in B_0, i.e. that there is $x \in B$ such that $B_0 \models \sigma[x]$ holds for every $\sigma \in \Sigma$. Let $\{c_\alpha : \alpha < \varkappa\}$ be a set of new individual constants and let L_C denotes the language of the theory of Boolean algebras extended by $\{c_\alpha : \alpha < \lambda\}$. Moreover let $\{\varphi_\alpha : \alpha < \lambda\}$ be an enumeration of all formulas of $L(B_0)$ with one free variable v. By the Weinstein version of Feferman-Vaught theorem ([1], [3],[7]) there is a function $\psi : \lambda \to L_C$ and a function ϑ, from λ into the set of all finite increasing sequences of ordinals less than λ, such that for every $\alpha < \lambda$ and every $x \in A^I|G$

(23) $\quad B_0 \models \sigma_\alpha [x/F]$ if and only if

$$2^I_F|G \models (\psi(\alpha))[K(\varphi_{\xi_1}, x), \ldots, K(\varphi_{\xi_k}, x)]$$

where $(\xi_1, \ldots, \xi_k) = \vartheta(\alpha)$ and $K(\varphi, x) = \{i \in I : A_i \models \varphi[x(i)]\}$. If $s,t \in S_\omega(\lambda)$ then we put

$$p(s,t) = \{i \in I: A_i \models (\exists v)(\bigwedge\{\varphi_\alpha : \alpha \in s\} \wedge \bigwedge\{\neg \varphi_\alpha : \alpha \in t\})\} .$$

Only finitely many constants b_ξ appear in the definition of $p(s,t)$ and every such constant belongs to $2^I|G$, therefore (3) holds. The conditions (4) and (5) are also easy to verify.

Now let

$$\Pi_0 = \{\bigcap\{c_\alpha : \alpha \in s\} \cap \bigcap\{-c_\alpha : \alpha \in t\} \subseteq p(s,t)/F : s,t \in S_\omega(\lambda)\} .$$

and

$$\Pi_1 = \{(\psi(\alpha))(c_{\xi_1},\dots,c_{\xi_k}) : (\xi_1,\dots,\xi_k) = \vartheta(\alpha) , \alpha < \lambda\} .$$

We claim that

(24) $\Pi = \Pi_0 \cup \Pi_1$ is consistent with

$$Th((2^I_F|G , p(s,t)/F)_{s,t \in S_\omega(\lambda)}) .$$

In fact let u be a finite subset of λ and let

$$\Pi_u = \{\bigcap\{c_\alpha : \alpha \in s\} \cap \bigcap\{-c_\alpha : \alpha \in t\} \subseteq p(s,t)/F : s,t \subseteq u\}$$

$$\cup\{(\psi(\alpha))(c_{\xi_1},\dots,c_{\xi_k}) : (\xi_1,\dots,\xi_k) = \vartheta(\alpha) , \alpha \in u\} .$$

Moreover let $\Sigma_u = \{\sigma_\alpha : \alpha \in u\}$. By (22) ther is $b \in A^I|G$ such that

(25) $B_0 \models \bigwedge \Sigma_u [b/F]$.

For $\xi \in u \cup \bigcup\{\vartheta(\alpha) : \alpha \in u\}$ we put $X_\xi = \{i \in I: A_i \models \varphi_\xi[b(i)]\}$. Since $b \in A^I|G$ and only a finite number of constants, each of them in $A^I|G$, appear in φ_ξ , we have $X_\xi \in 2^I|G$. Now (23) and (25) imply that if $\vartheta(\alpha) = (\xi_1,\dots,\xi_k)$ then $(\psi(\alpha))(X_{\xi_1}/F,\dots,X_{\xi_k}/F)$ holds in $2^I_F|G$ for every $\alpha \in u$. Moreover if $s,t \subseteq u$, then by the definition of function p we have $\bigcap\{X_\alpha : \alpha \in s\} \cap \bigcap\{I - X_\alpha : \alpha \in t\} \subseteq p(s,t)$. This implies that $\{X_\alpha/F : \alpha \in u\}$ satisfies Π_u in $(2^I_F|G , p(s,t)/F)_{s,t \in S_\omega(\lambda)}$.

Since $2^I_F|G$ is χ-saturated it follows from (24) that there is a function $C : \lambda \to 2^I$ such that

(26) $C(\alpha) \in 2^I|G$ for every $\alpha < \lambda$

and if $c_\alpha = C(\alpha)/F$, then

(27) $((2^I_F|G, p(s,t)/F)_{s,t\in S_\omega(\lambda)}, c_\alpha)_{\alpha<\lambda}$ is a model of Π.

This by $\Pi_0 \subseteq \Pi$ implies (7), whence by (26) p and C satisfy the hypotheses of Lemma 1.4. Now let ρ be an equivalence relation and D^+, D^- be functions such that $\rho \in G$, $D^+ : \lambda \to 2^I|\rho$, $D^- : \lambda \to 2^I|\rho$, $D^+ \cap D^- = 0$ and (8) - (10) hold.

We shall modify D^+ and D^- to assure that an additional condition is satisfied. Let for $s \in S_\omega(\lambda)$

$$H_0(s) = \bigcap\{eq(b_\xi) : b_\xi/F \text{ appears in } \varphi_\alpha \text{ for some } \alpha \in s\}$$

and let $f(s) = I \times I$. Since $b_\xi \in 2^I|G$ for every $\xi < \lambda$, we have $H_0^* S_\omega(\lambda) \subseteq 2^I|G$, Henceforth, since (F,G) is χ-good, there is $\rho_4 \in G$ and an additive function $g_2 : S_\omega(\lambda) \to F$ such that (g.1)-(g.3) hold (with suitable substitutions). We put $\rho_0 = \rho \cap \rho_4$ and for $\alpha < \lambda$ let $E^+(\alpha) = D^+(\alpha) \cap g_2(\{\alpha\})$, $E^-(\alpha) = D^-(\alpha) \cap g_2(\{\alpha\})$. Let us list some properties of E^+ and E^-.

(28) $(E^+)^*\lambda \cup (E^-)^*\lambda \subseteq 2^I|\rho_0$,

(29) $(E^+(\alpha) \triangle C^+(\alpha)) \cup (E^-(\alpha) \triangle C^-(\alpha)) \in I_F$ for every $\alpha < \lambda$

(30) $E^+(\alpha) \cap E^-(\alpha) = 0$ and $(E^+(\alpha) \cup E^-(\alpha)) \in F$,

(31) if $s,t \in S_\omega(\lambda)$ then $\mathbb{E}(s,t) \subseteq p(s,t)$,

(32) if X is an infinite subset of λ then

$$\bigcap\{E^+(\alpha) \cup E^-(\alpha) : \alpha \in X\} = 0 ,$$

and finally

(33) if $x \in I/\rho_0$ and $x \subseteq \mathbb{E}(s,t)$ then there is $y \in I/F_0(s \cup t)$ such that $x \subseteq y$.

To check that E^+ and E^- have the properties listed above let us notice that (33) follows from the fact that ρ_4 and g_0 satisfy (g.3). To prove (32) we apply (10), and finally (28)-(31) easily follows from (8) and (9).

Now we are ready to complete the proof of Theorem 1. We shall define a function $d \in A^I|G$ such that

(34) $\quad B_0 \models \sigma\,[d/F] \quad$ for every $\quad \sigma \in \Sigma$.

Let $i\colon I/\rho_0 \to I$ be a fixed function such that $i(x) \in x$ for every $x \in I/\rho_0$.
Further let for $x \in I/\rho_0$

$$s(x) = \{\alpha < \lambda : i(x) \in E^+(\alpha)\}$$

and

$$t(x) = \{\alpha < \lambda : i(x) \in E^-(\alpha)\} .$$

By (32) $s(x)$ and $t(x)$ are finite for every $x \in I/\rho_0$. Of course $i(x) \in \mathbb{E}(s(x), t(x))$ and hence by (33) $x \subseteq \mathbb{E}(s(x), t(x))$. Moreover by (31) $x \subseteq p(s(x), t(x))$. Consequently for every $j \in x$ we have

(35) $\quad A_j \models \exists v(\bigwedge\{\varphi_\alpha\colon \alpha \in s(x)\} \wedge \bigwedge\{\neg\, \varphi_\alpha\colon \alpha \in t(x)\})$.

Since $i(x) \in \mathbb{E}(s(x),t(x))$ it follows by (33) that if b_ξ/F appears in φ_α for some $\alpha \in s(x) \cup t(x)$ then b_ξ is constant on x . Thus by (35) there is $d_0(x) \in A$ such that for every $j \in x$

(36) $\quad A_j \models (\bigwedge\{\varphi_\alpha\colon \alpha \in s(x)\} \wedge \bigwedge\{\neg\, \varphi_\alpha\colon \alpha \in t(x)\})[d_0(x)]$.

To define d we put $d(j) = d_0(j/\rho_0)$ for every $j \in I$.

It remains to verify that d has all the required properties. First of all $d \in A^I|G$ because d is constant on all equivalence classes of ρ_0 . Now let $\xi < \lambda$. We have

$$K(\varphi_\xi,\, d) = \{j \in I\colon A_j \models \varphi_\xi[d(j)]\} .$$

But if $j \in E^+(\xi)$ then by the definition of d we have $A_j \models \varphi_\xi[d(j)]$, similarly if $j \in E^-(\xi)$, then $A_j \models \neg\, \varphi_\xi[d(j)]$. Therefore $E^+(\xi) \subseteq K(\varphi_\xi,\, d)$ and $E^-(\xi) \subseteq K(\neg\, \varphi_\xi,\, d) = I - K(\varphi_\xi,\, d)$ whence by

(30) $K(\varphi_\xi, d) \triangle E^+(\xi) \in I_F$ and by (29)

$$K(\varphi_\xi, d) \triangle C(\xi) \in I_F \qquad \text{(recall that } C(\xi) = C^+(\xi)) .$$

This shows that

(37) $K(\varphi_\xi, d)/F = c_\xi$.

Now we can prove that d/F satisfies Σ in B_0 . By (23) it suffices to check that, for a given $\alpha < \lambda$,

$$2^I_F|G \models (\phi(\alpha))[K(\varphi_{\xi_1}, d), \ldots, K(\varphi_{\xi_k}, d)]$$

where $(\xi_1, \ldots, \xi_k) = \vartheta(\alpha)$. The last formula is by (37) equivalent to $2^I_F|G \models (\phi(\alpha))[c_{\xi_0}, \ldots, c_{\xi_k}]$ but this, in view of $\Pi_1 \subseteq \Pi$, is a consequence of (27).

2. Universality.

It was stated without any proof in [11] (Theorem 3.7) that if 2^I_F is $\varkappa^+$-universal and F is $\varkappa$-regular, then for every relational structure A such that $|L(A)| \leq \varkappa$ the reduced power A^I_F is $\varkappa^+$-universal. The author could not reconstruct the proof of this fact but was more successful in looking for a counterexample. We give one in this section.

LEMMA 2.1. Assume that 2^I_F is infinite and for every relational structure A such that $|L(A)| \leq \varkappa$ the reduced power is $\varkappa^+$-universal. Then there is a family $\langle X_\xi : \xi < \varkappa \rangle$ of subsets of I such that $X_\xi \cap X_\eta = 0$ and $X_\xi \notin I_F$ for $\xi < \eta < \varkappa$.

Proof. Let $A = \langle \varkappa, \neq, \xi \rangle_{\xi < \varkappa}$ (here ξ is an element of $\varkappa$) and let Σ denotes the following set of sentences

$$\{\neg (x \neq \xi) : \xi < \varkappa\}$$

Of course for a given $\xi < \varkappa$ and $f \in A^I$ we have

(1) $\quad A_F^I \models (\neg\,(x \neq \xi))[f/F]$ if and only if $\{i \in I\colon f(i) = \xi\} \notin I_F$.

Since 2_F^I is infinite, it follows from (1) that $\Sigma \cup Th(A_F^I)$ is consistent. But by the assumption A_F^I is $\varkappa^+$-universal, whence there is a function $f \in A^I$ such that $(A_F^I, f/F)$ is a model of Σ .
Let for $\xi < \varkappa$

$$X_\xi = \{i \in I\colon f(i) = \xi\} .$$

Of course $X_\xi \cap X_\eta = 0$ for every $\xi < \eta < \varkappa$ and by (1) $X_\xi \notin I_F$ holds for every $\xi < \varkappa$.

LEMMA 2.2. Let $\varkappa$ be an infinite cardinal and let B be an infinite Boolean algebra. Then there is a filter F on $\varkappa^+$ such that

(i) $\quad 2_F^{\varkappa^+} \equiv B$

(ii) $\quad$ F is $\varkappa^+$-regular

(iii) $\quad 2_F^{\varkappa^+}$ is $\varkappa^{++}$-universal

and

(iv) $\quad$ If $\langle X_\xi\colon \xi < \eta\rangle$ is a family of disjoint subsets of $\varkappa^+$ and $X_\xi \notin I_F$ for every $\xi < \eta$, then $\eta \leq \varkappa$.

Proof. Let E^0 be an arbitrary $\varkappa^+$-regular filter on $\varkappa^+$. Let $E = \{E_\xi\colon \xi < \varkappa^+\}$ be a family of elements of E^0 such that every infinite subfamily of E has empty intersection. Assume moreover, that E is closed under finite intersections. By a lemma of Keisler ([4], see also [1], Lemma 6.1.6) there is a family $\{D_\xi\colon \xi < \varkappa^+\}$ such that for $\xi < \eta < \varkappa^+$ we have $|D_\xi| = \varkappa^+$, $D_\xi \subseteq E_\xi$ and $D_\xi \cap D_\eta = 0$.
Let for $\xi < \varkappa^+$ $\{D_{\xi,\eta}\colon \eta < \varkappa\}$ be a partition of D_ξ into $\varkappa$ sets, each of power $\varkappa^+$. We put $I_\alpha = \bigcup\{D_{\xi,\alpha}\colon \xi < \varkappa^+\}$ for every $\alpha < \varkappa$. Since for $\alpha < \varkappa$ and $\xi < \varkappa^+$ $I_\alpha \cap E_\xi \neq 0$ we have

(2) $\quad I_\alpha \notin I_E$

Moreover, since E is closed under intersections, the family $E_\alpha = \{X \cap I_\alpha\colon X \in E\}$ is a filter on I_α for every $\alpha < \varkappa$. Now let F_α be an arbitrary ultrafilter on I_α such that $E_\alpha \subseteq F_\alpha$ and let F^0 be a filter on $\varkappa$ such that

(3) $\quad 2^{\varkappa}_{F^0} \equiv B$

and

(4) $\quad 2^{\varkappa}_{F^0}$ is $\varkappa^+$-saturated.

The existence of such a filter is an easy consequence of the existence of $\varkappa^+$-good ultrafilter on $\varkappa$ (see [1]) and a theorem of Ershov ([2], cf also [7] and [11]) . Let F denotes the filter on $\varkappa^+$ defined by the condition

$$X \in F \leftrightarrow \{\alpha : X \cap I_\alpha \in F_\alpha\} \in F^0$$

We claim that F has the desired properties. To prove (i) notice that

(5) $$2^{\varkappa^+}_F = \prod_{\alpha<\varkappa} (2^{I_\alpha}_{F_\alpha})/F^0$$

This in view of maximality of the filters F_α $(\alpha < \varkappa)$ implies $2^{\varkappa^+}_F = 2^{\varkappa}_{F^0}$, whence by (3) $2^{\varkappa^+}_F \equiv B$. It is easy to check that $E \subseteq F$, whence (ii) holds. By (4) and (5) $2^{\varkappa}_F$ is $\varkappa^+$-saturated. From this it follows easily $2^{\varkappa}_F$ is $\varkappa^{++}$-universal. It remains to prove (iv). Let $\langle X_\xi : \xi < \eta\rangle$ be a family as in (iv) and let $\xi < \eta$. We put

$$H_\xi = \{\alpha : X_\xi \cap I_\alpha \in F_\alpha\} .$$

We claim that $H_\xi \neq 0$. In fact, if $H_\xi = 0$, then by maximality of the filters F_α we get $\{\alpha : (I - X_\xi) \cap I_\alpha \in F\} = \varkappa \in F^0$, whence $I - X_\xi \in F$, which is impossible since $X_\xi \notin I_F$. To prove (iv) assume to the contrary that $\eta > \varkappa$. Since $H_\xi \neq 0$ and $H_\xi \subseteq \varkappa$, there are $\xi_1, \xi_2 < \eta$ such that $H_{\xi_1} \cap H_{\xi_2} \neq 0$. Let $\alpha \in H_{\xi_1} \cap H_{\xi_2}$. Then by the definition of H_ξ we have $I_\alpha \cap X_{\xi_1} \cap X_{\xi_2} \neq 0$ which is impossible since we assumed that the sets X_ξ are disjoint.

THEOREM 2.3. If $\varkappa = \lambda^+$ and B is an infinite Boolean algebra, then there is a filter F and a relational structure A with $|L(A)| \leq \varkappa$ such that

1. $2^I_F \equiv B$

2. F is $\varkappa$-regular

3. 2^I_F is $\varkappa^+$-universal

4. A^I_F is not $\varkappa^+$-universal .

Theorem 2.3 is an immediate consequence of Lemma 2.1 and Lemma 2.2.

The assumption that B is infinite is necessary. This follows from a theorem of Keisler [6]. Also Theorem 2.3 does not hold for $\varkappa = \omega$ (see [7]).

The filter defined in Lemma 2.2 has the property that there is no family of $\varkappa^+$ disjoint subsets of $\varkappa^+$ with complements not in F, but, on the other hand there is a family of $\varkappa^+$ sets with complements not in F, which are F-almost disjoint. This can not be the case if F is $\varkappa^+$-separatistic (see [8]).

DEFINITION 2.4. A filter F is $\varkappa$-separatistic if for every set Σ of sentences of $L_B(\varkappa)$ which is satisfiable in 2^I_F there is a sequence $X = \langle X_\xi : \xi < \varkappa \rangle$ of subset of I such that $(2^I_F, X_\xi/F)_{\xi<\varkappa} \models \Sigma$ and moreover

$$\mathrm{Sep}(X) \begin{cases} \text{for any } s \in S_\omega(\varkappa) \text{ and any function } h\colon s \to 2 \\ \bigcap\{X_\xi^{h(\xi)} : \xi \in \mathrm{dom}(h)\} = 0 \text{ if and only if} \\ \bigcap\{X_\xi^{h(\xi)} : \xi \in \mathrm{dom}(h)\}/F = 0 \ . \end{cases}$$

Using the notion of a $\varkappa$-separatistic filter we can give a characterisation of filters F which have the property that the reduced power A^I_F is always $\varkappa$-universal.

THEOREM 2.5. Let F be a filter of subsets of I. The reduced power A^I_F is $\varkappa^+$-universal for any relational structure A with $|L(A)| \leq \varkappa$ if and only if

1. 2^I_F is $\varkappa^+$-universal

2. F is $\varkappa$-regular

and

3. F is $\varkappa$-separatistic.

Theorem 2.5 is a special case of Theorem 2.8 below. It easily follows from Lemma 2.2 that there are $\varkappa$-regular filters with 2^I_F being $\varkappa^+$-universal which are not separatistic. Below we give a characterisation of $\varkappa$-separatistic filters such that 2^I_F is $\varkappa^+$-universal. A similar chracterisation can be given for the case of limit reduced powers.

PROPOSITION 2.6. Let F be a filter over I. Then 2^I_F is $\varkappa^+$-universal and F is $\varkappa$-separatistic if and only if for every Boolean algebra B such that $B \equiv 2^I_F$ and $|B| \leq \varkappa$ there is an isomorphic embedding f of B into 2^I such that f/F defined by $f/F(x) = (f(x))/F$ is an elementary embedding of B into 2^I_F.

Now we go back to limit reduced powers.

DEFINITION 2.7. A filter pair (F,G) is $\varkappa$-separatistic if and only if for every set Σ of sentences of $L_B(\varkappa)$ which is satisfiable in $2^I_F|G$ there is $\rho \in G$ and a sequence $X = \langle X_\xi : \xi < \varkappa \rangle$ of elements of $2^I|\rho$ such that $(2^I_F|G, X_\xi/F) \models \Sigma$ and $Sep(X)$ holds.

THEOREM 2.8. Let (F,G) be a filter pair. Then for every relational structure A whose similarity type has power $\leq \varkappa$ the limit reduced power $A^I_F|G$ is $\varkappa^+$-universal if and only if

1. $2^I_F|G$ is $\varkappa^+$-universal

2. (F,G) is $\varkappa$-regular

and

3. (F,G) is $\varkappa$-separatistic.

<u>Proof</u>. We omit the proof that 1. 2. and 3. are sufficient. It is also clear that 1. is necessary. To prove that 2. is necessary it suffices to consider the structure

$$(S_\omega(\varkappa) \, , \subseteq \, , \{\alpha\})_{\alpha<\varkappa} \, ,$$

the set

$$\Sigma = \{\{\alpha\} \subseteq x : \alpha < \varkappa\}$$

and apply the routine technique.

Now we shall prove the necessity of 3. Let Σ be a complete set of sentences of $L_B(\varkappa)$ which is finitely satisfiable in $2^I_F|G$. Let

$$S = \{x \subseteq \varkappa : \Sigma \vdash \bigcap_{\alpha\in s} c_\alpha \cap \bigcap_{\alpha\in t} -c_\alpha = 0 \quad \text{for all} \quad s,t \in S_\omega(\varkappa)$$
$$\text{such that} \quad s \subseteq x \quad \text{and} \quad t \subseteq -x\} .$$

Since Σ is consistent S is non-empty and moreover if $\Sigma \vdash \bigcap_{\alpha\in s} c_\alpha \cap \bigcap_{\alpha\in t} -c_\alpha = 0$, then there is $x \in S$ such that $s \subseteq x$ and $x \subseteq -t$.

Now let $A = \langle S \cup 2 ; B, C, \subseteq, f_\alpha\rangle_{\alpha<\varkappa}$ where 2 is the two-element Boolean algebra, $B(x) \leftrightarrow x \in 2$, $C(x) \leftrightarrow x \in S$, $x \subseteq y \leftrightarrow B(x) \wedge B(y) \wedge x \subseteq y$ and

$$f_\alpha(x) = \begin{cases} 1 & \text{if} \quad \alpha \in x \quad \text{and} \quad x \in S \\ 0 & \text{otherwise} \end{cases}$$

We put

$$\Pi = \{C(x)\} \cup \{\varphi^B(f_{\alpha_1}(x), \ldots, f_{\alpha_n}(x)) : \varphi(c_{\alpha_1}, \ldots, c_{\alpha_n}) \in \Sigma\}$$

where φ^B is the relativisation of φ to B . We claim that Π is finitely satisfiable in $A^I_F|G$. In fact let

$$\Pi_0 = \{C(x)\} \cup \{\varphi_i^B(f_{\alpha_1}(x), \ldots, f_{\alpha_n}(x)) : i \leq k\}$$

be a finite subset of Π and let

$$\Sigma_0 = \{\varphi_i(c_{\alpha_1}, \ldots, c_{\alpha_n}) : i \leq k\} \cup$$

$$\{\bigcap\{c_{\alpha_i}^{h(i)} : i \in s\} = 0 : s \subseteq n,\ h : s \to 2,\ \Sigma \vdash (\bigcap\{c_{\alpha_i}^{h(i)} : i \in n\} = 0)\}.$$

Since Σ is satisfiable in $2^I_F|G$ there are $X'_{\alpha_1}, \ldots, X'_{\alpha_n} \in 2^I|G$ such that $(2^I_F|G, X'_{\alpha_1}/F, \ldots, X'_{\alpha_n}/F) \models \Sigma_0$. It is an easy exercise to check that there are $X_{\alpha_1}, \ldots, X_{\alpha_n}$ such that $X_{\alpha_i} \triangle X'_{\alpha_i} \in I_F$ for

$i \leqslant n$ and moreover for every $s \subseteq n$ and every $h : s \to 2$ $\bigcap\{X_{\alpha_i}^{h(i)} : i \in s\} = 0$ if and only if $\bigcap\{X_{\alpha_i}^{h(i)} : i \in s\}/F = 0$. Let $g \in {}^n2$ be a function such that $\bigcap\{X_{\alpha_i}^{g(i)} : i \in n\} \neq 0$. Then by the definition of Σ_0 there is $x_g \in S$ such that $\{\alpha_i : i \leqslant n,\ g(i) = 0\} \cap x = 0$ and $\{\alpha_i : i \leqslant n,\ g(i) = 1\} \subseteq x$. For $\xi \in I$ we put $a(\xi) = x_g$ where $g \in {}^n2$ is the (unique) function such that $\xi \in \bigcap\{X_{\alpha_i}^{g(i)} : i \in n\}$. It is obvious that the definition is correct and that $(A_F^I|G\ ,\ a/F) \models \Pi_0$

Now, since $A_F^I|G$ is $\varkappa^+$-universal there is $b \in A^I|G$ such that $(A_F^I|G\ ,\ b/F) \models \Pi$. For $\alpha < \varkappa$ we put

$$Y_\alpha = \{\xi \in I : (f_\alpha(b))(\xi) = 1\} .$$

It is left to the reader to check that $\{Y_\alpha : \alpha < \varkappa\}$ forms a sequence with the desired properties.

References

[1] C.C. Chang, H.J. Keisler, Model theory, North-Holland, Amsterdam 1973.

[2] Yu.L. Ershov, Decidability of the elementary theory of relatively complemented distributive lattices and the theory of filters, Algebra i Logika Sem. 3 No. 3 (1964), p.17.

[3] S.Feferman, R.L. Vaught, The first order properties of algebraic systems, Fundamenta Mathematicae, 47 (1959), pp. 57-103.

[4] H.J. Keisler, Good ideals in fields of sets, Annals of Mathematics 79(1964), pp. 338-359,

[5] H.J. Keisler, Ultraproducts and saturated models, Indagationes Mathematicae 26(1964), pp.178-186.

[6] H.J. Keisler, Ultraproducts which are not saturated, Journal of Symbolic Logic 32(1967), pp. 23-46.

[7] L. Pacholski, On countably compact reduced products III, Colloquium Mathematicum 23(1971), pp. 5-15.

[8] L. Pacholski, On countably universal Boolean algebras and compact classes of models, Fundamenta Mathematicae 78(1973),pp.43-60.

[9] L. Pacholski, Homogenity, universality and saturatedness of limit reduced powers III, to appear in Fundamenta Mathematicae.

[10] L. Pacholski, Homogeneous limit reduced powers, in preparation.

[11] S. Shelah, For which filters is every reduced product saturated, Israel Journal of Mathematics 12(1972), pp.23-31.

EXTENSIONS OF MODELS FOR ZFC TO MODELS FOR ZF + V = HOD WITH APPLICATIONS

by S. Roguski

ABSTRACT:

In this paper we investigate the extendability of models for ZFC to models for ZF + V = HOD , and the relative consistency of sentences of L_{ZF} with the theory ZF + V = HOD .

INTRODUCTION:

The first results in this domain were obtained by Mc Aloon. In the paper [1] he showed that the theories ZF + V = HOD $\neq$ L and ZF + V = HOD + $\neg$ GCH are consistent if ZF theory is consistent. Kunen in [4] has shown that if the theory ZFC + MC($\varkappa$) is consistent then the theory ZF + V = HOD + MC($\varkappa$) is consistent too. Clearly, the formulas $V \neq L$, $\neg$GCH , MC($\varkappa$) are Σ_2^{ZFC}-sentences. Thses fact gave us the idea that if Φ is a Σ_2^{ZFC}-sentence such, that ZFC + Φ is consistent then the theory ZF + V = HOD + Φ is consistent too. In this the assumption the axiom of choice is essential, because $\neg$ AC is Σ_2-sentence and ZF + V = HOD $\vdash$ AC . We cannot replace Σ_2 by any anöther stronger degree in Levy's hierarchy, because the sentence $(\exists x)(x \subseteq \omega \wedge x \notin OD)$ is consistent with ZFC (see [6]) and it is a Π_2^{ZFC}-sentence.
This paper is divided into three parts. In the first one we give fundamental definitions and facts about extended structures and forcing. In the second part we prove

---

This paper is a fragment of doctoral dissertation of author.

Theorem 1. Let $\langle M,\in,X_1,\dots,X_n\rangle$ be a c.s.m. for ZFC and $\alpha \in On^M$. Then there are formulas $\Phi_0,\Phi_1,\dots,\Phi_n$ of L_{ZF} with one free variable and with only ordinal parameters and there exist a c.s.m. N for ZF + V = HOD such, that $M \subseteq N$, $On^M = On^N$, $R^M_\alpha = R^N_\alpha$ and $x \in M \iff N \models \Phi_0(x)$ $x \in X_i \iff N \models \Phi_i(x)$ for $i = 1,\dots,n$.
In the third part we prove the

Theorem 2 (Schema). If Φ is a Σ_2^{ZFC}-sentence then
Con (ZFC + Φ) $\iff$ Con (ZF + V = HOD + Φ)

PART 1.

$U_1,\dots,U_n$ are fixed unary predicates. Then ZF $(U_1,\dots,U_n)$ is ZF theory plus all instances of the replacement schema for formulas from the language $L_{ZF}(U_1,\dots,U_n)$. We say that $\mathfrak{M} = \langle M,\in,X_1,\dots,X_n\rangle$ is a model for ZF if $X_i \subseteq M$, $x \in X_i \iff \mathfrak{M} \models U_i(x)$ for $i = 1,\dots,n$ and $\mathfrak{M} \models ZF(U_1,\dots,U_n)$.

Now we give some facts about forcing which we shall use in the next part of our paper. The fundamental definitions and denotations are the same as in [7].
From this moment we shall be working in ZFC theory.

Firstly we define functionals W and B.

Definition 1.1. $W(o) = \omega$, $W(\lambda) = \bigcup_{\alpha<\lambda} W(\alpha)$ for limit $\lambda > 0$
$W(\lambda + 1) = [W(\lambda)]^+$ for limit λ, $W(\beta + 2) = [2^{W(\beta+1)}]^+$
$(B(\alpha) = \omega_\beta) \overset{df}{\iff} (W(\alpha + 1) = \omega_{\beta+1})$

Definition 1.2. $\langle C, \leq\rangle$ is Jensen's notion of forcing if
(i) the elements of C are functions p such, that for any α,β,γ
$\langle\langle\alpha,\beta\rangle,\gamma\rangle \in p \iff \beta \in W(\alpha) \wedge \gamma \in B(\alpha)$ and $|p\restriction\{\alpha\}\times W(\alpha)| < W(\alpha)$
(ii) $\leq$ denotes reversed inclusion.
Let $p_{(\xi)} = \{\langle\langle\alpha,\beta\rangle,\gamma\rangle \in p : \alpha \leq \xi\}$, $p^{(\xi)} = p - p_{(\xi)}$.

Definition 1.3. $\langle Q, \leq\rangle$ is a notion of forcing such, that
(i) the elements of Q are functions p which satysfy the following condition $\langle\alpha,E\rangle \in p \iff E$ is a well-ordering of $R_{\omega+\alpha}$ of order type $|R_{\omega+\alpha}|$
(ii) $\leq$ denotes reversed inclusion.
Let $p_{(\xi)} = \{\langle\alpha,E\rangle \in p : \alpha \leq \xi\}$, $p^{(\xi)} = p - p_{(\xi)}$

Definition 1.4. Let λ be such cardinal that $\lambda = \omega_\lambda = \beth_\lambda$ and $2^{\omega_\alpha} = \omega_{\alpha+1}$ for $\alpha \geq \lambda$. $\langle P, \leq \rangle$ is Easton's notion of forcing if

(i) the elements of P are functions p such, that $\langle\langle\alpha,\beta,\gamma\rangle,i\rangle \in p \Leftrightarrow \alpha > \lambda \wedge \omega_\alpha$ is regular $\wedge$ $\beta \in \omega_\alpha \wedge \gamma \in \omega_{\alpha+2}$ $i = 0,1$ and $|\{\langle\langle\xi,\mu,\eta\rangle,i\rangle \in p : \xi \leq \alpha\}| < \omega_\alpha$ for ω_α regular,

(ii) $\leq$ denotes reversed inclusion.

Let $p_{(\xi)} = \{\langle\langle\alpha,\beta,\gamma\rangle,i\rangle \in p : \alpha \leq \xi\}$, $p^{(\xi)} = p - p_{(\xi)}$

We reserve symbols C,Q,P for notions of forcing defined in definitions 1.2, 1.3, 1.4.
For previous notions of forcing we define C_α, C^α, Q_α, Q^α, P_α, P^α in the usual way.
Clearly C,Q,P defined above are coherent notions of forcing.

Definition 1.5. A notion of forcing is set-closed if is m-closed for any cardinal m.

FACT 1.1. $|C_{\alpha+1}| < W(\alpha + 1)$ and $C^{\alpha+1}$ is $W(\alpha + 1)$ closed.

FACT 1.2. $P_{\alpha+1}$ satisfies $\omega_{\alpha+1}$ - c.c. and $P^{\alpha+1}$ is $\omega_{\alpha+1}$ closed.

FACT 1.3. Q is set-closed.

In the next part of our paper we shall need the following

MAIN LEMMA. If $\mathfrak{M} = \langle M,\in,X_1,\dots,X_n\rangle$ is a c.s.m. for ZFC, $\mathcal{P}$ is a coherent notion of forcing in $\mathfrak{M}$, $\mathcal{P}$ satisfies the m_ξ, β_ξ-density condition for $\xi \in On^M$ and $\bigcup\{m_\xi : \xi \in On^M\} = On^M$, then for every G which is $\mathcal{P}$-generic over $\mathfrak{M}$, $\mathfrak{M}[G] = \langle M[G],\in,X_1,\dots,X_n,M,G\rangle$ is a c.s.m. for ZFC.

The proof of Main Lemma is a simply modification of the proof of another version of this Lemma given in [7].

PART 2.

In this part we prove

Theorem 1. Let $\langle M,\in, X_1,\dots,X_n\rangle$ be a c.s.m. for ZFC and $\alpha \in On^M$. Then there are formulas $\Phi_0,\Phi_1,\dots,\Phi_n$ of L_{ZF} with one free variable and with only ordinal parameters and there exists a c.s.m. N for ZF + V = HOD such that $M \subseteq N$, $On^M = On^N$, $R^M_\alpha = R^N_\alpha$ and $x \in M \iff N \models \Phi_0(x)$ $x \in X_i \iff N \models \Phi_i(x)$ for $i = 1,\dots,n$.

We now pass to the proof of Theorem 1. Let α be a fixed ordinal in M and let λ be the smallest cardinal μ greater then α such that $\mu = \omega^M_\mu = \beth^M_\mu$.

Let C be Jensen's notion of forcing in M (see def. 1.2). Let us take $R = C^\lambda$. From Fact 1.1 it follows that R is an ω^+_λ-closed notion of forcing in M. Let G_1 be R-generic over $\mathfrak{M}$. Using the Main Lemma we obtain

$$\mathfrak{M}[G_1] \models ZFC + R^M_\lambda = R_\lambda + 2^{\omega_\alpha} = \omega_{\alpha+1} \quad \text{for} \quad \alpha \geq \lambda .$$

Now we take in the model $\mathfrak{M}[G_1]$ the notion of forcing Q defined in definition 1.3. Let G_2 be Q-generic over $\mathfrak{M}[G_1]$. Then using the Main Lemma $\mathfrak{M}[G_1][G_2] \models ZFC$

Next, we take in $\mathfrak{M}[G_1,G_2]$ Easton's notion of forcing P (see definition 1.4). Let G_3 be P-generic over $\mathfrak{M}[G_1,G_2]$. Then

$$\mathfrak{N} = \mathfrak{M}[G_1,G_2][G_3] \models ZFC + R^M_\lambda = R_\lambda + cf = cf^{M[G_1]} + 2^{\omega_{\alpha+1}} = \omega_{\alpha+3}$$

for $\alpha > \lambda$

In the model $\mathfrak{N}$ we define a model N satisfying the conditions of Theorem 1.

For this we introduce the following definitions and descriptions.

a. Let $x,y \in M[G_1]$. We define

$x \prec y \iff \mathrm{rank}(x) < \mathrm{rank}(y) \vee (\mathrm{rank}(x) = \mathrm{rank}(y) \wedge \langle x,y\rangle \in \bigcup G_2(\mathrm{rank}(x+1)$

FACT 2.1. The relation $\prec$ defines a well-ordering of $M[G_1]$ of type On.

Let x_α be the α-th element in $\prec$.

FACT 2.2. $\mathrm{rank}(x_\alpha) < \mathrm{rank}(x_\beta) \to \alpha < \beta$.

b. Let J be the pairing function for ordinals defined in [5]. we define

$\alpha \in Y_0 \overset{df}{\iff} x_\alpha \in M$, $\alpha \in Y_i \overset{df}{\iff} x_\alpha \in X_i$ for $i = 1,\dots,n$,

$\lambda + 3J(\alpha,i) + 2 \in Y \overset{df}{=} \alpha \in Y_i$ for $i = 0,1,\ldots,n$,

$K(\beta,\alpha) \overset{df}{=} \lambda + 3J(\beta,\alpha) + 1$.

K is a 1-1 functional because J is 1-1 .

Let $t_\alpha = \{K(\beta,\alpha) : x_\beta \in x_\alpha\}$

FACT 2.3. (i) $K(\beta,\alpha) \in t_\alpha \to \beta < \alpha$.

(ii) $\alpha \neq \beta \to t_\alpha \cap t_\beta = \emptyset$

(iii) $x_\alpha = \{x_\beta : \lambda + 3J(\beta,\alpha) + 1 \in t_\alpha\}$

Let $F = Y \cup \bigcup_{\alpha \in On} t_\alpha$

FACT 2.4. (i) $\gamma \in t_\alpha \Leftrightarrow \gamma \in F \wedge (\exists\beta)_\alpha(\lambda + 3J(\beta,\alpha) + 1 = \gamma)$

(ii) $\gamma \in Y_i = \lambda + 3J(\alpha,i) + 2 \in F$ for $i = 0,1,\ldots,n$.

From the Facts above we clearly obtain

COROLLARY 2.5. The clases $\langle t_\alpha : \alpha \in On\rangle$, $\langle x_\alpha : \alpha \in On\rangle$, and Y_i for $i = 0,1,\ldots,n$ are definable with parameters F and λ , and thus $M, X_1,\ldots,X_n, M[G_1]$ are definable with parameters F and λ. The elements of conditions of Easton's notion of forcing are four-tuples $\langle\alpha,\beta,\gamma,i\rangle$ where $\alpha > \lambda$ and ω_α is regular, $\beta \in \omega_\alpha$, $\gamma \in \omega_{\alpha+2}$, $i \in 2$.

We define $A = \{\langle\alpha,\beta,\gamma\rangle : (\bigcup G_3)(\langle\alpha,\beta,\gamma\rangle) = 0\}$

Now we are going to define a class $H \subseteq A$ such, that H and F will be definable in $L[H]$ by some formulas of L_{ZF} with λ as its only parameter.

Let $J^*(\alpha,\beta,\gamma) = \lambda + 3J(\alpha, J(\beta,\gamma)) + 3$ where $\alpha > \lambda$, ω_α is regular, $\beta \in \omega_\alpha$ and $\gamma \in \omega_{\alpha+2}$

Let

$$\bar{E}(\delta) \overset{df}{=} \begin{cases} \{\langle\alpha,\beta,\gamma\rangle\} \cup \bar{E}(\alpha) & \text{for } \delta = J^*(\alpha,\beta,\gamma) \\ \emptyset & \text{if there is no such } \alpha,\beta,\gamma \end{cases}$$

Now we give a definition of the class H :

$$\langle\alpha,\beta,\gamma\rangle \in H \overset{df}{\Longleftrightarrow} \langle\alpha,\beta,\gamma\rangle \in A \wedge (\alpha \in F \vee (\bar{E}(\alpha) \subseteq A \wedge$$
$$\wedge (\exists\alpha',\beta',\gamma')(\alpha' \in F \wedge \langle\alpha',\beta',\gamma'\rangle \in \bar{E}(\alpha))))$$

Let $H_\delta = \{\langle\beta,\gamma,\alpha\rangle \in H : \beta \leq \delta\}$ and $N = L[H]$.

FACT 2.6. The class F can be defined from class H in the following way:

$\alpha \in F \Leftrightarrow (\exists \beta,\gamma,\delta)(\langle\alpha,\beta,\gamma\rangle \in H \wedge (\alpha = 3\delta + 2 \vee \alpha = 3\delta + 1))$

Now we prove

LEMMA 2.1. $N \models ZFC + V = HOD + R_\lambda = R_\lambda^M$

We divide the proof into some steps

a) $N \models ZF$ This is obvious because the class H is defined in $\mathcal{N}$

b) $N \models AC$ This follows from the fact that $H \subseteq On \times On \times On$

c) $M[G_1] \subseteq N$ This follows from Corollary 2.5 and Fact 2.6

d) $cf^{M[G_1]} = cf^N = cf^{\mathcal{N}}$ This is implied by c)

e) $R_\lambda^M = R_\lambda^N$ This follows from $R_\lambda^M \subseteq R_\lambda^N \subseteq R_\lambda^{\mathcal{N}} = R_\lambda^M$

f) If $\delta = J^*(\alpha,\beta,\gamma)$ then $\langle\alpha,\beta,\gamma\rangle \in H \Longleftrightarrow N \models 2^{\omega_\delta} = \omega_{\delta+2}$

The proof is analogous to proofs of similar facts in [1].

Using f) and Fact 2.6 we obtain

FACT 2.7. If $\delta = \lambda + 3J(\alpha,i) + j \quad (j = 1,2)$ then $\delta \in F \Longleftrightarrow N \models 2^{\omega_\delta} = \omega_{\delta+2}$

And this Fact together with Lemma 2.1 and Corollary 2.5 gives us Theorem 1.

PART 3

We prove the following

THEOREM 2 (Schema). If Φ is Σ_2^{ZFC}-sentence then $Con(ZFC + \Phi) \Longleftrightarrow Con(ZF + V = HOD + \Phi)$

We need

LEMMA 3.1. (Schema). Let $\Phi(\overline{p}) \in \Sigma_2$. Then $\Phi(\overline{p}) \Longleftrightarrow (\exists\lambda)(\lambda = \omega_\lambda = \beth_\lambda \wedge \overline{p} \in R_\lambda \wedge (R_\lambda \models \Phi(\overline{p})))$

This Lemma follows from Levy's Theorem about Σ_1-formulas and the fact that $\lambda = \omega_\lambda = \beth_\lambda$ iff $y \in R_\lambda \Longleftrightarrow HC(y) < \lambda$

Now let Φ be Σ_2^{ZFC}-sentence and let M be c.s.m. for a large enough fragment of $ZFC + \Phi$

Using Lemma 3.1 we can find $\lambda \in On^M$ such that $\lambda = \omega_\lambda^M = \beth_\lambda^M$ and $R_\lambda^M \models \Phi$. If M_1 is a c.s.m. for the corresponding fragment of ZFC such that $R_\lambda^M = R_\lambda^{M_1}$ then $R_\lambda^{M_1} \models \Phi$ and thus from Lemma 3.1 we have

$M_1 \models \Phi$.

We must show that the theory $ZF + V = HOD + \Phi$ is consistent. For this it is enough to show that if $\alpha \in On^M$ then there is an N which is a c.s.m. for a large enough finite fragment of $ZF + V = HOD$ such that $On^M = On^N$ and $R_\alpha^M = R_\alpha^N$. This fact is an immediate corollary of Theorem 1.

INSTITUTE OF MATHEMATICS , TECHNICAL UNIVERSITY, WROCŁAW .

References

[1] K. Mc Aloon, Consistency results about ordinal definability, Ann. Math. Logic 2 1971 pp. 449-467.

[2] W. B. Easton, Powers of regular cardinals, Ann. Math. Logic 1 1970 pp. 139-178.

[3] R. Jensen, The Generalized Continuum Hypotesis and Measurable Cardinals, Summer Institute on Set Theory, Univ. of California, Los Angeles 1967.

[4] K. Kunen, Inaccessibility properties of cardinals, Ph. D. Dissertation, Stanford Univ.

[5] K. Kuratowski, A. Mostowski, Teoria Mnogości, Warszawa 1967.

[6] A. Levy, Definability in axiomatic set theory I , Logic Methodology and Philosophy of Science, North Holland Amsterdam 1965 pp. 137-154.

[7] A. Zarach, Forcing with proper classes, Fund. Math. LXXXI (1973), pp. 1-27.

DECISION PROBLEMS FOR GENERALIZED QUANTIFIERS - A SURVEY

by Alan Slomson

§ 1. Introduction.

In this survey by a generalized quantifier is meant a cardinality quantifier of the kind first introduced by Mostowski [1957]. We adopt the following notation. If φ is a formula with one free variable, then $\varphi^{\underline{\underline{A}}}$ denotes the subset of the universe of $\underline{\underline{A}}$ which φ defines in $\underline{\underline{A}}$. If φ has two free variables, then $\varphi^{\underline{\underline{A}}}$ denotes the subset of $A \times A$ defined by φ in $\underline{\underline{A}}$, and so on. Using this notation the existential quantifier $\exists$ can be defined by the satisfaction clause

$$\underline{\underline{A}} \models (\exists v)\, \varphi \iff \mathrm{card}(\varphi^{\underline{\underline{A}}}) \geq 1 .$$

For each ordinal α there is a quantifier Q_α which behaves syntactically like the existential quantifier and whose meaning is given by the satisfaction clause

$$\underline{\underline{A}} \models (Q_\alpha v)\, \varphi \iff \mathrm{card}(\varphi^{\underline{\underline{A}}}) \geq \aleph_\alpha .$$

If L is a (countable) first order language we denote by L_α the language obtained when we add the new quantifier symbol Q_α. In addition to these quantifiers there are three other cardinality quantifiers which merit attention in the present context. First there is the Chang or equi-cardinal quantifier, Q_C. This also binds one variable and is defined by

$$\underline{\underline{A}} \models (Q_C v)\, \varphi \iff \mathrm{card}(\varphi^{\underline{\underline{A}}}) = \mathrm{card}(\underline{\underline{A}}) .$$

For each positive integer n and each ordinal α, the Malitz quantifier Q_α^n is one which binds n variables and whose interpretation is given by

$$\underline{\underline{A}} \models (Q_\alpha^n v_1 \ldots v_n)\varphi \iff \text{for some } X \subseteq A, \text{ with } \mathrm{card}(X) \geq \aleph_\alpha , \quad (X^n)' \subseteq \varphi^{\underline{\underline{A}}} ,$$

where $(X^n)'$ denotes the set of all n-tuples of distinct elements of X . Thus when $n = 1$, Q_α^1 is just the quantifier Q_α already mentioned. Finally there is the Hartig quantifier H . This binds one variable in each of a pair of formulas and its interpretation is given by

$$\underline{\underline{A}} \models (Hv)(\varphi;\psi) \iff \mathrm{card}(\varphi^{\underline{\underline{A}}}) = \mathrm{card}(\psi^{\underline{\underline{A}}}) .$$

In terms of the Hartig quantifier we can define both the Chang quantifier and the Q_0 quantifier (which says "there exist infinitely many). Thus $(Q_C v)\varphi$ is equivalent to $(Hv)(\varphi;\ v = v)$ and $(Q_0 v)\ \varphi(v)$ is equivalent to $(\exists w)(\varphi(w) \wedge (Hv)(\varphi(v);\ \varphi(v) \wedge v \neq w))$.

In a structure of cardinality $< \aleph_\alpha$ the Q_α quantifier acts vacuously and therefore when we are dealing with this quantifier it is technically convenient to assume that all the structures we consider have cardinality at least $\aleph_\alpha$. Similarly when dealing with the Chang quantifier it is convenient to assume that all structures are infinite.

For each class K of structures we let Th(K) be the first order theory of K and $Th_\alpha(K)$ be the theory of K in the language L_α. Thus

$$Th_\alpha(K) = \{\sigma : \sigma \text{ is a sentence of } L_\alpha \text{ and for all } \underline{\underline{A}} \in K , \text{ with } \mathrm{card}(\underline{\underline{A}}) \geq \aleph_\alpha ,\ \underline{\underline{A}} \models \sigma \} .$$

Similarly $Th_C(K)$ denotes the theory of K in the language L_C with the Chang quantifier.

We shall be concerned with problems about the decidability of $Th_\alpha(K)$ for various classes K and ordinals α. Clearly $Th_\alpha(K)$ can only be decidable if Th(K) is decidable. It is easy to provide artificial examples of classes K such that $Th(K)$ is decidable while $Th_\alpha(K)$ is not decidable for certain (or all) α. So we only want to consider classes K which are "natural" in some sense. Especially we consider the case where K is a first order elementary class. If K is the class of all models of the set Δ of first order sentences, we denote $Th_\alpha(K)$ by $Th_\alpha(\Delta)$ and $Th(K)$ by $Th(\Delta)$. Herre and Wolter [1975] have given an example of a theory Δ such that $Th_0(\Delta)$ is decidable while $Th_1(\Delta)$ is undecidable. It is not yet known whether an example of the converse situation exists. Herre and Wolter's example exploits the fact that while $Th(\Delta)$ is decidable, if a new unary predicate is added to the language in this extended language in this language the theory of Δ is undecidable.

§ 2. Basic results and methods

In this section we list some basic results and techniques which underlie the decidability results mentioned below. In connection with decision problems, and for other reasons, it is interesting to know, for fixed K, how $Th_\alpha(K)$ varies as α varies. The key method here is the reduction technique due to Fuhrken [1964,1965] (see also chapter 13 of Bell and Slomson [1971]). This shows how problems about the existence of models of sentences of L_α can be reduced to the existence of models of first order sentences with special properties, i.e. cardinal-like and two-cardinal models. Known results about these models then enable us to obtain the following comparison theorems. (Note that the assertion $Th_\alpha(\Delta) \subseteq Th_\beta(\Delta)$ must be interpreted as meaning that if in $Th_\alpha(\Delta)$ each occurrence of Q_α is replaced by an occurrence of Q_β, then we obtain a subset of $Th_\beta(\Delta)$. We adopt this convention throughout.)

<u>Comparison Theorems</u>

(1) For all α, $Th_\alpha(\Delta) \subseteq Th_0(\Delta)$.

(2) For all $\alpha > 0$ with $\aleph_\alpha$ regular, $Th_1(\Delta) \subseteq Th_\alpha(\Delta)$.

(3) (G.C.H.) For all α with $\aleph_\alpha$ regular $Th_{\alpha+1}(\Delta) \subseteq Th_1(\Delta)$.

(4) (V = L) For all α, $Th_{\alpha+1}(\Delta) \subseteq Th_1(\Delta)$.

(5) For all α,β with $\aleph_\alpha$ a strong limit cardinal and $\aleph_\beta$ singular, $Th_\alpha(\Delta) \subseteq Th_\beta(\Delta)$.

These results depend on theorems of MacDowell and Specker [1961], Morley and Vaught [1962], Chang [1965], Jensen (unpublished, see Chang and Keisler [1973]), and Keisler [1968], respectively. It follows from them that if we make some strong assumption such as $V = L$ + "there are no inaccessible cardinals" then for a given first order theory Δ there are only three distinct theories $Th_\alpha(\Delta)$ at most, namely $Th_0(\Delta)$, $Th_1(\Delta)$ and $Th_\omega(\Delta)$. Although in general these three theories are distinct in some special cases it is known these are equal. Some of these cases are noted below.

Of the three Q_α quantifiers Q_0, Q_1 and Q_ω that give rise to these three theories Q_0 is somewhat different in character from the other two. With this quantifier we can express a categorical recursive set of exioms for the standard model of arithmetic. It therefore follows from Gödel's Incompleteness theorem that L_0 is not axiomatizable. On the other hand the powerful theorem of Rabin [1969] on the decidability of the second order theory of two successor functions enables us to obtain the decidability of many theries in the language L_0 . Examples are given below. In contrast to the non-axiomatizability of L_0 we have the following:

<u>Axiomatizability results for</u> L_1 <u>and</u> L_ω .

(1) If Δ is recursively enumerable then so is $Th_1(\Delta)$.

(2) If Δ is recursively enumerable then so is $Th_\omega(\Delta)$.

(1) is an observation due to Vaught [1964]. An explicit axiomatization for L_1 has been given by Keisler [1970], and (2) is also due to Keisler [1968].

It follows that in cases where Δ is recursively enumerable to prove that $Th_1(\Delta)$ (or $Th_\omega(\Delta)$) is decidable it is sufficient to show that the sentences of L_1 (or L_ω) consistent with Δ form a recursively enumerable set.

<u>Ehrenfeucht's Game</u>

Ehrenfeucht [1961], extending the work of Fraissé [1954], showed that elementary equivalence of structures with respect to a first order language can be characterized in terms of a game played with these structures. Lipner [1970] and Brown [1971] independently showed how this game could be extended to cope with elementary equivalence in the languages L_α . Vinner [1972], also independently, gave a similar characterization but expressed in terms of partial isomorphisms and observed that it could be used to compare the L_α theory of one structure with the L_β theory of another. For an account of the game see Slomson [1972].

Badger [1975] has shown that this game can be generalized further to deal with the Malitz quantifiers.

Apart from the use of Ehrenfeucht's game the chief technique used in proving the results listed below is that of elimination of quantifiers.

§ 3. Decidability results.

(a) Monadic predicates

The theory of monadic predicates without equality was shown by Mostowski [1957] to be decidable in each of the languages L_α , and to be the same for each α . In Slomson [1968] this is extended to a language with equality. The argument here is given in terms of the Chang quantifier but is easily seen to work also for each language L_α . Vinner [1972] gives a more direct proof. Slomson [1968] also shows, using a theorem of Löb [1967] that the theory of monadic predicates, without equality, but with one unary function, is decidable in the language with the Chang quantifier.

(b) One equivalence relation

Rabin [1969] proved that 2S2 , the second order theory of two successor functions, is decidable. Vinner [1972] showed that the L_0 theory of one equivalence relation is interpretable in 2S2 and hence is decidable. He also proved that for all α the L_α theory of one equivalence relation is the same as the L_0 theory, and hence is decidable. Since the theory of two equivalence relations is undecidable these results cannot be improved.

(c) Trees and one unary function

A tree is a relational structure with a single symmetric binary relation and in which there are no circuits. The theory of trees in interpretable in the theory 2S2 and so it follows that the L_0 theory of trees is decidable. The same applies to the L_0 theory of one unary function. Vinner [1972] observed that the L_1 theory of one unary function is not the same as the L_0 theory and he proved that for all $\alpha > 0$, with $\aleph_\alpha$ regular, the L_α theory is the same as the L_1 theory. Herre [1975] proved that for $\alpha > 0$ and $\aleph_\alpha$ regular the L_α theories of trees and one unary function are decidable.

(d) Abelian groups

Baudisch [1975] has proved that the L_α theory of Abelian groups is the same for all α and is decidable. His method is to extend the basis for the theory of Abelian groups given by Szmielew [1955] in her proof of the decidability of the first order theory of Abelian groups.

(e) Arithmetic with + and <

In Wolter [1973] it is proved that the L_0 theory of the natural numbers with addition and the usual ordering is decidable, and in Wolter [1975] this is extended to the same theory in the language $L_{\alpha,\beta}$ which comes from L by adding the two quantifiers Q_α and Q_β, with $0 < \alpha, \beta$.

(f) p-adic numbers

Weese [1975] showed that a certain theory of p-adic number fields is decidable in the language L_α , for all α . The class of structures he considers is not "natural" in the sense mentioned above since the theory he works with includes some non-first-order axioms, for example, the axiom $(\forall x)[x \neq 0 \rightarrow ((Q_\alpha y)(x < y) \wedge (Q_\alpha y)(y < x))]$.

(g) Well-ordered sets

We identify each ordinal ξ with the well-ordered structure (ξ, ϵ) , and we denote the class of all ordinals by On . Of course On is not a first-order elementary class, but in a good sense it is a "natural" class of structures.

Lipner [1970] proved that for each ordinal ξ , and for all α with $\aleph_\alpha$ regular, $Th_\alpha(\xi)$ is decidable. From the decidability of the theory 2S2, Rabin [1969], it follows that $Th_0(On)$ is decidable. In Slomson [1972] it is proved that $Th_1(On)$ is decidable and a proof is also given of the result due to Vinner that for all $\alpha > 0$, $Th_\alpha(On) = Th_1(On)$, and hence is also decidable. These proofs use Ehrenfeucht's game. In his thesis Badger [1975] raises the question as to whether these results can be extended to the theory of ordinals in the language $L_\alpha^{<\omega}$, which contains all the Malitz quantifiers Q_α^n , for $n < \omega$ .It is not too difficult to see that the techniques of Slomson [1972] can be extended to give a positive answer to this question, and indeed to prove that if two ordinals are elementarily equivalent with respect to the language L_α , then they are also elementarily equivalent with respect to the language $L_\alpha^{<\omega}$.

Herre and Wolter [1975], using quantifier elimination arguments, show that the theory of well-ordered sets in the language with the two quantifiers Q_0 and Q_α is decidable. In contrast to these decidability results Weese [1975i] has proved that the theory of well-ordered sets in the language with the Hartig quantifier is undecidable.

In most of the examples above we also have the decidability of the corresponding theory in the language L_C with the Chang quantifier. This is because for any set of first order sentences Δ,

$$\mathrm{Th}_C(\Delta) = \bigcap_{\alpha \in On} \mathrm{Th}_\alpha(\Delta) \quad .$$

§ 4. A Remark on Dense Linear Orderings

Perhaps the most notable omission from the list of theories given above is that of linear orderings. We let LO denote this theory. Again it follows from Rabin [1969] that $\mathrm{Th}_0(\mathrm{LO})$ is decidable, but the question as to the decidability of, for example, $\mathrm{Th}_1(\mathrm{LO})$ remains open. Rabin's method applies essentially to countable sets and so is not capable of immediate generalization to the language L_1. Similarly, the original proof of the decidability of the first order theory of linear orderings, due to Läuchli and Leonard [1966] makes essential use of Ramsey's theorem and so cannot be easily extended from the countable case to the uncountable case.

The difficulty of settling the decidability of $\mathrm{Th}_1(\mathrm{LO})$ is also seen if we look at the theory DLO of dense linear orderings without endpoints. As is well known, a famous theorem due to Cantor says that DLO is $\aleph_0$-categorical, hence by Vaught's test DLO is complete, and so being recursively axiomatizable it is decidable. In contrast the L_1 theory of dense linear orderings, i.e. $\mathrm{Th}_1(\mathrm{DLO})$ is not $\aleph_1$-categorical, but has $2^{\aleph_1}$ isomorphism types among its models of cardinal $\aleph_1$, and is far from complete, but has $2^{\aleph_0}$ complete extensions. Furthermore $\mathrm{Th}_1(\mathrm{LO})$ can be interpreted in $\mathrm{Th}_1(\mathrm{DLO})$. This can be seen as follows.

Let η be the order type of the rationals and let ϑ be the order type of a dense linear ordering without endpoints of cardinal $\aleph_1$ and with $\aleph_1$ points between any two distinct points (i.e. in case the G.C.H. holds ϑ is the order type of $(R,<)$ and otherwise of an L_1 elementary substructure of this structure of cardinal $\aleph_1$). Let λ be the order type $\eta + 1 + \vartheta$, and let $\varphi(x)$ be the formula

$$(\exists y)(y < x \wedge "(y,x) \models \sigma_1") \wedge (\exists y)(x < y \wedge "(x,y) \models \sigma_2") ,$$

where σ_1, σ_2 are finite axiomatizations of $Th_1(\eta)$ and $Th_1(\vartheta)$ respectively, and $"(y,x) \models \sigma_1"$, $"(x,y) \models \sigma_2"$ denote their relativizations to the formulas $(y < v \wedge v < x)$ and $(x < v \wedge v < y)$ respectively. Then in an ordered set of order type λ the formula $\varphi(x)$ defines a unique element ("the 1 in the middle"). Further if γ is any order type, then in a structure of order type $\lambda\cdot\gamma$ $\varphi(x)$ defines a subset of order type γ . Also $\lambda\cdot\gamma$ is a dense linear order type without endpoints. It therefore follows that for any sentence σ of the language L_1 for linear orderings with the Q_1 quantifier,

$$LO \models \sigma \iff DLO \models \sigma^{(\varphi)} \wedge (\exists x)\, \varphi(x) ,$$

where $\sigma^{(\varphi)}$ denotes the relativization of σ to the formula $\varphi(x)$. Thus, in striking contrast to the first order case, in the language L_1 the decidability of the theory of dense linear orderings is no easier than of the theory of all linear orderings.

References

Lee W. Badger [1975] ,
The Malitz quantifier meets its Ehrenfeucht game, Ph.D.Thessis, University of Colerado.

Andreas Baudisch [1975] ,
Elimination of the quantifier Q_α in the theory of Abelian groups, typescript.

J.L.Bell and A.B.Slomson [1971] ,
Models and Ultraproducts: an Introduction, North-Holland, Amsterdam, second revised printing.

W.Brown [1971] ,
Infinitary languages, generalized quantifiers and generalized products, Ph.D.Thesis, Dartmouth.

C.C.Chang [1965] ,
A note on the two cardinal problem, Proc. Amer. Math. Soc. 16, pp. 1148-1155.

C.C.Chang and H.J.Keisler [1973] ,
Model Theory, North-Holland, Amsterdam.

A.Ehrenfeucht [1961] ,
An application of games to the completeness problem for formalized theories, Fund. Math. 49, 129-141.

R.Fraïssé [1954] ,
Sur le classification des systems de relations, Pub. Sci. de l'Universite d'Alger I, no I .

G.Fuhrken [1964],
Skolem-type normal forms for first order languages with a generalized quantifier, Fund. Math. 54, 291-302.

[1965],
Languages with the added quantifier "there exist at least $\aleph_\alpha$" in The Theory of Models, edited by J.Addison, L.Henkin and A.Tarski, North-Holland, Amsterdam, 121-131.

H.Herre [1975],
Decidability of the theory of one unary function with the additional quantifier "there exist $\aleph_\alpha$ many", preprint.

H.Herre and H.Wolter [1975],
Entscheidbarkeit von Theorien in Logiken mit verallgemeinerten Quantoren, Z. Math. Logik, 21, 229-246.

H.J.Keisler [1968],
Models with orderings, in Logic, Methodology and Philosophy of Science III, edited by B.van Rotselaar and J.F.Staal, North-Holland, Amsterdam, 35-62.

[1970],
Logic with the quantifier "there exist uncountably many", Annals Math. Logic, 1, 1-94.

H.Läuchli and J.Leonard [1966],
On the elementary theory of linear order, Fund. Math., 49, 109-116.

L.D.Lipner [1970],
Some aspects of generalized quantifiers, Ph. D. Thesis, Berkeley.

M.H.Löb [1967],
Decidability of the monadic predicate calculus with unary function symbols, J.Symbolic Logic, 32, 563.

R.MacDowell and E.Specker [1961],
Modelle der Arithmetik, in Infinitistic Methods, Pergamon Press, Oxford, 257-263.

M.Morley and R.L.Vaught [1962],
Homogeneous universal models, Math. Scand., 11, 37-57.

A.Mostowski [1957],
On a generalization of quantifiers, Fund. Math., 44, 12-36.

M.O.Rabin [1969],
Decidability of second-order theories and automata on infinite trees, Trans. Amer. Math. Soc., 141, 1-35.

A.B.Slomson [1968],
The monadic fragment of predicate calculus with the Chang quantifier in Proceeding of the Summer School in Logic Leeds 1967, edited by M.H.Löb, Springer Lecture Notes, 70, 279-301.

[1972],
Generalized quantifiers and well orderings, Archiv.Math.Logik, 15, 57-73.

W.Szmielew [1955],
Elementary properties of Abelian groups, Fund.Math., 41, 203-271.

R.L.Vaught [1964],
The completeness of logic with the added quantifier "there are uncountably many", Fund. Math., 54, 303-304.

S.Vinner [1972],
A generalization of Ehrenfeucht's game and some applications, Israel J. Math., 12, 279-298.

M.Weese [1975] ,
Zur Entscheidbarkeit der Topologie der p-adischen Zähkorper in Sprach mit Machtigkeitsquantoren, Thesis, Berlin.

[1975i] ,
The undecidability of the theory of well-ordering with the quantifier I, preprint.

H.Wolter [1975] ,
Eine Erweiterung der elementaren Prädikatenlogik anwendungen in der Arithmetik und anderen mathematischen Theorien, Z. Math. Logik, 19, 181-190.

[1975] ,
Entscheidbarkeit der Arithmetik mit Addition und Ordnung in Logiken mit verallgemeinerten Quantoren, Z. Math. Logik, 21, 321-330.

School of Mathematics,
University of Leeds,
Leeds, LS2 9JT,
England.

THE ALTERNATIVE SET THEORY

by Antonin Sochor , Prague

The aim of this paper is to give a brief outline of Alternative Set Theory (AST) . This theory makes possible the synthesis of a number of mathematical disciplines using new methods, and these new approaches are natural from the point of view of AST .

Alternative Set Theory was created by P. Vopěnka and he presented its first version in his seminar in 1973. After the investigation of the consistency of that axiomatic system (by the author [2]) the original system was modified (by P. Vopěnka) and is now called AST . P. Vopěnka developed in AST such basic notions as e.g. natural numbers, infinite powers and real numbers, and proved a large number of fundamental statements and proposed the conception of topology. During the last two years the foundations for the development of mathematics in AST have been laid. Besides P. Vopěnka other members of his seminar, in particular J. Mlček, K. Čuda, J. Chudáček and the author of the present paper also participated by their results in the creation of mathematics in AST . At the same time the metamathematical problems of AST were investigated by the author.

This paper includes only some mathematical and metamathematical results concerning AST selected to show the possibilities of the theory and to explain its relation to the usual set theory. The results concerning model theory in AST (obtained by J. Mlček and the author) are not included at all. The first comprehensive text about AST , including most mathematical results about AST , was written by P. Vopěnka (in Czech). A similar text about the metamathematics of AST is also being prepared.

First let us explain some reasons why we started to deal with AST . At the end of the last century, Cantor developed set theory. Although his theory was inconsistent, it influenced the whole of mathematics in a decisive way. Very soon theories (consistent, we hope) based on Cantor's ideas were constructed - now we have e.g. the Zermelo-Fraenkel, Gödel-Bernays, Morse and New Foundations set theories. We shall speak about all these theories as Cantor's set theories. We can ask whether there existed possibilities to build up another theory that could replace Cantor's set theory and, consequently, whether there were other possibilities to develop mathematics in our century. At first let us mention at least the following two reasons why Cantor's theory was so important and so fruitful:

1) Cantor's theory became the world of mathematics. All theories investigated up to Cantor's time can be considered as parts of set theory. More precisely they have models in Cantor's theory. For some theories (e.g. for the theory of real numbers) their creation was finished only after this modelling. We have an interpretation of infinitesimal calculus in Cantor's theory, too, but Leibniz's and Newton's original ideas had to be reformulated before this modelling. This was necessary since the notion "infinitely small" cannot be naturally modelled in Cantor's theory.

2) Cantor's theory is a theory of infinity. In Cantor's set theory we have actual infinities and moreover Cantor's theory made possible a general investigation and classification of the notion of infinity.

A theory which wants to be an alternative to Cantor's set theory must satisfy these two requirements at least. Our AST is a theory of infinity and contrary to Cantor's is as poor as possible - there are only two infinite powers. Another difference between AST and Cantor's theory consists in the fact that Cantor's set theory places infinity "behind" finite sets and AST places it "among" finite sets. Infinity is represented in our theory by indeterminate (by a set formula), vague parts of finite sets (see the definition of the "countable" class An further in the text).

The problem whether AST fulfils the first requirement is much more complicated. To show that AST could be the world of mathematics in Cantor's time we have to interpret all the theories in question in AST . We hope that this is possible, up to now we have modelled real numbers (more generally we have developed topology in AST). Moreover we are trying not only to model all these theories in AST, but are looking for their natural interpretations (this concerns mainly infinitesimal calculus). And this is the main reason why we started to build up AST.

In AST there are means which are not available in Cantor's theory. For example we have "inaccessible" natural numbers and therefore we can model in AST the notion "infinitely small". This enables us to interpret directly Leibniz's and Newton's ideas. Further we are able to investigate in AST the connection between the continuous and the discrete. From one point of view we can consider a space (and therefore a motion) as discrete and from the second point of view the same space appears as continuous.

Now, what is the connection between AST and nonstandard methods? In some aspects, they are similar e.g. models showing consistency of AST with respect to Cantor's set theory are particular non-well-founded models. On the other hand there are the following two differences at least: At first nonstandard methods deal with models in Cantor's theory and AST is a new axiomatic theory (which can hardly be considered as a precise axiomatization of nonstandard methods). The second difference is even more important. We want to use means which are available in AST to obtain new approaches and new formalizations of notions in an immediate and natural way (and without intermediate steps such as Cantor's set theory and nonwell-founded models as in the case of nonstandard methods).

For every set theory, T, the theory T for finite sets (T_{Fin}) denotes the theory T where we replace the axiom of infinity by its negation.

AST is similar to the theory of semisets (see [3]) in the sense that both admit classes which are subclasses of sets and which are not sets. It is possible to say that AST is some strengthening of the theory of semisets for finite sets (without the axiom C2). But the main difference is again in <u>what</u> we want to do in AST; from this point of view, the theory of semisets is very near to Cantor's theory.

Now we shall describe the construction of AST. At first it is a theory of sets, because we want to keep the useful procedures and notions of Cantor's set theory. Our theory is similar to Gödel-Bernays or Morse for we have classes and sets. Sets can be thought of as objects of our investigation and classes can be thought of as our view (approach) to these objects or, in other words, classes can be considered as idealizations of some properties. Our theory has only finite sets, but classes can be infinite. This approach corresponds with one's idea of the real world - all sets as sets of people, houses and so on are finite and only our generalizations and idealizations are infinite, as e.g. the class of all natural numbers, the class of all real numbers and so on. On the other hand there are possibilities to treat some sets (formally finite) as infinite. We have precedents in real life for this, too. For example the number of all atoms on our globe is considered as finite, but it is also possible to consider it as inaccessible.

AST is a theory with one sort of variables - class variables - denoted by $X, \ldots$ and two binary predicates - relationship $\in$ and equality $=$. Sets are defined as members of classes and are denoted by $x, \ldots$

§ 1 The axioms of AST

1) Axiom of extensionality for classes i.e.

$$(\forall X)(X \in Y \equiv X \in Z) \equiv Y = Z$$

2) All axioms of Zermelo-Fraenkel set theory for finite sets.

3) Morse's class existence scheme i.e. for every (including non-normal) formula $\varphi(X)$ we have the axiom

$$(\exists X)(\forall x)(x \in X \equiv \varphi(x))$$

Up to now we have formulated only axioms which are either axioms or are provable in Morse's set theory for finite sets. The following axiom is inconsistent with $\mathrm{Morse}_{\mathrm{Fin}}$ and therefore by accepting our fourth axiom we depart from Cantor's set theory. In $\mathrm{Morse}_{\mathrm{Fin}}$ the statement

$$(*) \qquad X \subseteq x \rightarrow M(X)$$

is provable, on the other hand its negation is provable in AST . Each mathematician is accustomed to the statement (*) and the question is if there are reasons to assume its negation. Vopěnka's argument must be repeated here:

Ch. Darwin teaches us that there is a finite sequence with monkey Charlie as the first element, with Mr. Charles Darwin as the last element and such that each element of the sequence is the father of the following one. Of course the first element is a monkey and the last element is not a monkey since it is a man. Moreover if some element is a monkey then the following one is a monkey, too. If there existed a set of all the monkeys in our sequence, we would have trivially a contradiction with the statement that Mr. Ch. Darwin is not a monkey (every set of natural numbers has a first element). It is natural therefore to assume that the property "to be a monkey" describes only a class (in this case we do not obtain a contradiction because we do not require that every class of natural numbers has a first element).

Our example is not artificial, such situations are very frequent in real life. Moreover the existence of proper classes which are subclasses of sets enables us to assume that every set is finite and simultaneously to have infinite powers.

We are now going to formulate our fourth axiom. Using axioms 1) - 3) we can define the natural numbers as usual, N denoting the class of all natural numbers. We define the class of all absolute natural numbers by

$$An = \{n \in N \; ; \; (\forall X)(X \subseteq n \rightarrow M(X))\}$$

(a natural number is absolute if all its subclasses are sets). Let us recall that in Morse_{Fin} we have trivially $An = N$. On the other hand in our theory we accept the axiom

4) Axiom of extension

$$Fnc(F) \wedge D(F) = An \rightarrow (\exists f)(Fnc(f) \wedge F \subseteq f)$$

(every function defined on An is a subclass of a function which is a set).

$An \neq N$ follows from this axiom and therefore we get the negation of the statement (*) . The axiom of extension is very strong and one can say that it is the most important axiom of our theory. It enables us to grasp the notion of limit very naturally. Moreover natural numbers which are not absolute can be considered as inaccessible i.e. in some sense infinite. The existence of such natural numbers enables us to model the notion of "infinite small".

Our fifth axiom is the axiom of choice:

5) For every equivalence relation there is a selector.

Since all sets are finite we can prove the existence of a selector for every set equivalence relation from the other axioms. Therefore our axiom of choice gives something new only for proper classes.

The last axiom says how many infinite powers we have. Let $X \mathrel{\hat{\leq}} Y$ denote that there is a 1-1 mapping (possibly a proper class!) of X into Y and let $X \mathrel{\hat{\approx}} Y$ stand for $X \mathrel{\hat{\leq}} Y \wedge Y \mathrel{\hat{\leq}} X$. We can prove using the axiom of extension that $\neg\, An \mathrel{\hat{\approx}} N$. The cardinality of An can be considered as the infinity of "real" natural numbers and the cardinality of N can be considered as the infinity of real numbers (continuum; in the axiomatic system 1) - 5) we can code all subclasses of An by some natural numbers). The last axiom of AST postulates that there are no other infinite cardinalities.

6) <u>Axiom of cardinalities</u>

$$\neg\, X \mathrel{\hat{\leq}} An \rightarrow X \mathrel{\hat{\approx}} N$$

We have thence in AST two kinds of proper classes - countable ("small") and the others, the cardinality of which is that of the continuum ("large"). Therefore the axiom of extension can be reformulated in the following form:
Every countable ("small") function is a subclass of a set function. Trivially this statement cannot be true for "large" functions.

§ 2 <u>Metamathematics of AST</u>

We have now described all the axioms of AST. Before we describe <u>what</u> we can do <u>in</u> AST we are going to discuss the consistency of AST and more generally the <u>connection between AST and Cantor's set theory</u>.
We have the following diagram:

$$\text{Morse}_{Fin} \;\underset{\not\longleftarrow}{\longrightarrow}\; \text{AST} \;\underset{\not\longleftarrow}{\longrightarrow}\; \text{ZF}$$

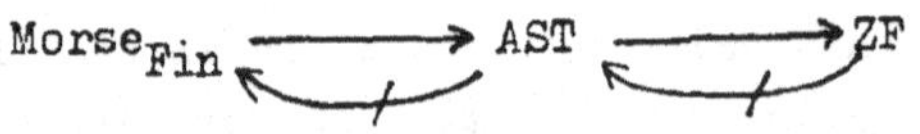

where $\longrightarrow$ means that there is an interpretation of the first theory in the second one and $\not\longleftarrow$ means that the interpretation in question does not exist. In this paper we restrict ourselves to sketching a proof of the existence of an interpretation of AST in Zermelo-Fraenkel set theory, a fact which is almost obvious. The following construction is done in Zermelo-Fraenkel.

Let $\underline{M} = \langle M, E \restriction M \rangle$ be the model of all hereditarily finite sets and let Z be a non-trivial ultrafilter on ω_0. Let $\underline{\tilde{N}} = \langle \tilde{N}, \tilde{E} \rangle = \underline{M}^{\omega}/Z$ be the usual ultrapower. $\underline{\tilde{N}}$ is a model of ZF_{Fin} and to obtain a model containing classes we add "all subclasses of $\underline{\tilde{N}}$" i.e.

$\underline{\tilde{N}}' = \langle \tilde{N} \cup Q, \tilde{E} \cup (E \upharpoonright Q)\rangle$ where $Q = \{x \subseteq \tilde{N} ; \neg(\exists f)(x = \{g ; \underline{\tilde{N}} \models g \in f\})\}$

then $\underline{\tilde{N}}' \models$ AST is provable ; in the following two paragraphs we are going to prove particularly that the axiom of extension and the axiom of cardinalities hold in $\underline{\tilde{N}}'$.

Let k_x be the constant function the value of which is x . The class of all absolute natural numbers in our model is the set of all constants the values of which are natural numbers i.e. $An^{\underline{\tilde{N}}'} = \{k_n : n \in \omega_0\}$. To prove the axiom of extension in the model let us suppose $\underline{\tilde{N}}' \models Fnc(F) \wedge D(F) = An$, then for every $n \in \omega_0$ there is a function f_n such that $\underline{\tilde{N}}' \models \langle f_n, k_n\rangle \in F$. Let us define a function f on ω_0 by

$$f(n) = \{\langle f_1(n), 1\rangle, \ldots, \langle f_n(n), n\rangle\} .$$

We can suppose $f \in \tilde{N}$ and moreover we have $\underline{\tilde{N}}' \models Fnc(f) \wedge F \subseteq f$.

To prove the axiom of cardinalities suppose that we had started in $ZF + 2^{\aleph_0} = \aleph_1$. In this case we have $card(An^{\underline{\tilde{N}}'}) = \aleph_0$ and $card(N^{\underline{\tilde{N}}'}) = card(\tilde{N}) = \aleph_1$. Therefore for every infinite $X \subseteq \tilde{N}$ there is 1-1 mapping between X and either $An^{\underline{\tilde{N}}'}$ or $N^{\underline{\tilde{N}}'}$. Hence for $X \in Q$ we have $\underline{\tilde{N}}' \models X \hat{\approx} An \vee X \hat{\approx} N$. If $X = \{g : \underline{\tilde{N}} \models g \in f\}$ then $\underline{\tilde{N}}' \models f \hat{\approx} An \vee f \hat{\approx} N$ (in fact only the second case can happen). If $X \subseteq \tilde{N}$ is finite then there are f and $n \in \omega_0$ such that $X = \{g : \underline{\tilde{N}} \models g \in f\} \wedge \underline{\tilde{N}}' \models f \hat{\approx} k_n$.

The following metamathematical results concern <u>independence</u> of the axioms. Our attention is directed mainly to the problems relating to the last three axioms. The results concerning the last axiom are satisfactory: We can prove that the theories obtained, from AST , by the substitution of the last axiom by one of the axioms "there are there (four,... resp.) cardinalities"
"there are cofinally many cardinalities"
are consistent with respect to Zermelo-Fraenkel set theory.

We can also prove that the theory obtained from AST by the substitution of the axiom of extension by its negation and by adding the axiom (which seems to be only a slight modification of the axiom in question)

$$F \subseteq An^2 \wedge Fnc(F) \,.\rightarrow (\exists f)(Fnc(f) \wedge F \subseteq f)$$

is consistent with respect to Zermelo-Fraenkel set theory.

The question concerning the independence of the axiom of choice is open up to now.

It is well known that Zermelo-Fraenkel and Gödel-Bernays set theories are equiconsistent. In AST the situation is not so simple. Since An plays in AST the role of ω_0 in Cantor's theory, we suppose that all formalizations of theories are parts of An . In AST we can define the notions of "formula" and "proof" either as usual (i.e. with respect to all natural numbers) or we can substitute in the usual definition the words "natural number" by the words "absolute natural number". Therefore we have in AST proofs - the length of which can be an arbitrary natural number, and absolute proofs - the length of which must be an absolute natural number. It seems better to restrict ourselves to absolute proofs. If we do not do so we can prove e.g. the following strange result:

The theory

$$\text{AST} + \text{Con}\ (\text{ZF}_{\text{Fin}}) + \neg\ \text{Con}\ (\text{GB}_{\text{Fin}})$$

is consistent with respect to Zermelo-Fraenkel (of course the length of the proof of inconsistency of GB cannot be an absolute natural number).

§ 3 Topology in AST

In this section we want to show how it is possible to define topology in AST and furtemore roughly how AST makes it possible to grasp the connection between the continuous and the discrete.

A pair $(a, \doteq)$ is called a topological space if a is a <u>set</u> and if $\doteq$ is an equivalence relation on it (possibly a proper class). We can interpret the relation $\doteq$ as a relation of infinitesimal nearness. First we need some definitions in which x,y denote elements of a and X,u,v denote subclasses of a .

$$\text{Mon}(x) = \{y : y \doteq x\}$$

(The monad of x is the class of all points infinitely near to x)

$$\text{Fig}(X) = \{y : (\exists x \in X)(y \doteq x)\} = \bigcup_{x \in X} \text{Mon}(x)$$

(The figure of X is the class of all points infinitely near to some point of X).

$$Sep(y,X) \equiv (\exists u,v)(Mon(y) \subseteq u \wedge Fig(X) \subseteq v \wedge u \cap v = 0)$$

(We can separate a point y from a class X if there are two disjoint <u>sets</u> one containing the monad of y and the other containing the figure of X).

We have the following axioms of separation:

S1 $\quad Mon(x) \cap Mon(y) = 0 \rightarrow Sep(x, \{y\})$

S2 $\quad Mon(x) \cap Fig(u) = 0 \rightarrow Sep(x,u)$

Of course there is the natural question as to the connection between this notion of topological space and the classical one. Now we shall define the closure operation which constitutes the classical topological space corresponding to our topological space. Let a pair $(a,\doteq)$ be a topological space and let $A \subseteq a$ be a selector with respect to $\doteq$. For every $Y \subseteq A$ we define $U(Y)$ by

$$U(Y) = \{y : y \in A \wedge \neg Sep(y,Y)\}$$

(the "classical closure" of Y is the class of all elements of A which cannot by separated from Y).
The class A with the closure operation U is called the skeleton of $(a,\doteq)$. For $Y,Z \subseteq A$ we have

$$U(0) = 0$$

$$U(Y \cup Z) = U(Y) \cup U(Z)$$

$$Y \subseteq U(Y)$$

and therefore the skeleton of a topological space is a classical topological space in a weak sense - Čech's closure space (see [4]).
The closure of a S1-space is a semi-separated closure space since we have

$$U(\{x\}) = \{x\} .$$

If a pair $(a,\doteq)$ is a S2-space then the skeleton of it is a topological space because we have moreover

$$U(U(Y)) = U(Y) .$$

Constructing the skeletons we create classical topological spaces and we can ask whether we obtain enough classical topological spaces in this way. Theorem 2 gives a positive answer showing that we can obtain in this way every compact metric space.

To define the notion of metric space we need real numbers. The class of real numbers can be constructed in AST otherwise than in the classical case. As we have noted the class An plays in AST the same role as ω_0 in Cantor's theory. Therefore we define the rational numbers as pairs $\frac{n}{m}$ where $n,m \in An$ i.e.

$$Rac = \{ \pm \frac{n}{m} : n,m \in An \wedge m \neq 0\} .$$

Moreover all pairs $\frac{n}{m}$ where n,m run over all natural numbers are called hyperrational numbers i.e.

$$HRac = \{ \pm \frac{n}{m} : n,m \in N \wedge m \neq 0\} .$$

We define (the idea is the same as in nonstandard analysis) two hyperrational numbers x,y to be infinitely near iff their distance apart is less than $\frac{1}{n}$ for every <u>absolute</u> natural number or if both x,y are infinitely great i.e. greater than every <u>absolute</u> natural number:

$$x \doteq y \equiv (\forall n \in An)(|x - y| < \frac{1}{n} \vee (n < x \wedge n < y)) .$$

If we choose a selector with respect to $\doteq$ we obtain a class which has some of the properties of the real numbers e.g. for which the theorem about supremum holds. But there is one disadvantage - there is no x with $x^2 = 2$, we have only x with $x^2 \doteq 2$ (and similarly for the other irrational numbers). Therefore it is better to construct at first a real closed field containing HRac and to extend the equality $\doteq$ to these new elements (we add new elements to old monads e.g. $\sqrt{2}$). Then it is possible to choose a selector Real having the properties which are required from the class of real number (this construction is due to P. Vopěnka).

A function ρ is called hypermetric if

(1) $W(\rho) \subseteq HReal$

(2) $\rho(x,y) = 0 \equiv x = y$

(3) $\rho(x,y) = \rho(y,x)$

(4) $\rho(x,y) + \rho(y,z) \geq \rho(x,z) \geq 0$

(5) $M(\rho)$

A function ρ (possibly a proper class) satisfying (2)-(4) and

(1′) $W(\rho) \subseteq \text{Real}$

(5′) $(\exists b)\, \rho \subseteq b$

is called metric. A pair (a,ρ) is called hypermetric space if $D(\rho) = a^2$ and if ρ is a hypermetric(a is a <u>set</u>). A pair (A,ρ) is called metric space if $D(\rho) = A^2$ and if ρ is a metric (viz. the notion of "classical metric space", the only difference is that A,ρ need not be sets). The class of real numbers with the metric $\rho_1(x,y) = |x - y|$ becomes a metric space. Every hypermetric induces a topology if we define

$$x \doteq y \equiv (\forall n \in An)\, \rho(x,y) < \frac{1}{n}$$

We have the following metrization theorem:

<u>Theorem</u> 1.(Mlček). A topology $\doteq$ is induced by a hypermetric iff $\doteq$ is an intersection of countably many sets i.e. iff there is a class $\{d_n : n \in An\}$ such that $\doteq$ is equal to $\bigcap_{n \in An} d_n$.

This theorem has a nice history. We looked for a long time for a metrization theorem. One day J. Mlček came up with a theorem the formulation of which was rather complicated, but when P. Vopěnka formulated the above theorem, we saw that Mlček's proof worked. Therefore the theorem in question was proved before it was formulated. The names given with the following theorems indicate only the person who brought the main idea ; the other members of the seminar also participated in the creation of the results. Due to the method of work of the seminar it is very difficult to attribute a result to only one person. The proof of the last theorem was essentially simplified by K. Čuda.

<u>Theorem</u> 2. (the author). If a pair (X,ρ) is a metric space compact in the classical sense then there is a topological space

(which is moreover induced by a hypermetric space) such that its skeleton is isomorphic with (X,ρ) .

Now, we come to the crucial point of topology in AST . We shall explain the connection between the continuous and the discrete and the notion of motion. Our construction of skeletons makes it possible to view one space from two different angles and therefore to have space simultaneously discrete and continuous. The field of every hypermetric space is a set (hence formally finite) and therefore every hypermetric space is discrete, on the other hand its skeleton can be continuous (see Theorem 2).

Let a pair (a,ρ) be a hypermetric space. We call a function f (it is a set) a motion of a point if

(1) $D(f) \in N \wedge W(f) \subseteq a$

(2) $(\alpha + 1) \in D(f) \rightarrow f(\alpha) \doteq f(\alpha + 1)$

Note. The set $D(f)$ is (formally) finite, but the interesting cases are only those for which $D(f) \notin An$ holds, i.e. for which $D(f)$ is in some sense infinite.

The explanation of why we can speak about such a function as about motion, is again connected with the skeleton of the hypermetric space. For example, let (a,ρ) be a hypermetric space the skeleton of which is $(Real, \rho_1)$. Let f be the function numbering all elements between 0 and 1 and at the same time preserving the ordering (such a function exists since a itself is finite). Then, turning to the skeleton, we obtain "classical" continuous motion starting with 0 and finishing with 1.

We call a function d (it is a set) a motion of a set in the hypermetric space (a,ρ) if

(1) $D(d) \in N \wedge W(d) \subseteq P(a)$

(2) $(\alpha + 1) \in D(d) \rightarrow Mon(x) \cap d(\alpha) \hat{\approx} Mon(x) \cap d(\alpha + 1)$

(2) of the above definition demands that the cardinality of the class of all elements of $d(\alpha)$ infinitely near to x is the same as the cardinality of the class of all elements of $d(\alpha + 1)$ infinitely near to x . Therefore in $Mon(x) \cap d(\alpha)$ has n elements and $n \in An$

then $Mon(x) \cap d(\alpha + 1)$ must again have n elements. But if $Mon(x) \cap d(\alpha)$ has n elements and $n \notin An$, then $Mon(x) \cap d(\alpha + 1)$ can have m elements for every $m \notin An$ since $n \mathrel{\hat{\approx}} m$ for every non-absolute n and m.

The definition of motion of a set is so weak that we can doubt if this definition in fact expresses the notion of "real" motion. The following theorem shows that this is so.

Theorem 3. (Vopěnka). Let d be a motion of a set in a hypermetric space. Then there is a system T of motions of points in this hypermetric space such that

(1) $\alpha \in D(d) \rightarrow d(\alpha) = \{f(\alpha) : f \in T\}$

(2) $f,g \in T \wedge \alpha \in D(d) \wedge f \neq g \,.\rightarrow f(\alpha) \neq g(\alpha)$

(3) $\alpha \in D(d) \rightarrow M(\bigcup\{f \in T : f(\alpha) \in u\})$

The first statement implies that every point of a given set $(d(0))$ has its motion in T. (2) conveys that two motions of points in T cannot go through one point. The third statement expresses the fact that the system T determines moreover the motion of every subset of $d(0)$.

References

[1] P. Vopěnka, Matematika v alternativni teorii množin, (Mathematics in the Alernative Set Theory), manuscript.

[2] A. Sochor, Real classes in ultrapower of hereditarily finite sets, to appear in CMUC.

[3] P. Vopěnka and P. Hájek, The theory of semisets, North Holland P. C. and Academia, Prague, 1972.

[4] E. Čech, Topological spaces, Academia, Prague 1966

SOME CASES OF KÖNIG'S LEMMA

by J. Truss (Oxford)

§ 1 . "König's Infinity Lemma" is the statement that any well-founded tree of length ω with finite branching has an infinite branch. It is easy to see that this is equivalent to the axiom of choice for countable families of non-empty finite sets, $C^{\omega}_{<\omega}$. Thus in this case we have the equivalence between a countable axiom of choice and an axiom of dependent choices. This contrasts with the situation for the full countable axiom of choice C^{ω} , and the axiom of dependent choices, DC . Jensen showed that
Consis ZF $\rightarrow$ Consis (ZF + C^{ω} + $\neg$ DC) . (See [1 p.151] for a proof). The question as to what happens when restrictions are placed on the degree of branching of the finitary tree was raised by W.Guzicki. I shall show that here, as in Jensen's case, the appropriate axiom of dependent choice is stronger than corresponding countable axiom of choice.

The notation used is as follows. If n is a natural number (≥ 2 to exclude trivial cases) and α is an ordinal, C^{α}_{n} asserts that any family of n-element sets indexed by α has a choice function. C^{*}_{n} is $(\forall\alpha)\ C^{\alpha}_{n}$. C_n is the (ordinary) axiom of choice for families of n-element sets. I shall chiefly be interested here in C^{ω}_{n} , the countable axiom of choice for n-element sets. DC_n is the statement that any well-founded tree of length ω in which every element has exactly n immediate successors has an infinite branch. It is also useful to introduce C^{ω}_{Z} , DC_Z etc. for a set Z of natural numbers. DC_Z asserts that any well-founded tree of length ω in which every element has exactly n immediate successors for some $n \in Z$ has an infinite branch. C^{ω}_{Z} etc. are defined similarly. Note carefully that although when $Z = \{n_1, \ldots, n_k\}$ is finite, C_Z^ω is obviously equivalent to $C_{n_1}^\omega \wedge \ldots \wedge C_{n_k}^\omega$, the same is probably not true of DC_Z. For example we shall show that $DC_{\{2,3\}} \to DC_4$, but presumably $DC_2 \wedge DC_3 \not\to DC_4$. (This has not been proved yet; but Theorem 4.2 will give a possible method). When Z is infinite it is not even true that $(\forall n \in Z)\ C_n^\omega$ is equivalent to C_Z^ω (as was shown, essentially, by Levy in [4]).

The main idea in studying finite versions of the axiom of choice is to reduce questions about them to questions about groups, usually finite ones. It was Mostowski in [5] who first realized the intimate connection between the problem of choosing effectively an element of a finite set and the structure of the group of symmetries of that set. Because of this the most natural setting for independence proofs about finite axioms of choice is the Fraenkel-Mostowski method, where the group of symmetries appears very explicitly. Of course all the proofs can be carried out using forcing, but the extra complications tend to obscure what is really going on. Besides, if a ZF consistency is desired, the metatheorems of [7] can be appealed to. Also Pincus has shown [6] that the use of forcing in [3] and [8] is unnecessary. We shall therefore work officially in FM (for Fraenkel-Mostowski) which is the theory obtained from ZF, Zermelo-Fraenkel set theory, by deleting the axiom of foundation. In practice it will be more convenient to use urelemente in place of "reflexive" sets. However the two approaches are well known to be equivalent.

Our principal results are as follows.

(i) Consis FM $\to$ Consis (FM + $(\forall n)\ C_n^* + (\forall n > 1) \neg DC_n$) .

(ii) $\mathrm{FM} \vdash DC_Z \to DC_n \iff \mathrm{FM} \vdash DC_Z \to C_n^\omega \iff$

$$\mathrm{FM} \vdash (\forall m \in Z)\ DC_m \to C_n^\omega \iff L(n,Z)\ .$$

Here $L(n,Z)$ is the following condition formulated in [8], which is due to Gauntt : For any fixed point free group G of permutations of an n-element set, there are proper subgroups H_i of G such that $\Sigma\ |G: H_i| \in Z$. In [2] Gauntt announced that $L(n,Z) \iff \mathrm{FM} \vdash C_Z^* \to C_n^*$ for finite Z. This works also for infinite Z, and a similar proof shows that $L(n,Z) \iff \mathrm{FM} \vdash C_Z^\omega \to C_n^\omega \iff \mathrm{FM} \vdash C_Z^\alpha \to C_n^\alpha$ for any infinite α.

Unfortunately the question of when $(\forall m \in Z)\ DC_m \to DC_n$ holds seems to be more complicated, so the following at present is only a conjecture.

(iii) $FM \vdash (\forall m \in Z)\ DC_m \to DC_n \iff$ for some $m \in Z$,

$FM \vdash DC_m \to DC_n \iff$ for some $m \in Z$ (the same m), $L(n,\{m\})$.

To complete the picture, we quote the following result of Levy [4] .

(iv) Consis $FM \to$ Consis $(FM + (\forall n)\ C_n + \neg DC_{<\omega})$.

Of course for each n DC_n follows from C_n .

The negative answer to Guzicki's original question, "does $C_2^\omega \to$ $\to DC_2$?" follows from (i). We also show, as an easy consequence of the simplicity of the alternating group A_n $(n \geq 5)$ that $L(n, \{2,3,\ldots,n-1\})$ is false unless $n = 4$, so that $C_2^\omega \wedge \ldots \wedge C_{n-1}^\omega \not\to$ $\not\to C_n^\omega$ and $DC_{\{2,\ldots,n-1\}} \not\to DC_n$ for $n \neq 4$.

§ 2 . We use standard notation for Fraenkel-Mostowski models. Proofs of the basic facts about these models can be found in [1 pp. 52-57]. $\mathfrak{M}$ is the ground model, which satisfies FM + AC , U is the class of urelemente in $\mathfrak{M}$, which in all the cases considered here may be taken countable and infinite. G is a group of permutations of U , and J a filter of subgroups of G closed under conjugacy. $\mathfrak{N}$ is the resulting Franekel-Mostowski model. $\mathfrak{N}$ is obtained from U , G and J as follows. Firstly the action of G on any member x of $\mathfrak{M}$ is defined by transfinite induction on rankx ; $\sigma x = \{\sigma y : y \in x\}$. Then for any $x \in \mathfrak{M}$ we let $H(x) = \{\sigma \in G: \sigma x = x\}$ and $K(x) = \{\sigma \in G: (\forall y \in x)\ \sigma y = y\}$. $H(x)$ is the stabilizer of x and $K(x)$ the pointwise stabilizer. Finally $\mathfrak{N} = \{x \in \mathfrak{M}: x \subseteq \mathfrak{N}\ \ H(x) \in J\}$. In other words $\mathfrak{N}$ is the class of those members of $\mathfrak{M}$ which are hereditarily symmetric with respect to the filter J . We make the following additional assumption on J ; that J is generated by subgroups of the form $H(x)$ with $x \in \mathfrak{N}$. This results in no loss of generality, since J can be replaced by $\{H \in J : (\exists x \in \mathfrak{N})\ H(x) \subseteq H\}$ without altering $\mathfrak{N}$. Also it is very easily checked in all the cases we cover.

LEMMA 2.1. For any $H \in J$ there is $y \in \mathfrak{N}$ such that $H(y) = H$.

<u>Proof.</u> By our assumption on J there is $x \in \mathfrak{N}$ such that $H(x) \subseteq H$. Let $y = \{\sigma x : \sigma \in H\}$. Now it is easily checked that $H(\sigma x) = \sigma H(x)\sigma^{-1}$ and so, as J is closed under conjugacy, $y \subseteq \mathfrak{N}$. As $H \subseteq H(y)$, $y \in \mathfrak{N}$. It remains to show that $H(y) \subseteq H$. Let

$\sigma \in H(y)$. Then $\sigma x \in y$, so $\sigma x = \tau x$ for some $\tau \in H$. Hence $\tau^{-1}\sigma \in H(x) \subseteq H$ giving $\sigma \in \tau H = H$.

In order to lead up to the construction of the models we give here necessary and sufficient conditions on G and J for $\mathfrak{N}$ to satisfy C_n^α and DC_n . The proof of (i) is then just a matter of the appropriate choice of G and J . For $n > 2$ these conditions are rather involved, so to aid comprehensibility we give first the versions for $n = 2$. It will be shown that

(v) C_2^α holds in $\mathfrak{N}$ $\Longleftrightarrow$ for any $H \in J$ there is $K \leq H$ in J such that if $\{K_\beta : \beta < \alpha\}$ are subgroups of K each of index 2 in K , lying in J , then $\bigcap\{K_\beta : \beta < \alpha\} \in J$,

and

(vi) DC_2 holds in $\mathfrak{N}$ $\Longleftrightarrow$ for any $H \in J$ there is $K \leq H$ such that for any family $\langle K_n : n \in \omega\rangle$ of subgroups of K lying in J satisfying $K_0 = K$ and for each n , K_{n+1} is a subgroup of K_n of index 2 , $\bigcap\{K_n : n \in \omega\} \in J$.

THEOREM 2.2. Let $n \geq 2$. Then C_n^α holds in $\mathfrak{N}$ $\Longleftrightarrow$ for any $H \in J$ there is $K \leq H$ lying in J such that for any family $\langle K_\beta^i : i < n_\beta, \beta < \alpha\rangle$ of subgroups of K lying in J satisfying $(\forall\beta) \sum_{i<n_\beta} |K : K_\beta^i| = n$ there are functions ϑ , φ with domain α such that $\bigcap\{\varphi(\beta). K_\beta^{\vartheta(\beta)}. \varphi(\beta)^{-1} : \beta < \alpha\} \in J$ (each $\varphi(\beta) \in K$) .

<u>Remark</u>. When $n = 2$ this condition is equivalent to that in (v), since $n_\beta = 1$ or 2 for each β , and any subgroup of index 2 is normal.

<u>Proof</u>. $\Longleftarrow$; Let $\langle X_\beta : \beta < \alpha\rangle$ be a family of n-element sets in $\mathfrak{N}$ indexed by α . Then there must be $H \in J$ such that $H(X_\beta) \supseteq H$, each β . Take $K \leq H$ in J as given by the hypothesis. For each β the action of K on X_β allows us to express it as a union of K-orbits. Let n_β be the number of such orbits, and $\{x_\beta^0, x_\beta^1, \ldots, x_\beta^{n_\beta - 1}\}$ be representatives. Let $K_\beta^i = K \cap H(x_\beta^i)$. Then the orbit containing x_β^i has $|K : K_\beta^i|$ members, so $\sum_{i<n_\beta} |K : K_\beta^i| = n$. Let ϑ , φ be functions with domain α as given by the hypothesis, and let $K' = \bigcap\{\varphi(\beta). K_\beta^{\vartheta(\beta)}. \varphi(\beta)^{-1} : \beta < \alpha\}$. It is then clear that the function f given by $f(X_\beta) = (\varphi(\beta))(x_\beta^{\vartheta(\beta)})$ lies in $\mathfrak{N}$, since $H(f) \supseteq K'$.

$\Longrightarrow$; Suppose C_n^α holds in $\mathfrak{N}$, and let $H \in J$. We take $K = H$. Let $\langle K_\beta^i : i < n_\beta, \beta < \alpha \rangle$ be a family of subgroups of K lying in J satisfying $(\forall\beta) \sum_{i<n_\beta} |K : K_\beta^i| = n$. By Lemma 2.1 there are elements x_β^i of $\mathfrak{N}$ such that $H(x_\beta^i) = K_\beta^i$. Let $X_\beta = \{\langle \sigma x_\beta^i, i \rangle : \sigma \in K, i < n_\beta\}$. Then $|X_\alpha| = n$ for each α. By C_n^α in $\mathfrak{N}$ there is a function f such that $K' = H(f) \in J$ and such that for each β, $f(X_\beta) = \langle \sigma x_\beta^i, i \rangle$, some i, σ. Let $\vartheta(\beta)$ be the unique i such that for some $\sigma \in K$, $f(X_\beta) = \langle \sigma x_\beta^i, i \rangle$, and let $\varphi(\beta)$ be some such σ.

THEOREM 2.3. Let $n \geq 2$. Then DC_n holds in $\mathfrak{N}$ $\Longleftrightarrow$ for any $H \in J$ there is $K \leq H$ in J such that for any family $\langle K_\pi : \pi \in n^{<\omega} \rangle$ of members of J satisfying $K_{()} = K$, for each π $\{K_{\pi^\frown\langle i \rangle} : i < n\}$ is a family of subgroups of K_π closed under conjugacy (in K_π) and if $\langle K_{\pi^\frown\langle i \rangle} : i < n_\pi \}$ are representatives of the conjugacy classes, $\sum_{i<n_\pi} |K_\pi : K_{\pi^\frown\langle i \rangle}| = n$, then there is a function $\vartheta \in n^\omega$ such that $\bigcap\{K_{\vartheta \restriction i} : i \in \omega\} \in J$.

<u>Proof</u>. $\Longleftarrow$; Let $\langle T, < \rangle$ be an n-branching tree in $\mathfrak{N}$, and let H be its stabilizer. Let $K \leq H$ be given by the hypothesis. Let $T = \{t_\pi : \pi \in n^{<\omega}\}$ be an indexing of T in $\mathfrak{M}$, and define K_π by induction on the length of $\pi \in n^{<\omega}$. Let $K_{()} = K$. Otherwise suppose K_π has been defined. Let $\{x_\pi^0, \dots, x_\pi^{n_\pi - 1}\}$ be representatives of the orbits of $\{t_{\pi^\frown\langle i \rangle} : i < n\}$ under (the group induced by) K_π. By relabling the tree beyond t_π we suppose that $x_\pi^i = t_{\pi}{}_{\langle i \rangle}$ for $i < n_\pi$. Let $K_{\pi^\frown\langle i \rangle} = K_\pi \cap H(\langle T_{\pi^\frown\langle i \rangle}, < \rangle)$ for each $i < n$, where $T_{\pi^\frown\langle i \rangle} = \{t \in T : t_{\pi^\frown\langle i \rangle} \leq t\}$, and where we are identifying $n^{<\omega}$ with T via the indexing. It is clear that the appropriate hypotheses on $\langle K_\pi : \pi \in n^{<\omega} \rangle$ are now satisfied. Hence there is $\vartheta \in n^\omega$ such that $\bigcap\{K_{\vartheta \restriction i} : i \in \omega\} = K' \in J$. Then $\{t_{\vartheta \restriction i} : i \in \omega\}$ is an infinite branch of T lying in $\mathfrak{N}$.

$\Longrightarrow$; The converse is proved in a similar way to the converse for the previous theorem.

<u>Remark</u> 2.4. By allowing n to vary over a set Z in Theorems 2.2 and 2.3 we obtain necessary and sufficient conditions for C_Z^α and DC_Z to hold in $\mathfrak{N}$. The proofs, which are similar to those just given, are omitted.

§ 3 . Although the following lemma is a special case of Theorem 4.1 we give its proof here as it simplifies the proof of the next theorem.

LEMMA 3.1. If $n \geq 2$ and $k \geq 1$, $DC_{nk} \rightarrow DC_n$.

Proof. Let $\langle T, <\rangle$ be an n-branching tree of length ω , with m-th level T_m . Form another tree T' thus. The domain of T' is $\{\langle t,\pi\rangle : t \in T_m \wedge \pi \in k^m$, some $m\}$. T' is partially ordered by $\langle t,\pi\rangle \leq \langle t',\pi'\rangle$ if $t \leq t'$ and $\pi \leq \pi'$. Clearly each member of T' has exactly nk immediate successors in T' . By DC_{nk} there is an infinite branch of T' , and this clearly provides an infinite branch of T .

THEOREM 3.2. Consis FM $\rightarrow$ Consis (FM + $(\forall n)C_n^*$ + $(\forall n>1)\neg DC_n$) .

Proof. By Lemma 3.1. it is enough to make DC_p false in our model for each prime p . The problem is that it is not enough to use finite supports, for by Theorem 2.2 the filter J has to be closed under some countable intersections. However to make DC_n false it cannot be closed under all countable intersections, (since when J is closed under countable intersections, DC holds in $\mathfrak{N}$). The key idea is to use a Frattini-like operation on subgroups of the symmetries of $p^{<\omega}$ to generate an appropriate proper filter. If G is a group and p a prime we denote by $\Phi_p(G)$ the intersection of all subgroups of G of index p .

LEMMA 3.3. Let G be the group of permutations of the tree $p^{<\omega}$ generated by the action of the cyclic group on each set of immediate successors, i.e. the group generated by permutations of the form $\sigma(\pi,\tau)$ for $\pi \in p^{<\omega}$ and $\tau \in \mathbb{Z}_p$, where $\sigma(\pi,\tau)(\pi^\frown\langle i\rangle^\frown\pi') = \pi^\frown\langle\tau(i)\rangle^\frown\pi'$ and $\sigma(\pi,\tau)$ fixes every element not of this form. Let $G_0 = G$ and $G_{n+1} = \Phi_p(G_n)$. Then for all n , $G_n \neq 1$.

Remark. The group G is isomorphic to the direct limit of the syste $\langle \mathbb{Z}_p, \mathbb{Z}_p \wr \mathbb{Z}_p, \mathbb{Z}_p \wr \mathbb{Z}_p \wr \mathbb{Z}_p, \ldots\rangle$ where $\mathbb{Z}_p$ is the cyclic group of order p and $\wr$ denotes wreath product.

Proof of Lemma. Let G^n be the subgroup of G generated by those $\sigma(\pi,\tau)$ such that π has length $\leq n$. Then $G^0 \leq G^1 \leq G^2 \leq \ldots$ and G is the union of the G^n . Now if $H \leq K$ are both p-groups and $|K: M| = p$ then $|H: H \cap M| \leq p$. It follows that $\Phi_p(H) \leq \Phi_p(K)$. So to show that $G_n \neq 1$ it is enough to show that $G_n^{n+1} \neq 1$, and this is done by induction on n . For $n = 0$ it is immediate. Otherwise we show that $\Phi_p(G^{n+1})$ contains an isomorphic copy of G^n , thus establishing the induction step.

G^{n+1} may be expressed as the group generated by $\{\langle x_1,\dots,x_p\rangle : x_i \in G^n\}$ and y where $\langle x_1,\dots,x_p\rangle \cdot \langle x_1',\dots,x_p'\rangle = \langle x_1\cdot x_1',\dots,x_p\cdot x_p'\rangle$ and y is an element of order p satisfying $y^{-1}\cdot\langle x_1,\dots,x_p\rangle\cdot y = \langle x_p,x_1,\dots,x_{p-1}\rangle$. Let $H = \{\langle x,x,\dots,x\rangle : x \in G^n\}$. Then H is a subgroup of G^{n+1} isomorphic to G^n , and we show that $H \leq \Phi_p(G^{n+1})$. For this we use the following two facts. Firstly for any finite p-group K , $\Phi_p(K)$ is the Frattini subgroup of K . Secondly, the Frattini subgroup of a finite p-group is generated by the set of commutators and pth powers.

Let $x \in G^n$ be arbitrary and let $x_i = y^{-i+1}\langle x^1,1,\dots,1\rangle\cdot y^{i-1}$. We have $\prod_{i=1}^{p-1} (y^{-1}x_i^{-1}yx_i)\cdot\langle 1,1,\dots,1,x\rangle^p =$
$= \langle x,x^{-1},1,\dots,1\rangle\langle 1,x^2,x^{-2},1,\dots,1\rangle\dots\langle 1,\dots,1,x^{p-1},x^{1-p}\rangle\cdot\langle 1,\dots,1,x^p\rangle$
$= \langle x,x,\dots,x\rangle$.
Since by the above remarks $\Phi_p(G^{n+1})$ contains all commutators and pth powers, $\langle x,x,\dots,x\rangle$ lies in $\Phi_p(G^{n+1})$, as desired.

<u>Proof of Theorem</u> 3.2, <u>continued</u> : Let $U_n = \{u_\pi : \pi \in p_n^{<\omega}\}$, where p_n is the nth prime, and let $U = \bigcup\{U_n : n \in \omega\}$. Let the group G of Lemma 3.3 now be called G(p) . G is the group of all permutations of U fixing each U_n and generated by the action of $G(p_n)$ on U_n. Thus G is isomorphic to the restricted direct product of $\{G(p_n) : n \in \omega\}$. J is the filter of subgroups of G generated by $\{G_m^{p_n} : m,n \in \omega\}$, where again the subscript indicates iteration of the Φ_{p_n} operation. Thus $G_0^{p_n} = G$, and $G_{m+1}^{p_n} = \Phi_{p_n}(G_m^{p_n})$. $\mathfrak{N}$ is the resulting model.

Firstly we have to check that $U \in \mathfrak{N}$, i.e. that for any $u \in U$, $H(u) \in J$. However if $u \in U_n$, and u is in the mth level of U_n , it is clear that $G_m^{p_n} \leq H(u)$, so this is immediate. Next to see that DC_{p_n} is false in $\mathfrak{N}$, we show that U_n has no infinite branch in $\mathfrak{N}$. If b is an infinite branch of U_n in $\mathfrak{N}$ then there must be $m \in \omega$ such that $G_m^{p_n} \leq H(b)$. Now the set of branches, B , which are images of b under some member of G , is dense, and as $G_m^{p_n}$ is a normal subgroup of G , $G_m^{p_n} \leq H(b')$ for every $b' \in B$. It follows that $G_m^{p_n}$ fixes any well-ordering of B lying in $\mathfrak{M}$, and hence also U_n can be well-ordered in $\mathfrak{N}$. Thus the action of $G_m^{p_n}$ on U_n is trivial, contrary to Lemma 3.3.

We now just have to verify C_n^* in $\mathfrak{N}$ for each $n \geq 2$. We use Theorem 2.2. Let $H \in J$. Then for some m,

$$K = G_m^{p_0} \cap G_m^{p_1} \cap \ldots \cap G_m^{p_m} \leqslant H .$$

We show by induction on n that the intersection of any family of normal subgroups of K lying in J of index n is in J. It will then folow at once that for each n the intersection of any family of subgroups of K lying in J of index n also lies in J (since J is closed under conjugacy). This is clearly enough. We therefore let $\langle K_\beta : \beta < \alpha \rangle$ be such a family of normal subgroups of K. As K acts on U_k as a p_k-group, for each k, K/K_β is nilpotent, and so there is a normal chain $K_\beta \lhd K_\beta^{(1)} \lhd \ldots \lhd K_\beta^{(r_\beta)} = K$ of subgroups of K containing K_β in which each factor is cyclic of prime order. This chain is of length $\leqslant n$, and hence

$$\bigcap \{K_\beta : \beta < \alpha\} \supseteq G_{m+n}^{p_0} \cap G_{m+n}^{p_1} \cap \ldots \cap G_{m+n}^{p_{m+n}}$$

as desired.

<u>Remarks</u> 3.4. It would make no difference to the proof if we used the inverse limit of the system $\langle \mathbb{Z}_p, \mathbb{Z}_p \wr \mathbb{Z}_p, \mathbb{Z}_p \wr \mathbb{Z}_p \wr \mathbb{Z}_p, \ldots \rangle$ instead of the direct limit, or used the full direct product of the $G(p_n)$ instead of the restricted direct product. It is a result of Jensen that C^* (the axiom of choice for well-ordered families of non-empty sets) implies DC (see [1 pp.148-149] for a proof.) It follows trivially from the remarks at the beginning of the paper that $C^*_{<\omega}$ implies $DC_{<\omega}$. However for finite n the situation is different. The theorem shows that C_n^* does not imply DC_n.

§ 4 In this section we collect together various results which are connected with the condition $L(n,Z)$: If G is a fixed point free group of permutations of an n-element set there are proper subgroups H_i of G such that $\Sigma|G: H_i| \in Z$. This condition is due to Gauntt [2], and the main facts known about it are as follows (see [8])

$$L(n,Z) \iff C^*_Z \to C^*_n \iff C^o_Z \to C^o_n \iff C^o_Z \to C^*_n \ .$$

(here C^o_n is the axiom of choice for linearly ordered families of n-element sets). The same proof shows that for any infinite $\alpha \leq \beta$, $L(n,Z) \iff C^\beta_Z \to C^\alpha_n$.

THEOREM 4.1.

$$FM \vdash DC_Z \to DC_n \iff FM \vdash DC_Z \to C^\omega_n \iff FM \vdash (\forall m \in Z)DC_m \to C^\omega_n \iff L(n,Z).$$

Proof. That each of $FM \vdash DC_Z \to DC_n$ and $FM \vdash (\forall m \in Z)DC_m \to C^\omega_n$ implies $FM \vdash DC_Z \to C_n$ is obvious. Since $L(n,Z)$ implies that $FM \vdash (\forall m \in Z)C^\omega_m \to C^\omega_n$ it also implies that $FM \vdash (\forall m \in Z)DC_m \to C^\omega_n$.

It remains to show that $FM \vdash DC_Z \to C^\omega_n \Rightarrow L(n,Z) \Rightarrow FM \vdash DC_Z \to DC_n$.

Suppose $L(n,Z)$ is false. Then there is a group G_o of permutations of $\{0,\dots,n-1\}$ without fixed points such that for any proper subgroups H_i of G_o, $\Sigma|G_o : H_i| \notin Z$. We form a Franekel-Mostowski model as follows, in the manner of Mostowski [5].
$U = \{u_{ij} : i \in \omega, j < n\}$, $U_i = \{u_{ij} : j < n\}$. G is the group of all permutations of U fixing all but finitely many points, each U_i, and acting on each U_i like G_o. J is the filter of finite supports, and $\mathfrak{N}$ the resulting model.

Since G_o has no fixed points C^ω_n fails in $\mathfrak{N}$. It remains to verify DC_Z in $\mathfrak{N}$. By the appropriate modification of Theorem 2.3 it is enough to consider the following situation. Let $K = K(\bigcup_{i<m} U_i)$ some m, and let $\langle K_i : i < k\rangle$ be a family of k members of J such that $\Sigma_{i<k} |K : K_i| \in Z$. It is enough to show that $K = K_i$ for some i (and then continue inductively through the tree. Suppose not. Then for some m', $K_i \supsetneqq K(\bigcup_{j<m+m'} U_j)$ all i. Hence there are proper subgroups H_i of $G^{m'}_o$ such that $\Sigma|G^{m'}_o : H_i| \in Z$. It is then easy to find proper subgroups H'_i of G_o such that $\Sigma|G_o : H'_i| \in Z$, contrary to the choice of G_o.

Now assume $L(n,Z)$ and DC_Z. We shall show that DC_n follows. So let $\langle T,<\rangle$ be an n-branching tree with mth level T_m. We define from T another tree $\langle T',<\rangle$ such that each point in T' has m immediate successors for some $m \in Z$. There will be an order-preserving map ϑ (i.e. $t < t' \to \vartheta(t) < \vartheta(t')$) from a cofinal subset of T' onto T. Hence any infinite branch of T' defines an infinite branch of T.

T' is defined in stages corresponding to the levels T_m of T.

It is enough to show how to define a fragment of T' extending t' which will correspond to the part of T consisting of $\vartheta(t')$ and its immediate successors $\{t_1,\ldots,t_n\}$, and to define the mapping ϑ from the tips of this fragment onto $\{t_1,\ldots,t_n\}$.

The immediate successors of t' in T' are determined thus. As $L(n,Z)$ holds there are proper subgroups H_i $(1 \leq i \leq k)$ of S_n such that $\Sigma|S_n : H_i| \in Z$. Let $\underline{t} = \langle t_1,\ldots,t_n\rangle$ and $\underline{t}_i = \{\sigma\underline{t} : \sigma \in H_i\}$. Then the immediate successors of t' in T' are $\{\langle\sigma\underline{t}_i, i, t'\rangle : \sigma \in S_n, 1 \leq i \leq k\}$. There are $\Sigma|S_n : H_i| \in Z$ of these as desired. Moreover their definition is independent of the particular well-ordering of $\{t_1,\ldots,t_n\}$ chosen. Now repeat this for each point $\langle\sigma\underline{t}_i, i, t'\rangle$ such that the corresponding group H_i is fixed point free. This can only happen finitely often (at most $n!$ times) and then each tip t'' of the fragment corresponds to a group H of permutations of $\{t_1,\ldots,t_n\}$ with a fixed point. By a standard argument due to Mostowski [5 pp. 142-143] we may choose effectively one such fixed point t_i, and let $\vartheta(t'') = t_i$. ϑ is clearly order-preserving, and since the whole construction from t' is symmetric in $\{t_1,\ldots,t_n\}$, it must also be "onto".

To clarify this rather involved argument we give a diagram to illustrate the proof that $DC_6 \rightarrow DC_4$.

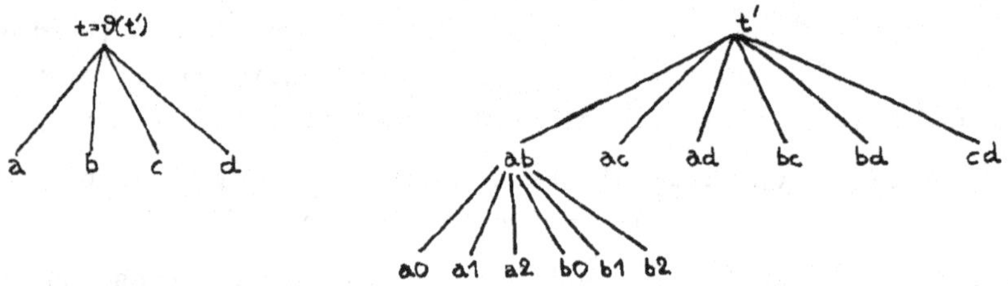

The fragment of T on the left is replaced by the fragment of T' on the right.

We now sketch a possible method for proving the conjecture (iii) mentioned in § 1. This is that $FM \vdash (\forall m \in Z)DC_m \rightarrow DC_n \Leftrightarrow$ for some $m \in Z$, $FM \vdash DC_m \rightarrow DC_n \Leftrightarrow$ for some $m \in Z$ (the same m), $L(n,\{m\})$. The only part of this causing difficulty is to show that if $L(n,\{m\})$ is false for every member m of Z, then so is $FM \vdash (\forall m \in Z)DC_m \rightarrow DC_n$.

The general case is certainly simplified by the observation that $\{m : L(n,\{m\})\}$ is closed under addition, so that if

$Z = \{m: \neg L(n,\{m\})\}$ and $\Sigma|G: K_i| \in Z$ then for some i, $|G: K_i| \in Z$. However, even so it is quite involved, and we consider only the case $n = 4$, $Z = \{2,3\}$.

Naturally one takes U, the set of urelemente, to be indexed by $4^{<\omega}$, so $U = \{u_\pi : \pi \in 4^{<\omega}\}$, G is the group $S_4 \wr S_4 \wr S_4 \quad \ldots$ acting on U in the natural way; i.e. the group whose action on U is generated by the action of S_4 on each set of immediate successors. The problem then is to make the correct choice of J. As in § 3 let $\Phi_m(H)$ be the intersection of all subgroups of H of index m. Let $\Phi_m^0(H) = H$, and $\Phi_m^{k+1}(H) = \Phi_m(\Phi_m^k(H))$. Finally let $\Phi_m^\omega(H) = \bigcap\{\Phi_m^k(H) : k \in \omega\}$. J is the filter of subgroups of G generated by $\{\Phi_2^\omega(G), \Phi_3^\omega \Phi_2^\omega(G), \Phi_2^\omega \Phi_3^\omega \Phi_2^\omega(G), \ldots\}$.

THEOREM 4.2. If J is a proper filter, then $FM \vdash DC_2 \wedge DC_3 \rightarrow DC_4$ is false.

The proof is omitted. It clearly uses Theorem 2.3. The feature which makes this case particularly simple is the fact that $\Sigma k_i = 2$ implies some $k_i = 1$ or 2, and similarly for $\Sigma k_i = 3$.

We now turn to some numerical cases. In particular we consider the question as to when $C_2^\omega \wedge \ldots \wedge C_{n-1}^\omega \rightarrow C_n^\omega$, and similarly for the other variants. Part (ii) of the following theorem appeared (implicitly) in Mostowski's paper [5].

THEOREM 4.3. (i) For every $n \geq 2$ except 4, $(\forall m < n)DC_m \not\rightarrow DC_n$, $DC_{\{2,\ldots,n-1\}} \not\rightarrow DC_n$, $(\forall m < n)C_m^\omega \not\rightarrow C_n^\omega$, etc. but $DC_{\{2,3\}} \rightarrow DC_4$ and $C_2^\omega \wedge C_3^\omega \rightarrow C_4^\omega$.

(ii) For every prime p, $(\forall m < p)C_m \not\rightarrow C_p$, and for every $n \geq 2$ which is not a prime, $(\forall m < n)C_m \rightarrow C_n$.

<u>Remark</u>. The only case left open is the question as to whether $DC_2 \wedge DC_3 \rightarrow DC_4$. As indicated above, the answer is probably "no".

<u>Proof</u>. (i) It is sufficient to show that $L(n,\{2,\ldots,n-1\})$ is true only for $n = 4$. That $L(4,\{2,3\})$ is true is an immediate consequence of the solubility of S_4. $L(2,\emptyset)$ and $L(3,\{2\})$ are trivially false. So we let $n \geqslant 5$ and show that $G = A_n$ has no proper subgroup of index less than n.

Suppose not, and let H be a maximal proper subgroup of G of index $m < n$. As G is simple, $H = N_G(H)$. Hence H has exactly m conjugates in G. G defines as action on these by conjugation, and this determines a non-trivial homomorphism from G into S_m. As G is simple this homomorphism is $1-1$, and so $|G| \leq |S_m|$, a contradiction.

References

[1] U.Felgner, Models of ZF - Set Theory, Springer Lecture Notes, vol.223.

[2] R.J.Gauntt, Some Restricted Versions of the Axiom of Choice, Notices A.M.S. 68T - 176 (1968) 351.

[3] R.J.Gauntt, Axiom of Choice for Finite Sets, Notices A.M.S. 70T - E12 (1970) 454.

[4] A.Levy, Axioms of Multiple Choice, Fund.Math. 50(1962), 475-483.

[5] A.Mostowski, Axiom of Choice for Finite Sets, Fund.Math. 33(1945), 137-168.

[6] D.Pincus, Two Model-Theoretic Ideas in Independence Proofs, to appear.

[7] D.Pincus, Zermelo-Frankel Consistency Results by Fraenkel -Mostowski Methods, J. Symbolic Logic 37(4) (1972),721-743.

[8] J.Truss, Finite Axioms of Choice, Annals of Mathematical Logic 6(1973), 147-176.

CATEGORICITY RELATIVE TO ORDINALS FOR MODELS OF SET THEORY AND THE NONABSOLUTENESS OF L

by Z. Vetulani

Investigating the consistency of AC and GCH with the axioms of ZF , Gödel introduced the so-called constructible sets and proved that they constitute the smallest subuniverse of the ZF-universe in which the axioms of ZF hold. Contructible sets are objects made from ordinal numbers by iterating some simple set theoretical operations (i.e. taking pairs, cartesian products etc.), so they are in some sense very effective. One of the most useful properties of constructible sets is their absoluteness with respect to standard models. Namely if we take any two standard models of ZF with height α then we observe that they have the same class of constructible sets L_α i.e. the class of constructible sets of a given standard model depends only on its height. That agrees well with the intuitive "effectiveness" of constructible sets. So the natural question to ask is whether we have this absoluteness also in the case of nonstandard models. Our claim is that this is not the case.

H. Friedman in [2] introduces the notion of "α-categoricity" of set theories, for $\alpha \in On$. We expand this to the notion of "τ-categoricity" where τ is an order type, but the only interesting case is when $\tau = \alpha \times (1 + Q)$, where $\alpha \in On$ and Q is a dense linear ordering without endpoints, because this is the general form of the order type of ordinal numbers in nonstandard ω-models of Kripke-Platek set theory having standard ordinal α (see [2]). In this paper we assume that Q is countable.

All the terminology of the present paper is standard (see also [2]).

DEFINITION . Let τ be any order type. The theory $T \supset KP$ is τ-categorical iff any two models of T with the ordinals ordered in the type τ are isomprphic.

For α-categoricity, $\alpha \in On$ we have the following theorem.

THEOREM 1. (Friedman [2]). If T is a recursive theory in $L(\varepsilon)$, α a countable admissible ordinal such that $R_\alpha \cap L_{\alpha+} = L_\alpha$ then

$$(*) \qquad KP + T \text{ is } \alpha\text{-categorical iff } KP + T \vdash_{\overline{\alpha}} V = L$$

where for a theory S and a formula φ :

$$S \vdash_{\overline{\alpha}} \varphi <-> (M)((On^M = \alpha \wedge M \models S) \rightarrow M \models \varphi)$$

and On^M is the order type of ordinals in the model M .

We will show that for some recursively inaccessible α the equivalence (*) with $\alpha \times (1 + Q)$ instead of α does not hold for some complete T .

Let PKP denote here Kripke-Platek set theory with the powerset axiom and with the new defining axiom for the functional symbol P : $(x)(y)(y \in P(x) <-> (v)_y \; v \in x)$ and with collection axioms which are Δ_0 with respect to the atomic formulas $P(x) = y$, $P(x) \in y$, $P(x) = P(y)$ and $P(x) \in P(y)$. Z denotes Zermelo set theory and KMC is Kelley-Morse theory of classes with the axiom of choice.

THEOREM 2. Let α be a countable ordinal number, $\alpha > \omega$ such that $(\exists\beta)$ $(L_\beta \models$ "α is inaccessible" $\wedge$ $L_\beta \models$ PKP + Z) and T be a theory in $\mathcal{L}_{L_\alpha}(\epsilon)$ such that $PKP \subset T \in L_\beta$, T having a model containing L_α as initial segment. Then there are models $\mathfrak{M}_1 = \langle M_1, E_1 \rangle$ and $\mathfrak{M}_2 = \langle M_2, E_2 \rangle$ of T such that $Sp(\mathfrak{M}_1) = L_\alpha$ and $\langle M_1, E_1 \rangle \not\cong \langle M_2, E_2 \rangle$. In particular the theory T is not $\alpha \times (1 + Q)$-categorical.

THEOREM 3. Let $\langle \mathcal{F}, L_\alpha, \epsilon \rangle$ be a countable model of KMC and T a theory in the language $\mathcal{L}_{L_\alpha}(\epsilon)$ such that $PKP \subset T \in \mathcal{F}$ and T has a model containing L_α as initial segment. Then there are models

$\mathfrak{M}_1 = \langle M_1,E_1\rangle$ and $\mathfrak{M}_2 = \langle M_2,E_2\rangle$ of T such that $Sp(\mathfrak{M}_i) = L_\alpha$ and $\langle M_1,E_1\rangle \not\equiv \langle M_2,E_2\rangle$. In particular T is not $\alpha \times (1 + Q)$ categorical.

Remark. A nice example of a theory being not $\alpha \times (1 + Q)$-categorical is then the theory of L_α, where L_α is expandable to a model of KMC. This theory contains $ZF + V = L$ (even $ZF^{KM} + V = L$) and is complete.

Proof of theorem 2. The lemma below shows that it is enough to find two nonstandard models of T with standard part L_α and different "standard systems".

DEFINITION. The standard system of the model $\mathfrak{N} = \langle N,E\rangle$ (denoted by $STS(\mathfrak{N})$) is the pair $\langle A,B\rangle$ where $A = Sp(\mathfrak{N})$ and (b) $(b \in B \Longleftrightarrow (\exists x \in N)(b = x_N \cap A))$ where $t \in x_N \Longleftrightarrow t \in N \wedge t E x$. Sometimes we write $STS(\mathfrak{N}) = B$.

LEMMA If $STS(\mathfrak{N}_1) = \langle A,B_1\rangle$ and $STS(\mathfrak{N}_2) = \langle A,B_2\rangle$ and $B_1 \neq B_2$ then $\mathfrak{N}_1 \not\cong \mathfrak{N}_2$.

Proof. Assume not. Let us take $I: \mathfrak{N}_1 \cong \mathfrak{N}_2$ and $b \in B_1 \setminus B_2$. Then $(\exists c)\, c \in N_1 \wedge b = c_{N_1} \cap A$. Take Ic. We have $b = (Ic)_{N_2} \cap A \in B_2$ because $(Ic)_{N_2} = \{Ia : a \in c_{N_1}\}$, $I \restriction A = id \restriction A$ and Id is nonstandard iff d is nonstandard.

The following result of Friedman permits us to confine ourselves to finding two standard systems of some poweradmissible sets, each containing T as an element.

LEMMA (Friedman [2]). If $\langle A,C\rangle$ is the countable standard system of some poweradmissible set, $PKP \subset T \subset L_A(\epsilon)$ for $T \in C$, and T has a model having $\langle A,\epsilon\rangle$ as initial segment, then there is N such that $N \models T$ and $STS(\mathfrak{N}) = \langle A,C\rangle$.

Now we will find two such standard systems. Let $\alpha > \omega$, $L_\beta \models PKP + Z + \alpha$ is inaccessible. We may assume that β is countable. Let us state here one more Lemma from Friedman's [2] :

LEMMA. Let $\langle A,C,\epsilon\rangle$ be a poweradmissible system and T be a theory in $\mathcal{L}_A$ such that $T \in C$, and there is a model of T with A

as initial segment. Then there is a model of T with standard part exactly equal A .

By poweradmissible systems we mean here the countable standard models for the following axioms in the language of GB : all axioms of PKP , the extensionality axiom for classes, Δ_0^c-separation axioms: $(\exists v)(w)(w \in v \Longleftrightarrow w \in a \wedge \varphi(w))$ for Δ_0^c-formulas φ , Δ_0^c-collection axioms : $(v)_a(\exists w)\varphi \to (\exists z)(v)_a(\exists w)_z \varphi$ for Δ_0^c-formulas φ , Δ^c-class existence axioms $(w)(\varphi \Longleftrightarrow \sim \psi) \to (\exists V)(w)(w \in V \Longleftrightarrow \varphi)$ where φ, ψ are Σ^c-formulas. Δ_0^c (Σ^c) formulas are formulas which are Δ_0 (Σ) with respect to atomic formulas of the form $a \in V$, where a is a set and V class variable.

COROLLARY. Let $\mathfrak{a}$ be a standard countable model for PKP + Z . Let $\mathfrak{a} \models$ "α is regular $\wedge$ $L_\alpha = R_\alpha$" . Let $a_0, a_1 \in \mathfrak{a}$ and $\mathfrak{a} \models$ "$a_i \subset L_\alpha \wedge a_i \notin L_\alpha$" . Then there is a model $\mathfrak{b}$ of PKP such that $Sp(\mathfrak{b}) = L_\alpha$ and $a_i \in STS(\mathfrak{b})$

Proof of the corollary. Let
$T = PKP \cup Z \cup \{\bar{t} \in \overline{a_0} : t \in a_0\} \cup \{\bar{t} \notin \overline{a_0} : t \in L_\alpha \wedge t \notin a_0\}$

For each element x of L_α we have the constant symbol $\bar{x}$ in L_{L_α} (also for a_0 and a_1) . With a suitable coding of elements from L_α in L_{L_α} the theory T is definable in $\mathfrak{a}$. Also $\mathfrak{a} \models T \subset L_\alpha$.

Let us observe that $\langle L_\alpha, P(L_\alpha)^{\mathfrak{a}}, \in\rangle$ is a poweradmissible system. This is the case because $\mathfrak{a} \models$"$\langle L_\alpha, P(L_\alpha), \in\rangle$ is a poweradmissible system" . $T \in (P(L_\alpha))^{\mathfrak{a}}$, so applying the Lemma to T and $\langle L_\alpha, P(L_\alpha)^{\mathfrak{a}}, \in\rangle$ we obtain a model $\mathfrak{b}$ of T such that $Sp(\mathfrak{b}) = L_\alpha$. In $\mathfrak{b}$ we have denotations for $\overline{a_0}$ and $\overline{a_1}$. The question which elements from L_α are in the extensions of these denotations is decided by the theory of T . These are exactly the elements of a_i . So $a_i \in STS(\mathfrak{b})$.

Now we can apply the above corollary to the theory T , L_β and α , because $L_\beta \models$ "α is inaccessible" and we obtain a countable standard system $\langle A,B\rangle$ such that $A = L_\alpha$ and $T \in B$.

To produce another standard system with the required property we first extend the model L_β to $L_\beta[G]$ by forcing. All we need is that this forcing step does not change the R_α and that there are $\aleph_1$ generic sets.

We can take the following notion of forcing : $\langle P,\leq\rangle$, where $P = \{f: \text{func}(f) \wedge rg(f) \subseteq \{0,1\} \wedge \text{dom } f \subseteq L_\alpha \wedge L_\beta \models \overline{\text{dom } f} < \alpha\}$ and $f \leq g \leftrightarrow g \subseteq f$. $\langle P,\leq\rangle$ satisfies the αdcc i.e. each descending chain of conditions with length $< \alpha$ has a lower bound in P . It is well known that for $\langle P,\leq\rangle$ satisfying the αdcc $L_\beta[G] \models$ "α is regular" and for $\xi < \alpha$ in $L_\beta[G]$ there are the same subsets of ξ as in L_β , for generic G . (See J.Jech [3], Lemma 57). So $L_\beta[G] \models$ "α is inaccessible".

Also $(R_\alpha)^{L_\beta} = (R_\alpha)^{L_\beta[G]}$: If not let γ be the least ξ such that $(R_\xi)^{L_\beta} \neq (R_\xi)^{L_\beta[G]}$. Then $\gamma = \delta + 1$ for some δ and $(R_\delta)^{L_\beta} = (R_\delta)^{L_\beta[G]}$. Let $L_\beta \models$ "$f: R_\delta \overset{1-1}{\longleftrightarrow} \eta$" , for some f and $\eta \in \alpha$. Then $L_\beta[G] \models$ "$f: R_\delta \overset{1-1}{\longleftrightarrow} \eta$. By collection for some $a \in (R_\gamma)^{L_\beta[G]} - (R_\gamma)^{L_\beta}$ there is in $L_\beta[G]$ the image of a by f - which is not in L_β . This is a contradiction.

So $L_\alpha = (R_\alpha)^{L_\beta[G]}$, because $L_\alpha = (R_\alpha)^{L_\beta}$. From this we get that $\mathfrak{A} \models L_\alpha = R_\alpha$ where $\mathfrak{A} = L_\beta[G]$. Note that G may be chosen in such a way that $a_1 = (\bigcup G)^{-1}(0) \notin B$ because B is countable and we have $\aleph_1$ generic sets at our disposal. Now we apply the corollary to $L_\beta[G]$, $a_0 = T$ and a_1 . We get a countable standard system $\langle L_\alpha, c\rangle$,where $T \in c$, $a_1 \in C$; so $C \neq B$.
This completes proof of theorem 2.

The proof of theorem 3 is quite similar. Let $\langle \mathcal{F}, L_\alpha, \in\rangle \models KMC$. By W. Marek's theorem [4] : $\langle \mathcal{F}, L_\alpha, \in\rangle \models KMC$ iff there exists a model N of $ZFC^- + V = HC$ such that $N \models$ "L_α is inaccessible family of sets" and $L_\alpha = Sp(\mathfrak{N})$ and $\mathcal{F} = P(L_\alpha) \cap N$. We now use the proof of theorem 2 with $\mathfrak{N}$ instead of L_β .

Let us note that some nonabsoluteness results about L were obtained by J. Rosenthal [5] and K.J. Barwise [1] but the models (witnesses of nonabsolutness of L) which they obtain are not elementarily equivalent.

So we claim that the present results are in that sense stronger.

We are greatly indebted to ours colleagues from Alistair Lachlan's seminars in Warsaw for many valuable discussions.

References

[1] Barwise, K.J., A preservation theorem for interpretations, Proceedings of the Cambridge Summer School in Mathematical Logic, Lectures Notes in Mathematics 337 (1973), Springer Verlag.

[2] **Friedman**,H., Countable Models of Set Theories, Proceedings of the Cambridge Summer School in Mathematical Logic, Lecture Notes in Mathematics 337 (1973), Springer Verlag, pp 539-573.

[3] Jech, T.J., Lectures in Set Theory with Particular Emphasi on the Method of Forcing, Lecture Notes in Mathematics 217 (1971), Springer Verlag.

[4] Marek W., On the metamathematics of impredicative set theory, Dissertationes Mathematicae XCVIII, PWN 1973, Warszawa.

[5] Rosenthal, J., Relations not determining the structure of L, Pacific Journal of Mathematics, 37 (1971), pp. 497-514.

Institute of Mathematics ,
Adam Mickiewicz University , Poznań , Poland ,

Institute of Mathematics ,
Warsaw University .

THE UNIVERSALITY OF BOOLEAN ALGEBRAS WITH THE HÄRTIG QUANTIFIER

by Martin Weese
(Humboldt University, Berlin)

Summary: By using the universality of the theory of irreflexive symmetric graphs the universality of the theory of Boolean algebras with the Härtig quantifier is shown. Let $\mathfrak{B}$ be any Boolean algebra with infinitely many atoms. Then the theory of $\mathfrak{B}$ with the Härtig quantifier is undecidable.

The quantifier I was introduced by Härtig [2]:
For any $\mathfrak{A}$,

$$\mathfrak{A} \models I\, x\, \varphi(x)\, \psi(x) \quad \text{iff} \quad \operatorname{card}\{a \in \mathfrak{A} : \mathfrak{A} \models \varphi(a)\}$$
$$= \operatorname{card}\{a \in \mathfrak{A} : \mathfrak{A} \models \psi(a)\} .$$

If T is an elementary theory, then T(I) denotes the theory of all models of T in the corresponding language with the added quantifier I . If T is any theory (an elementary theory or a theory with added quantifier I), then $T \vdash \varphi$ means, that for any $\mathfrak{A} \in \operatorname{Mod}(T)$ we have $\mathfrak{A} \models \varphi$.
It is possible to express the quantifier Q_0 with the quantifier I :

$$Q_0\, x\, \varphi(x) \longleftrightarrow \exists x(\varphi(x) \wedge I\, y\, (x \neq y \wedge \varphi(y))\, \varphi(y)) .$$

The theory T is <u>universal</u> iff any theory S can be interpreted in a suitable extension of T .

Let Ba be the theory of Boolean algebras. It is known (see Ers'hov [1] or Tarski [4]) that Ba is decidable.
Let $Ba' = Ba \cup \{\forall x(x \neq 0 \rightarrow \exists y(y \neq 0 \wedge y < x)\}$, that means, Ba' is the theory of atomless Boolean algebras.
Let Gis be the theory of irreflexive symmetric graphs. Hauschild and Rautenberg [3] proved the universality of Gis .
It is possible to interprete Gis in Ba'(I) . That means, there is an effective procedure attaching to any formula φ of the language of Gis a formula φ^* of the language of Ba'(I) such that

$$\text{Gis} \vdash \varphi \quad \text{iff} \quad \text{Ba}'(\text{I}) \vdash \varphi^* \ .$$

Thus we get the universality of Ba'(I) and also of Ba(I) . For any linearly ordered set τ with first element let $J(\tau)$ be the Boolean algebra generated by the left-closed right-open intervals. For any ordinal i let $\overline{\eta}_i$ be the set of all finite sequences of ordinals less than ω_i ordered in the following way: $\overline{\alpha} < \overline{\beta} \Leftrightarrow (lh(\overline{\alpha}) = 0 \wedge \wedge\, 0 < lh(\overline{\beta})) \vee (0 < lh(\overline{\beta}) \wedge lh(\overline{\beta}) < lh(\overline{\alpha}) \wedge (\forall i)_{lh(\overline{\beta})}(\alpha_i = \beta_i)) \vee \vee((\exists k)(k < lh(\overline{\alpha}) \wedge k < lh(\overline{\beta}) \wedge (\forall i)_k(\alpha_i = \beta_i) \wedge \alpha_k < \beta_k))$.
$J(\overline{\eta}_i)$ is an atomless Boolean algebra and every $a \in J(\overline{\eta}_i)$, $a \neq 0$ contains exactly ω_i elements smaller than a .
Let $\mathfrak{A} \in \text{Mod}(Ba')$, $a \in \mathfrak{A}$. The infinite cardinal λ is <u>determined by</u> $\mathfrak{A}$ iff there is $b \in \mathfrak{A}$ such that $card\{c \in \mathfrak{A} : c \leq b\} = \lambda$. The infinite cardinal λ is <u>determined by</u> a iff there is $b \leq a$ such that $card\{c \in \mathfrak{A} : c \leq b\} = \lambda$.
We use $C(\mathfrak{A})$ $(C(a))$ for the set of all cardinals determined by $\mathfrak{A}$ (by a).

$$c_1\, x \underset{df}{=} x \neq 0 \wedge \forall y(y \neq 0 \wedge y \leq x \rightarrow I\, z \quad z \leq x \quad z \leq y) \ ;$$

that means: $\mathfrak{A} \models c_1\, a$ iff $card(C(a)) = 1$.
If $\mathfrak{A} \models c_1\, a$, a is called <u>one-cardinal-like</u>.

$$c_2\, x \underset{df}{=} \exists y_1 y_2 (y_1 \leq x \wedge y_2 \leq x \wedge c_1\, y_1 \wedge c_1\, y_2 \wedge$$
$$\forall z(z \leq x \wedge z \neq 0 \rightarrow (I\, u \quad u \leq z \quad u \leq y_1 \vee I\, u \quad u \leq z \quad u \leq y_2))) \wedge$$
$$\neg \exists y_1 y_2 (c_1\, y_1 \wedge c_1\, y_2 \wedge x = y_1 \cup y_2) \ ;$$

that means: $\mathfrak{A} \models c_2\, a$ iff $card(C(a)) = 2$, there are one-cardinal-

-like elements $b_1, b_2 \leq a$ such that $C(a) = C(b_1) \cup C(b_2)$ and a is not the union of two one-cardinal-like elements.
If $\mathcal{A} \models c_2\, a$, a is called <u>two-cardinal-like</u>.
Let $\mathfrak{M} = \langle M, R\rangle$ be a model of Gis. Without loss of generality we assume that M is a set of ordinals. We construct a model $\mathcal{B}_{\mathfrak{M}}$ of Ba′ such that we can interprete $\mathfrak{M}$ in $\mathcal{B}_{\mathfrak{M}}$:

$$\mathcal{B}_{\mathfrak{M}} = \prod_{i \in M} J(\bar{\eta}_i) \times \prod_{(i,j)\in R} J((\bar{\eta}_i + \bar{\eta}_j)\cdot\omega) \quad .$$

Then we have:

$i \in M$ iff there is a one-cardinal-like element a such that $C(a) = \{\omega_i\}$;

$(i,j) \in R$ iff there is a two-cardinal-like element a such that $C(a) = \{\omega_i, \omega_j\}$.

Now we describe the construction of the formula φ^* for any given formula φ of the language of Gis :

$$(x = y)^* \underset{df}{=} c_1\, x \wedge c_1\, y \wedge I\, z\; z \leq x\; z \leq y \quad ;$$

$$(Rxy)^* \underset{df}{=} \exists z(c_2\, z \wedge \forall u(u \leq z \rightarrow (I\, v\; v \leq u\; v \leq x \vee I\, v\; v \leq u\; v \leq y))) \quad ;$$

$$(\psi \wedge \chi)^* \underset{df}{=} \psi^* \wedge \chi^* \quad ;$$

$$(\neg\, \psi)^* \underset{df}{=} \neg\, \psi^* \quad ;$$

$$(\exists x\, \psi)^* \underset{df}{=} \exists x(c_1 x \wedge \psi^*) \quad .$$

Now it is easy to see that:

$$\text{Gis} \vdash \varphi \quad \text{iff} \quad \text{Ba}'(I) \vdash \varphi^* \quad .$$

Thus we have shown:

THEOREM 1. The theory of atomless Boolean algebras with added quantifier I is universal.

COROLLARY 1. The theory of Boolean algebras with added quantifier I is universal.

If we replace $\bar{\eta}_i$ everywhere by $\bar{\eta}_i \cdot \omega$ in the construction of $\mathfrak{B}_m$ we get

COROLLARY 2. The theory of atomic Boolean algebras with added quantifier I is universal.

It follows from the universality that all these theories are undecidable.
Now we show, that for any Boolean algebra $\mathfrak{B}$ with infinitely many atoms, $Th(\mathfrak{B})(I)$ is undecidable.
Let N be the set of natural numbers, let + be the ternary predicate, defined by

$$+(k,m,n) \quad \text{iff} \quad k + m = n ,$$

and let P be the binary predicate defined by

$$P(m,n) \quad \text{iff} \quad \binom{m}{2} = n .$$

LEMMA. The theory of $\langle N, +, P\rangle$ is undecidable.

Proof. We have $n^2 = 2\binom{n}{2} + n$ and $k \cdot m = n$ iff $(k + m)^2 = k^2 + 2n + m^2$.
Thus it is possible to define multiplication in the system $\langle N, +, P\rangle$ and we get the undecidability of the system $\langle N, +, P\rangle$.

Let $\mathfrak{B}$ be a Boolean algebra with infinitely many atoms. We give an interpretation of $\langle N, +, P\rangle$ in $\mathfrak{B}$. That means, we describe an effective procedure attaching to any formula φ of the language of $\langle N, +, P\rangle$ a formula φ^* of the language of $\mathfrak{B}$ such that

$$\langle N, +, P\rangle \models \varphi \quad \text{iff} \quad \mathfrak{B} \models \varphi^* .$$

At first we give an abbreviation:

$$\text{at } x \underset{df}{=} x \neq 0 \wedge \forall y(y \cap x = x \vee y \cap x = 0) \quad ;$$

that means, $\mathfrak{A} \models$ at a iff a is an atom of $\mathfrak{A}$. Let n be a natural number. The element $a \in \mathfrak{A}$ is a <u>representation</u> of n iff a contains exactly n atoms. To describe the addition of the natural numbers m and n we choose disjoint representations a_m and a_n

of m and n . Then $a_m \cup a_n$ is a representation of $m + n$. To get a representation of $\binom{n}{2}$ we take a representation of a_n ; we consider the set $A = \{a \in \mathfrak{A} : a \leq a_n$ and a contains exactly two atoms$\}$ and choose an element $b \in \mathfrak{A}$ that contains as many atoms as A has elements. Then b is a representation of $\binom{n}{2}$.
Now we give the formal definition of :

$$(x = y)^* \underset{df}{=} I\, z(\text{at } z \wedge z \leq x)\,(\text{at } z \wedge z \leq y) \quad ;$$

$$\begin{aligned}(+(x,y,z))^* \underset{df}{=} \exists uv(u \cap v = 0 \wedge u \cup v = z \wedge \\ I\, z\,(\text{at } z \wedge z \leq x)\,(\text{at } z \wedge z \leq u) \wedge \\ I\, z\,(\text{at } z \wedge z \leq y)\,(\text{at } z \wedge z \leq v)) \quad ;\end{aligned}$$

$$\begin{aligned}(P(x,y))^* \underset{df}{=} I\, z\,(z \leq x \wedge \exists uv(\text{at } u \wedge \text{at } v \wedge u \neq v \wedge u \cup v = z)) \\ (\text{at } z \wedge z \leq y) \quad ;\end{aligned}$$

$$(\psi \wedge \chi)^* \underset{df}{=} \psi^* \wedge \chi^* \quad ;$$

$$(\neg\ \psi)^* \underset{df}{=} \neg\ \psi^* \quad ;$$

$$(\exists x\, \psi)^* \underset{df}{=} \exists x(\neg\, Q_0 y(\text{at } y \wedge y \leq x) \wedge \psi^*) \ .$$

Now it is easy to see that

$$\langle N, +, P\rangle \models \varphi \quad \text{iff} \quad \mathfrak{B} \models \varphi^* \ .$$

Thus we get

THEOREM 2. The theory $Ba(I) \cup \{Q_0 x(\text{at } x)\}$ is essential undecidable.

References.

[1] Ju.Eršov, Decidability of elementary theory of distributive lattices with the relative complements and theory of filters, Algebra i Logika, 3(1964), 17-38.

[2] K.Härtig, Über einen Quantifikator mit zwei Wirkungsbereichen; Kolloqium über die Grundlagen der Mathematik, mathematische Maschinen und ihre Anwendungen, Tihany (Ungarn), September 1962, Budapest 1965.

[3] K. Hauschild and W. Rautenberg, Interpretierbarkeit in der Gruppentheorie, Algebra Universalis 1(1971), 136-151.

[4] A. Tarski, Arithmetical classes and types of Boolean algebras, Bull. Amer. Math. Soc. 55(1949), 64.

VON NEUMANN'S HIERARCHY AND DEFINABLE REALS

by W. Zadrożny

Introduction. In this work we shall deal with ordinal-definable reals. There are natural connections between ordinal-definable sets and von Neumann's hierarchy (cf. Myhill, Scott [8]), but they have not yet been studied. We shall study the definability of real numbers on particular levels of von Neumann's hierarchy.

In 1963 Putnam [9] proved that $(\exists\alpha)_{\omega_1^L}\,[P(\omega)\cap(L_{\alpha+1}\setminus L_\alpha)=\emptyset]$ Srebrny [10] has generalized this result. A question arose, whether we can formulate a similar problem for ordinal-definable reals. Because there is no hierarchy for ordinal definable sets similar to the hierarchy of constructible sets we have defined our own as follows:

i) $a \in \mathrm{Def}\, R_\alpha \equiv a$ is definable in $\langle R_\alpha, \in\rangle$ by a formula involving ordinals as parameters.

ii) $a \in \mathrm{Df}\, R_\alpha \equiv a$ is definable in $\langle R_\alpha, \in\rangle$ by a formula without any parameters.

The αth level of these hierarchies is the set $\mathrm{Def}\, R_\alpha$ or $\mathrm{Df}\, R_\alpha$. From the reflection principle it follows that OD (the class of all ordinal-definable sets) is the union of all sets $\mathrm{Def}\, R_\alpha$ (or $\mathrm{Df}\, R_\alpha$).

These hierarchies may seem natural ; they divide the class OD into very "thin" sets. Notice that $|\mathrm{Df}\, R_\alpha| = \omega$ and $|\mathrm{Def}\, R_\alpha| = |\alpha|$. Unfortunately they are not cumulative i.e. $\mathrm{Def}\, R_\lambda \neq \bigcup_{\beta<\lambda} \mathrm{Def}\, R_\beta$ and $\mathrm{Df}\, R_\lambda \neq \bigcup_{\beta<\lambda} \mathrm{Df}\, R_\beta$.

For such defined hierarchies we have a negative answer to the question : is there $\alpha < \omega_1^L$ such that $P(\omega)\cap(\mathrm{Def}\, R_\alpha \setminus \bigcup_{\beta<\alpha} \mathrm{Def}\, R_\beta) = \emptyset$

or $P(\omega) \cap (Df\ R_\alpha \setminus \bigcup_{\beta<\alpha} Df\ R_\beta) = \emptyset$?

In [8] Myhill and Scott formulated the "question in which R_α a particular real number is definable?" We try to give in this paper a partial answer to this. The main result is the following:

Let M be a countable, standard model for ZF + V = L. Let $\alpha > \omega$ be an ordinal in M such that α is not the successor of any limit ordinal. Then there is a model $N \supseteq M$ for ZFC $\pm$ V = HOD such that

$$N \models (\exists a \subseteq \omega)[(a \in DfR_\alpha \setminus \bigcup_{\beta<\alpha} DfR_\beta) \wedge a \notin L] \wedge (\forall b)(\forall \beta)_\alpha [b \in DfR_\beta \rightarrow b \in L]$$

The above remains true when we replace "Df" by "Def".

To obtain this result we use forcing. In particular we use Mc Aloon's method introduced in [8] for making definable a Cohen subset of ω. We apply also, changing in some aspects, Jensen and Solovay's method of almost disjoint sets.

We do not examine properties of the proposed hierarchies, neither do we discuss any axiom assuming the existence of large cardinals. Perhaps these questions are worth to studying.

The problem, whether there is a limit ordinal $\lambda > \omega$ such that $(\exists a \subseteq \omega)[a \in Df\ R_{\lambda+1} \setminus \bigcup_{\beta\leq\lambda} Df\ R_\beta \wedge a \notin L] \wedge (\forall b)(\forall \beta)_{\lambda+1}[b \in DfR_\beta \rightarrow b \in L]$, remains open.

I would like to express my gratitude to M.Srebrny, A.Zarach and M. Krynicki for their moral and scientific support.

§ 0. Preliminary notions. The aim of this paper is to study sets definable in von Neumann's hierarchy. We start with some definitions.

Definition 0.1. (von Neumann's hierarchy). Let the power-set of a be denoted by P(a), then we define the following hierarchy: $R_0 = \emptyset$, $R_\alpha = \bigcup_{\beta<\alpha} P(R_\beta)$.

Definition 0.2. $\mu_\leq x[\Phi]$ means "the least x in the sense of the ordering $\leq$ such that x satisfies Φ".

$(\cdot)^N$ means that the notion in brackets must be relativized to the class N.

Definition 0.3. rank $x = \alpha$ iff $\alpha = (\mu\beta)[x \subseteq R_\beta]$.

Let $|a|$ denote the power of the set a . Then $|R_\alpha \times \ldots \times R_\alpha| = |R_\alpha|$. Of course $\mathrm{rank}\langle x_1 \ldots x_n\rangle \geqslant \alpha$ for $x_i \in R_\alpha$. Quine n-tuple allows us to identify an n-tuple belonging to $R_\alpha \times \ldots \times R_\alpha$ with an element of R_α i.e. it is a definable 1-1 function from $R_\alpha \times \ldots \times R_\alpha$ into R_α . Quine n-tuple has the following properties:

i) $(a_0, \ldots, a_{n-1}) = (b_0, \ldots, b_{m-1}) \equiv n = m \wedge (\forall i)_n\, [a_i = b_i]$

ii) $\alpha \geqslant \omega \wedge a_0, \ldots, a_{n-1} \in R_\alpha \rightarrow (a_0, \ldots, a_{n-1}) \in R_\alpha$

(For details see Drake [3]).

Remark. Starting from this section we shall not distinguish between Quine n-tuples $(a_0 \ldots a_{n-1})$ and ordinary n-tuples $\langle a_0 \ldots a_{n-1}\rangle$.

Definition 0.4. (The hierarchies for ordinal-definable sets.)

i) $a \in \mathrm{Def}\, R_\alpha \equiv$ there is a formula Φ and a sequence $(\beta_0 \ldots \beta_n) = \bar{\beta}$ of ordinal numbers less than α such that $R_\alpha \models (\exists! x)\Phi(\bar{\beta}, x) \wedge \Phi(\bar{\beta}, a)$.

ii) $a \in \mathrm{Df}\, R_\alpha \equiv$ there is a formula Φ without parameters such that $R_\alpha \models (\exists! x)\, \Phi(x) \wedge \Phi(a)$.

$a \in \mathrm{Df}\, R_\alpha$ means that a is definable in R_α without any parameters and $a \in \mathrm{Def}\, R_\alpha$ holds if a is definable in R_α with finitely many ordinal parameters.

Following Myhill and Scott [8] we introduce classes of ordinal-definable sets and hereditary-ordinal-definable sets (OD and HOD-classes), and we list some fundamental properties of them.

Definition 0.5. i) $x \in OD \equiv (\exists\alpha)[x \in \mathrm{Df}R_\alpha]\,(\equiv (\exists\alpha)[x \in \mathrm{Def}R_\alpha])$

ii) $x \in HOD \equiv x \in OD \wedge x \subseteq HOD$.

The following are true:

1° $N \models ZF \rightarrow HOD^N \models ZFC$,

2° $L \subseteq HOD \subseteq OD$,

3° $OD = HOD \equiv V = HOD \equiv V = OD$,

4° $\mathrm{Con}\, ZF \rightarrow \mathrm{Con}\, (ZF + V = HOD = L)$,

5° $\mathrm{Con}\, ZF \rightarrow \mathrm{Con}\, (ZF + V \neq HOD = L)$,

6° $\mathrm{Con}\, ZF \rightarrow \mathrm{Con}\, (ZF + V = HOD \neq L)$,

7° $\mathrm{Con}\, ZF \rightarrow \mathrm{Con}\, (ZF + V \neq HOD \neq L)$.

§ 1. Constructible sets in von Neumann's hierarchy. Addison [1] has shown that there is an analitical formula Constr( ) such that for every real x : Constr(x) iff $x \in L$. We generalize this result and then formulate some simple theorems about constructible sets.

<u>Theorem</u> 1.1. There is a formula $C(\cdot,\cdot)$ of the language of ZF-set theory such that for any model N of ZF , for every $a \in N$ and for every $\alpha > \max(\omega, \text{rank } a)$ we have:

$$(\forall x)\ [x \in L^N_\alpha \cap R^N_\alpha \equiv R^N_\alpha \models C(x,a)] \ .$$

Remark. If $a = \emptyset$ and $\alpha = \omega + 1$ this theorem is just Addison's result.

We omit proof of the above theorem. It is based on Addison's ideas and uses transfinite induction on α .

<u>Theorem</u> 1.2. There are formulae Φ_1, Φ_2, Φ_3 such that

a) Φ_1 defines in $R_{\omega+\alpha}$ a 1-1 function from $|\alpha|^L$ onto $\omega + \alpha$.

b) Φ_2 defines in $R_{\omega+\alpha}$ a 1-1 function from $|\alpha|^L$ into $\omega + \alpha$ whose range is cofinal with $\omega + \alpha$.

c) If $\alpha < \omega^L_\alpha$, Φ_3 defines in $R_{\omega+\alpha}$ a function from a well ordering definable in $R_{\omega+\alpha}$ without parameters, whose range is cofinal with $\omega + \alpha$.

Proof. We introduce the following symbols:

$\vec{x}$ - "an initial segment determined by x" .

$W.O(\varphi)$ - "φ is a well-ordering"

$<_L$ - "the canonical well ordering of L"

$lh(\vec{y}) = \beta$ - "there is an isomorphism between $\vec{y}$ and β".

Let $\varphi \approx On$ denote "$(\forall\beta)(\exists\vec{y})[y \in \varphi \wedge lh(\vec{y}) = \beta] \wedge \wedge (\forall\vec{y})(\exists\beta)[y \in \varphi \rightarrow lh(\vec{y}) = \beta]$. If $|\alpha|^L \leq \alpha < \omega^L_\alpha = |R_{\omega+\alpha} \cap L|$, then $(\exists\varphi_0)[\varphi_0 = \mu_{<_L} \varphi[W.O(\varphi) \wedge \varphi \approx (On)^{R_{\omega+\alpha}}]]$. By the same argument $(\exists f_0)[f_0 \in R_{\omega+\alpha} \cap L \wedge f_0 = \mu_{<_L} [f : |\alpha|^L \overset{1-1}{\underset{onto}{\rightarrow}} \varphi_0]]_{R_{\omega+\alpha}}$. Let us define $G(x,\beta) \equiv$ "$x \in \varphi_0 \wedge \beta \in On \wedge lh(\vec{x}) = \beta$" . $[G(x,\beta)]^{R_{\omega+\alpha}}$ determines an isomorphism of $\omega + \alpha$ and φ_0 , Let $\Phi^*_1(\gamma,\beta) \equiv (\forall x)[f_0(\gamma) = x \rightarrow G(x,\beta)]$ and $|On|^L < On$ means "$(\exists\varphi)_L[\varphi \approx On \wedge (\exists f)(\exists x)[x \in \varphi \wedge dom\, f = \vec{x} \wedge rg\, f = \varphi]]$

Now we can define Φ_1 as follows :

$$\Phi_1(\gamma,\beta) \equiv [|On|^L < On \rightarrow \Phi_1^*(\gamma,\beta)] \wedge [\neg|On|^L < On \rightarrow \gamma = \beta] .$$

b) and c) we obtain as a) . Q.E.D.

Theorem 1.3. There is a formula Φ such that, if $cf(\alpha) = \lambda < \alpha$ and $\alpha < \omega_\alpha^L$, then Φ defines in $R_{\omega+\alpha}$ a function from λ into $\omega + \alpha$ whose range is cofinal with $\omega + \alpha$.

Proof is based on the same idea as that of theorem 1.2.

§ 2. Definability. In this section we give some facts about definability in order to make precise some intuitive ideas about it. Moreover we shall occupy ourselves with "gaps" in the hierarchies Df and Def . First we cite results of Dowson [2] who studied ordinals definable in von Neumann's hierarchy. These results will be useful later.

Definition 2.1. (Dowson [2]) $\gamma_0 = (\mu\gamma)[(\exists\alpha)[\alpha < \gamma \wedge \alpha \notin DfR_\gamma]]$.

Theorem 2.2. (Dowson [2])

a) γ_0 is a constructible cardinal i.e. $\gamma_0 \in Card^L$.

b) $ZF \vdash \omega_1^L \leq \gamma_0 \leq \omega_1$

c) $V = L \rightarrow \gamma_0 = \omega_1$

d) Con ZF $\rightarrow$ Con (ZFC $\pm$ GCH + $\gamma_0 = \omega_1$)

e) Con ZF $\rightarrow$ Con (ZFC $\pm$ GCH + $\gamma_0 = \omega_1^L < \omega_1$)

f) Con ZF $\rightarrow$ Con (ZFC $\pm$ GCH + $\omega_1^L < \gamma_0 = \omega_1$)

g) Con ZF $\rightarrow$ Con (ZFC $\pm$ GCH + $\omega_1^L < \gamma_0 < \omega_1$)

The next theorem is a consequence of simple facts about the sets Def R_α and Df R_α .

Theorem 2.3.

a) $\alpha < \gamma_0 \rightarrow$ Def R_α = Df R_α

b) Df $R_\alpha \subseteq$ Def R_α for all α .

c) $|$Df $R_\alpha| = \omega$; $|$Def $R_\alpha| = |\alpha|$ for all $\alpha \geq \omega$.

d) Def $R_\alpha \in$ Df $R_{\alpha+1}$; Def $R_\alpha \in$ Def $R_{\alpha+1}$; Df $R_\alpha \in$ Df $R_{\alpha+1}$.

e) $V = L \rightarrow$ Def $R_{\omega+\alpha} \subseteq L_{\omega_\alpha}$

f) $V = L \rightarrow P(\omega) \subseteq \text{Def } R_{\omega_1}$

g) $V = L \rightarrow P(\omega) \subseteq \bigcup_{\alpha<\omega} \text{Df}_1 R_\alpha$

Proof. a), b), c) – obvious ; d) We have Quine n-tuples, so we can formalize in $R_{\alpha+1}$ the satisfaction relation for R_α ; e) – by "condensation" lemma ; f), g) $V = L$ implies that every $a \subseteq \omega$ is a value of Gödel's constructibility function F i.e. $a = F'\eta$ for some $\eta < \omega_1$. By Addison's argument $a \in \text{Df } R_{\eta+1}$. Q.E.D.

Corollary 2.4. i) $(\forall\gamma)(\exists\alpha)(\exists\beta)[\alpha,\beta > \gamma \wedge \text{Df } R_\alpha \equiv \text{Df } R_\beta]$
ii) $(\forall\gamma)(\exists\alpha)(\exists\beta)[\alpha,\beta > \gamma \wedge \text{Def } R_\alpha \equiv \text{Def } R_\beta]$.

Theorem 2.5. (about "gaps"). If $(\forall\alpha < \gamma_0)[|\alpha|^L = \omega]$, then $(\forall\alpha)_{\gamma_0}[(\text{Df } R_\alpha \setminus \bigcup_{\beta<\alpha} \text{Df } R_\beta) \cap P(\omega) \neq \emptyset]$.

Proof. By the theorem 1.2. a) there is a definable in $R_{\omega+\alpha}$ 1-1 function f from ω onto α. Let $a_{i\beta}$ denote the real definable in R_β by a formula whose Gödel's number is i. (If such a formula does not define any real, put $a_{i\beta} = \emptyset$.) Let $m_{i\beta}$ denote the mth element of $a_{i\beta}$. We define the element a of the set $\text{Df } R_\alpha \setminus \bigcup_{\beta<\alpha} \text{Df } R_\beta$ as follows : Let a_n denote the nth element of a, we put $a_n = m_{Kn, f'Ln} + 1$ where $(K,L) = J^{-1}$ and J is a pairing function. It is easy to see that this is a good definition. Q.E.D.

Remark. We suppose the above theorem is the strongest result about the existence of gaps which can be proved in ZF. For that reason we shall study the consistency with ZFC of some sentences about gaps.

Corollary 2.6. i) $V = L \rightarrow (\forall\alpha)_{\omega_1}[(\text{Df } R_\alpha \setminus \bigcup_{\beta<\alpha} \text{Df } R_\beta) \cap P(\omega) \neq \emptyset]$
ii) For any model N of ZF such that $\omega_1^N = \omega_1^L$ we have $N \models (\forall\alpha)_{\omega_1}[(\text{Df } R_\alpha \setminus \bigcup_{\beta<\alpha} \text{Df } R_\beta) \cap P(\omega) \neq \emptyset]$.

Corollary 2.7. If N is a model for ZF and $\alpha < \omega_\alpha^L$ then $\text{Df } R_\alpha$ is not a model for ZF or ZF^-.

Proof. By the theorem 1.2 a) there is a definable function from ω onto $\alpha = (\text{On})^{R_\alpha}$. This contradicts the replacement axiom.

We have not examined whether Def R_α and Df R_α can be models for weaker theories. This problem may be interesting.

§ 3. Outline of McAloon's method. In accordance with the above heading we present McAloon's method for obtaining models in which $ZF + V = HOD \neq L$ holds. This method will be used in the next section to obtain definable generic reals. All results and definitions are taken from McAloon's paper [7].

3.1. Forcing. Let M be a countable transitive model for ZFC and $\langle S,\leq\rangle \in M$, then the notion of a S-generic ultrafilter over M we define in the usual manner.

3.2. Good subsets. If ω_α is a regular cardinal, then X is a good subset of ω_α iff $\bigcup X = \omega_\alpha$ and there is no constructible subset Y of X such that $\bigcup Y = \omega_\alpha$. Let $\mathcal{F}(\alpha)$ denote that ω_α is regular and has a good subset. Let P_α be the following notion of forcing : $P_\alpha = \{p \in M : \text{func}(p) \wedge \text{dom}(p) \subseteq \omega_\alpha \wedge |\text{dom}(p)| < \omega_\alpha \wedge \wedge \text{rg}\, p \subseteq \{0,1\}\}$, $f \leq g$ iff $f \supseteq g$.

<u>Theorem</u> 3.2.1. If A is P_α-generic over M, then $M[A] \models \mathcal{F}(\alpha) \wedge (\forall\beta)[\beta \neq \alpha \rightarrow \neg\mathcal{F}(\beta)] \wedge \text{Card}^M = \text{Card}^{M[A]}$. The set $\{\beta : \langle\beta,0\rangle \in A\}$ is a good subset of ω_α. Moreover $M[A] \models (\forall\beta)[\beta < \alpha \rightarrow P(\omega_\beta) \subseteq L]$.

Remark. If $\alpha = 0$, then $\{n: \langle n,0\rangle \in A\}$ is a Cohen real.

Let $X \in M$ be a set of indices of regular cardinals in M. Let $Q_X = (\prod_{\alpha\in X} P_\alpha)^M$ and let $F \leq G$ iff $(\forall\alpha)_X[F'\alpha \leq G'\alpha]$.

<u>Theorem</u> 3.2.2. Let A be Q_X-generic. If $Y \subseteq X$ and $Y \in M$ then $A \upharpoonright Y = \{\langle\alpha,\beta,\varepsilon\rangle \in A : \alpha \in Y\}$ is a Q_Y-generic set over M.

<u>Theorem</u> 3.2.3. If $\beta < \bigcap X$ and A is Q_X-generic, then $M[A] \models {}^b a \subseteq L$ for all $a,b \in M$ such that $|b| \leq \omega_\beta$. (${}^b a$ denotes the set of all functions from b into a)

<u>Definition</u> 3.2.4. $X^\alpha = \{\beta \in X : \beta > \alpha\}$. $X_\alpha = X \setminus X^\alpha$.

3.3. Models for $V = HOD \neq L$. Let $\xi_0 = (\mu\xi)(\xi = \omega_\xi)$, $X_0 = \{\emptyset\} \cup \{\alpha + 2 : \alpha < \xi_0\}$ and A be Q_{X_0} -generic over M. Then we have $M[A] \models (\forall\alpha)_{X_0} \mathcal{F}(\alpha)$. Moreover, by the homogenity of Q_{X_0} (cf. Levy [6]) $M[A] \models V \neq HOD = L$. Mc Aloon's method consists in defining in $M[A]$ a certain submodel $M[B]$ in which B is definable.

Let $a_0 = A \restriction X_0$ and $A' = \{\langle\alpha,\beta,\varepsilon\rangle \in A : \alpha < \omega \to \alpha \in a_0\}$. Define $A(\hat{\gamma}) = \{\langle\alpha,\beta,\varepsilon\rangle \in A : \alpha \neq \gamma\}$. Then we have:

Theorem 3.3.1. i) $k \in a_0 \equiv M[A'] \models \mathcal{F}(k)$
ii) $k \in a_0 \to A' \in M[A(\hat{k})]$
iii) $\beta = \alpha + 1 \wedge |Q_{(X_0)_\alpha}| \leq \omega_\alpha \to M[A(\hat{\beta})] \models \neg \mathcal{F}(\beta)$

Now we introduce auxiliary function F, E, $\bar{\bar{E}}$:

<u>Definition</u> 3.3.2. i) $\mathrm{dom}\, F = \{\langle\alpha,\beta\rangle : \alpha \in X_0 \wedge \beta < \omega_\alpha\}$; for every $k<\omega$ $F'\langle 0,k\rangle = k$; $F'\langle\alpha+1,\beta\rangle = \omega_\alpha + \beta + 2$ for other α.
ii) $E = F^{-1} = \langle E_1, E_2\rangle$
iii) $\bar{\bar{E}}(\alpha) = \{E(\alpha)\} \cup \bar{\bar{E}}(E_1(\alpha))$.

In order to finish this section we define B as follows:

$$B = \{\langle\alpha,\beta,0\rangle \in A : (\forall\langle\gamma,\delta\rangle)[\langle\gamma,\delta\rangle \in \bar{\bar{E}}(\alpha) \to \langle\gamma,\delta,0\rangle \in A]\}$$

<u>Theorem</u> 3.3.3. $\langle\alpha,\beta,0\rangle \in B \equiv M[B] \models \mathcal{F}(F'\langle\alpha,\beta\rangle)$, $a_0 \in M[B]$, hence $M[B] \models V = HOD \neq L$.

§ 4. Definable generic reals I. Two theorems proved in this part give certain results about definability of generic reals in von Neumann's hierarchy.

<u>Theorem</u> 4.1. Let M be a countable, standard model for $ZF + V = L$. Let $\alpha_0 \in M$ be a limit ordinal of cofinality (in M) equal to ω. Then there is a model $N \supseteq M$ for $ZFC \pm V = HOD$ such that $N \models (\exists a)[a \subseteq \omega \wedge a \notin L \wedge a \in \mathrm{Df}\, R_{\alpha_0} \setminus \bigcup_{\beta<\alpha_0} \mathrm{Df}\, R_\beta] \pm (\bar{\forall} b)(\forall\beta)[b \in \mathrm{Df}\, R_\beta \wedge \beta < \alpha_0 \to b \in L]$.

<u>Proof</u>. We use McAloon's method. Case 1^0 : $\alpha_0 < \omega^M_{\alpha_0}$. By Theorem 1.3. there is a definable in R_{α_0} function $f : \omega \nearrow \alpha_0$, (the symbol

$\nearrow$ means that the range of f is cofinal with α_0.) We may assume for convenience that $\alpha_0 < \xi_0 = (\mu\xi)\ [\xi = \omega_\xi]$. We can suppose also that $f(0) = 0$ and $f(k) > \omega$ for $k > 0$. Following McAloon we look for some submodel $M[B]$ of a generic extension of M. Let $X_0 = \{\emptyset\} \cup \{\alpha + 2 : \alpha < \xi_0\}$ and $\lambda = (\mu\xi)[\omega_\xi > \alpha_0]$. As in § 3 we define the function F. Let for $\langle\alpha,\beta\rangle$ such that $\alpha \in X_0$ and $\beta < \omega_\alpha$ $F'\langle 0,0\rangle = 0$, $F'\langle 0,k\rangle = f(k) + 2$, $F'\langle\alpha,\beta\rangle = \omega_{\lambda+\alpha} + \beta + 2$ $(\alpha > 0)$. If $Q_{X_0} = (\prod_{\alpha\in X_0} \mathbf{P}_\alpha)^M$ and A is a Q_{X_0}-generic set over M, then we define B as in § 3. Of course, we have $M[B] \models \mathfrak{F}(F'\langle\alpha,\beta\rangle)$ iff $\langle\alpha,\beta,\ \rangle \in B$. Let $a = \{k: \langle 0,k,0\rangle \in B\}$. We must show that $a \in \mathrm{Df}\, R_{\alpha_0} \setminus \bigcup_{\beta<\alpha_0} \mathrm{Df}\, R_\beta$.

Lemma 1. $a \notin \mathrm{Df}\, R_\beta$ for $\beta < \alpha_0$.

Proof of Lemma 1. Let us suppose conversely. Let $a \in \mathrm{Df}\, R_\beta$ for some $\beta < \alpha_0$. There is k_0 such that $\beta < f(k_0)$. If $Y = \{\emptyset\} \cup \{f(k)+ 2 : k \leq k_0\}$ then $A \restriction Y$ is Q_Y-generic over M. It is easy to prove that $R_\beta^{M[A\restriction Y]} = R_\beta^{M[B]}$. But Q_Y is a homogeneous notion of forcing. Hence $M[A\restriction Y] \models V \neq HOD = L$ (cf. Levy [6] and Jech [4]). This means $a \notin \mathrm{Df}\, R_\beta^{M[A\restriction Y]} \subseteq OD^{M[A\restriction Y]}$. This contradiction completes the proof of Lemma 1. By the same argument $(\forall b)[b \in \mathrm{Df}\, R_\beta^{M[B]} \wedge \beta < \alpha_0 \rightarrow b \in L]$.

Lemma 2. $a \in \mathrm{Df}\, R_{\alpha_0}$.

Proof of Lemma 2. $M[B] \models k \in a \equiv \mathfrak{F}(F'\langle 0,k\rangle)$. There is a constructible function $f : \omega_\beta \xrightarrow[\text{onto}]{1-1} R^L_{\omega+\beta}$. If there is a good subset $\mathfrak{X}$ of ω_β, then there is $Y = f''\mathfrak{X} \subseteq R^L_{\omega+\beta}$ such that $|Y| = |R^L_{\omega+\beta}|$ and $(\forall y \subseteq Y)\,[|y| < |Y| \rightarrow y \in L]$. So we have $M[B] \models k \in a \equiv$
$\equiv (\exists Y)[Y \subseteq R^L_{F'\langle 0,k\rangle} \wedge |Y| = |R^L_{F'\langle 0,k\rangle}| \wedge Y \notin L \wedge (\forall y \subseteq Y)[|y|<|Y| \rightarrow y \in L]]$.
But the right hand of this equivalence can be expressed in $R^{M[B]}_{\alpha_0}$, since all notions used there are expressible in $R^{M[B]}_{\alpha_0}$ (cf.Thm.1.1.). This completes the proof of the theorem for the 1st case.

Case 2^0. $\alpha_0 = \omega^M_{\alpha_0}$. In this case a constructible function $f: \omega \nearrow \alpha_0$, definable in R_{α_0} can be found in a very technical manner. In order to simplify the proof we adopt a different method. First we extend M to $M[a]$ as follows: Let $\mathfrak{f} \in M$ and $\mathfrak{f}: \omega \nearrow \alpha_0$.

We put $P = \prod_{k\in\omega} P_{\underset{\sim}{f}(k)+1}$, where $P_{\underset{\sim}{f}(k)+1}$ is the notion of forcing introduced in § 3 to add a good subset to $\omega_{\underset{\sim}{f}(k)+1}$. If G is P-generic, then we have $M[G] \models \alpha = \underset{\sim}{f}(k) \equiv \mathcal{F}(\underset{\sim}{f}(k) + 1)$. By the same argument as in the 1st case, we can show that the right hand of the equivalence can be expressed in $R^{M[G]}_{\alpha_0}$. So we have a definable in $R^{M[G]}_{\alpha_0}$ function $\underset{\sim}{f}: \omega \nearrow \alpha_0$. If we put $f(0) = 0$, $f(k) = \underset{\sim}{f}(k) + 2$ and use the method introduced above, we obtain our model. So the proof of the theorem is complete.

In the model $M[B]$ of the 1st case we have $V = HOD$, in the 2nd case $V \neq HOD$, but this model can be extended in the usual way to a model in which $V = HOD$ holds.

It is easy to obtain "$\pm (\forall b)[b \in Df\, R_\beta \wedge \beta < \alpha_0 \rightarrow b \in L]$" . Namely in the models obtained in 1st and 2nd case this sectence holds. But, for example, if we start from Jensen and Solvay's [5] model, we can repeat the above procedure and obtain a model in which $\neg (\forall b)[b \in Df\, R_\beta \wedge \beta < \alpha_0 \rightarrow b \in L]$ holds.

<u>Corollary</u> 4.2. In the McAloon's [7] model the Cohen real is definable in $R^{M[B]}_{\omega+\omega}$.

<u>Proof</u>. McAloon uses the function $f: \omega \nearrow \omega + \omega$, $f(n) = \omega + n$.

Now, we shall present a rather technical proof of the following:

<u>Theorem</u> 4.3. Let M be a countable, standard model for $ZF + V = L$, $\alpha_0 = \beta_0 + 2$ for some $\beta_0 \in On^M$. Then there is a model $N \supseteq M$ for $ZFC \pm V = HOD$ such that

$$N \models (\exists a)[a \subseteq \omega \wedge a \notin L \wedge a \in DfR_{\alpha_0} \setminus \bigcup_{\beta<\alpha_0} DfR_\beta] \pm (\forall b)[b \in DfR_\beta \wedge \beta < \alpha_0 \rightarrow b \in L$$

<u>Proof</u>. Let $\alpha_0 = \beta_0 + 2$ and a be a Cohen real. We denote $M[a]$ by N_1 . Thus we have $N_1 \models HOD = L \neq V + Card^{N_1} = Card^M$. We define the following notion of forcing: $P_1 = \{f: f \in L \wedge dom\, f \subseteq \omega_{\beta_0+\omega} \times R^M_{\beta_0} \wedge |dom\, f| < \omega_{\beta_0} \wedge rg\, f \subseteq \{0,1\}\}$, $f \leqslant g$ iff $f \supseteq g$.

Let $N_2 = N_1[G_1]$, where G_1 is a P_1-generic set over N_1 . It is easy to see that $N_2 \models HOD = L \neq V + |R_{\beta_0+1}| \geqslant \omega_{\beta+\omega} + Card^{N_2} = Card^M$.

Remark. We supposed above that $|R^M_{\beta_0}| = \omega^M_{\beta_0}$ i.e. $\beta_0 \geqslant \omega \cdot \omega$. When $\beta_0 < \omega \cdot \omega$ the definition of P_1 must be replaced by $P_1 = \{f\colon f \in L \wedge \operatorname{dom} f \subseteq \omega_{\gamma+\omega} \times R^M_{\beta_0} \wedge \operatorname{rg} f \subseteq \{0,1\} \wedge |\operatorname{dom} f| < \omega_\gamma \wedge \omega_\gamma = |R^M_{\beta_0}|\}$. In the remaining part of the proof is also necessary to make certain minor changes. Of course the idea of the proof remains unchanged.

We proceed our proof. We will code the property $i \in a$ by $|\omega^M_{\beta_0+i+1}| = |\omega^M_{\beta_0+i+2}|$ in some extension N_3 of N_2. This equality will be expressed in $R^{N_3}_{\beta_0+2} = R^{N_3}_{\alpha_0}$. Hence a will be definable in $R^{N_3}_{\alpha_0}$. We shall show that a cannot be definable below.

Let $P^*_n = \{f\colon \operatorname{dom} f \subseteq \omega_{\beta_0+n+1} \wedge |\operatorname{dom} f| < \omega_{\beta_0+n+1} \wedge \operatorname{rg} f \subseteq \omega_{\beta_0+n+2}\}$ and $P = \prod_{n\in a} P_n$. For $F,G \in P$ let $F \leqslant G$ iff $(\forall n)[F'n \supseteq G'n]$. Then $\langle P,\leqslant\rangle \in N_2$. Let G be a P-generic ultrafilter. In the model $N_3 = N_2[G] = M[a][G_1][G]$ we have $i \in a$ iff $|\omega^M_{\beta_0+i+1}| = |\omega^M_{\beta_0+i+2}|$. In other words, some constructible alephs are collapsed. We shall prove that this can be expressed in R_{α_0}.

Since $|R^{N_3}_{\beta_0+1}| \geqslant \omega^{N_3}_{\beta_0+\omega}$, there are sets $y_i \subseteq R^{N_2}_{\beta_0+1}$ such that $|y_i| = \omega^M_{\beta_0+i+1}$. Because $\omega^{N_3}_{\beta_0+\omega} = \omega^{N_2}_{\beta_0+\omega}$ and $|R^{N_2}_{\beta_0+1}| \geqslant \omega^{N_2}_{\beta_0+\omega}$ we can suppose that $y_i \in N_2$. Hence $N_3 \models i \in a \equiv |y_i| = |y_{i+1}|$. We shall express this fact in $R^{N_3}_{\alpha_0}$. First we introduce some new symbols. Let $[|y_i| = |y_j|]^{L[X]}$ mean $(\exists f \in L[X])[f\colon y_i \xrightarrow[\text{onto}]{1-1} y_j]$. Let X_i denote the ith element of the sequence X. We identify the finite sequence $X = \langle X_0, X_1, \ldots, X_{n-1}\rangle$ with the Quine n-tuple $(X_0,\ldots,X_{n-1})$. Notice that, if $X = (y_0, y_1, \ldots, y_n, y_{n+1})$ belongs to N_2, then for all $i,j \leqslant n+1$ $[|y_i| \neq |y_j|]^{L[X]}$ (for $i \neq j$). Notice also that rank X = rank $y_i = \beta_0 + 1$, and if $f\colon y_i \to y_j$ then rank f = rank y_i = rank $y_j = \beta_0 + 1$ (i.e. $f \in R^{N_3}_{\beta_0+2}$). (Of course, we identity f with its code i.e. the set of Quine pairs of its elements.) Let us consider the following formula:

$$\begin{aligned} n \in a \equiv {} & (\exists X)[\text{"X is an } n+3\text{-tuple"} \wedge |X_0| = |R_{\beta_0}| \wedge (\forall i)_{\leqslant n+2}[|X_{i+1}| > \\ & > |X_i|]^{L[X]} \wedge \neg(\exists Y)[\text{"Y is an } n+3\text{-tuple"} \wedge |Y_0| = |R_{\beta_0}| \wedge \\ & (\forall i)_{\leqslant n+2}[|Y_{i+1}| > |Y_i|]^{L[Y]} \wedge (\forall i)_{\leqslant n+2}[|Y_i| = |X_i|]^{L[X,Y]} \wedge \\ & (\exists \underset{\sim}{Y})[\underset{\sim}{Y} \in L[Y] \wedge (\exists i)_{\leqslant n+1}[|Y_i| < |\underset{\sim}{Y}| < |Y_{i+1}|]^{L[Y]}]] \wedge |X_{n+2}| = |X_{n+1}|] \end{aligned}$$

Let us analyse this formula. The non-existence of Y in the above asserts that X is a collection of elements X_i, $i \leq n+2$ such that $L[X] \models |X_i| = \omega^M_{\beta_0+i}$. To prove this suppose conversely. Then there is $i_0 > 0$ such that $L[X] \models |X_i| > \omega^M_{\beta_0+i}$ i.e. $L[X] \models |X_{i_0}| \geq \omega_{\beta_0+i_0+k_0}$ for some $k_0 > 0$. We take the smallest such i_0. We shall construct Y and indicate $\underset{\sim}{Y}$. Let $Y = (Y_0,\dots,Y_{n+2})$ such that $Y_i \in R^{N_2}_{\beta_0+2}$ and $N_2 \models |Y_i| = \omega^M_{\delta_i}$ where $\omega^M_{\delta_i}$ is the power of X_i in $L[X]$. Because $\mathrm{Card}^M = \mathrm{Card}^{N_2}$, we have $L[Y] \models |Y_i| = \omega^M_{\delta_i}$ i.e. $L[X,Y] \models |Y_i| = |X_i|$. Moreover $L[Y] \models |Y_{i_0}| \geq \omega^M_{\beta_0+i_0+k_0}$. Let $f \in L[Y]$ be such that $f\colon \omega^M_{\beta_0+i_0+k_0} \xrightarrow[\text{onto}]{1-1} Y_{i_0}$. We put $\underset{\sim}{Y} = f''\omega^M_{\beta_0+i_0}$. Then $L[Y] \models |Y_{i_0-1}| < |\underset{\sim}{Y}| < |Y_{i_0}|$. This contradiction completes the proof of the above assertion.

It is easy to show that the formula holds in the model N_3. Now, we shall prove that $a \in \mathrm{Df}\, R_{\alpha_0}$. It suffices to see that the right hand of the equivalence can be expressed in $R^{N_3}_{\alpha_0}$. By the above arguments $X, Y, \underset{\sim}{Y}$ belong to $R^{N_3}_{\alpha_0}$. By Theorem 1 the notion "$f \in L[X]$" (for $X \in R^{N_3}_{\alpha_0}$) is expressible in $R^{N_3}_{\alpha_0}$.

To complete the proof we show that $a \notin \mathrm{Df}\, R_\beta$ for $\beta < \alpha_0$. Since $P = \prod_{n\in a} P^*_n$ satisfies $|R_{\beta_0}| = \omega_{\beta_0}$ d.c.c. (cf. Jech [4]) we have $R^{N_2}_{\beta_0+1} = R^{N_2[G]}_{\beta_0+1}$. But $\bigcup_{\beta\leq\beta_0+1} \mathrm{Df}\, R_\beta \cap P(\omega) \subseteq L$, because $N_2 \not\models \mathrm{HOD} = L$. Of course a cannot belong to any $\mathrm{Df}\, R^{N_2}_\beta = \mathrm{Df}\, R^{N_3}_\beta$ for $\beta < \beta_0 + 2 = \alpha_0$. Q.E.D.

Remark 4.4. In the above model $V \neq \mathrm{HOD}$. Of course, it can be extended to a model in which $V = \mathrm{HOD}$ holds together with the properties of reals expressed by the theorem.

The consequences of the above theorems for gaps will be drawn in the final section of this paper.

§ 5. Alomost disjoint sets. (see: Jensen and Solovay [5])

Definition 5.1. For any regular cardinal $\varkappa$ let ${}^{\varkappa}\varkappa$ denote the set of all functions from $\varkappa$ into $\varkappa$. If $f,g \in {}^{\varkappa}\varkappa$, then we say f and g are almost disjoint if $\{\alpha: f(\alpha) = g(\alpha)\} < \varkappa$.

Remark. The above condition is equivalent to $|\{\alpha: f(\alpha) = g(\alpha)\}| < \varkappa$ or to $|f \cap g| < \varkappa$.

Lemma 5.2. If $|{}^{\varkappa}\varkappa \cap L| = 2^{\varkappa}$, then there is a family of almost disjoint sets of power $2^{\varkappa}$.

Proof. Let $\langle s_\xi : \xi < \varkappa\rangle$ be a fixed, constructible enumeration of all constructible functions f such that $f: \eta \to \varkappa$ for some $\eta < \varkappa$. Let $S(f) = \{\xi: s_\xi$ is an initial segment of $f\}$. Then $f \neq g$ implies $|S(f) \cap S(g)| < \varkappa$. The set $\{S(f): f \in {}^{\varkappa}\varkappa\}$ has power $2^{\varkappa}$.

Definition 5.3. Let $Y \subseteq P(\varkappa)$. We define a notion of forcing $\langle P_Y, \leqslant\rangle$ to be a set satisfying the following conditions:

1^0 $p \in P_Y \to [p = \langle s,t\rangle \wedge s \subseteq \varkappa \wedge |s| < \varkappa \wedge t \subseteq Y \wedge |t| < \varkappa]$

2^0 $\langle s,t\rangle \leqslant \langle s',t'\rangle$ iff $s \supseteq s' \wedge t \supseteq t' \wedge (\ A)_{t'}[s \quad A = s' \quad A]$.

§ 6. Definable generic reals II. We are going to prove the following fact: for every limit ordinal α there is a model in which there is a real which is definable exactly on the level R_α of the rank hierarchy. First we shall do this for the case $\alpha = \omega_1$ and then generalize this method to prove the remaining cases. We proceed in this way in order not to introduce immediately the technical complications which are necessary in the general case. It seems that the proof is easier to understand in this way. Once one has understood the proof for the case $\alpha = \omega_1$, it is clear how to generalize it, and for this reason we confine ourselves to a sketch of the method of generalization. We shall use the method of almost disjoint sets.

We order $\omega_1 \times \omega_0$ lexicographically ; it is then order isomorphic to ω_1 . Let (α_ξ, n_ξ) be the ξth member of $\omega_1 \times \omega_0$. We shall write $\alpha_\xi, n_\xi < \beta$ instead of $\xi < \beta$, and $\alpha,n < \gamma,j$ instead of $(\alpha,n) < (\gamma,j)$.

Definition 6.1. (cf. Jensen and Solovay [5]) . We introduce a predicate $T_j(x)$: $T_j(x) \equiv (\forall \xi)_{\omega_1} (\forall i)_{\omega} [|x \cap S(a_{\xi,i})| < \omega \equiv i = j]$.

We now define the sequence $\langle a_{\xi,i} : \xi, i < \omega_1 \rangle$ of generic reals which appeared above. Let M be a countable, transitive model of $ZF + V = L$. Let $\{\lambda_\xi\}_{\xi < \omega_1}$ be the sequence of all successive limit ordinals less than ω_1 . Let $f_{\xi+1} : \omega \nearrow \lambda_{\xi+1}$ be the first (in the sense of $<_L$) function such that $\mathrm{rg}\, f_{\xi+1} \subseteq \lambda_{\xi+1} \setminus (\lambda_\xi + 1)$. Let P_1^* be the product of ω_1 Cohen notions of forcing i.e. $P_1^* = \prod_{\xi<\omega_1} P_\xi$ where $P_\xi = \{f : \mathrm{dom}\, f \subsetneq \omega \wedge |f| < \omega \wedge \mathrm{rg}\, f \subseteq \{0,1\}\}$. If G_1 is P_1^*-generic then in the model $N_1 = M[G_1]$ there are ω_1 non-constructible reals a_ξ , $\xi < \omega_1$ such that $M[a_\xi] \cap M[a_{\xi'}] = M$ for $\xi \neq \xi'$.

Of course, the sequence $\langle a_\xi, \xi < \omega_1 \rangle$ is not definable in N_1 . Now, using ω_1-times the method introduced in the proof of Theorem 4.1. we shall make this sequence definable in R_{ω_1} of a generic extension of N_1 .

Let $P_2^* = \prod_{\xi<\omega_1} P'_\xi$, where $P'_\xi = \prod_{n \in a_\xi} P_{f_{\xi+1}(n)}$. ($P_{f_{\xi+1}(n)}$ is the notion of forcing introduced in § 3 in order to add a good subset of $\omega_{f_{\xi+1}(n)}$). Hence, if G_2 is P_2^*-generic over N_1 , then $N_1[G_2] \models n \in a_\xi \equiv$ "$\omega_{f_{\xi+1}(n)}$ has a good subset" . Put $N_2 = N_1[G_2]$. As in § 4 we can prove that this equivalence can be expressed in $R_{\omega_1}^{N_2}$. Let a be a Cohen real, generic over N_2 . Then $a \notin \mathrm{HOD}^{N_3}$ (where $N_3 = N_2[a]$). This a we will make definable in R_{ω_1} .

Using the canonical bijection between $\omega_1 \times \omega_0$, we rename $\langle a_\xi : \xi < \omega_1 \rangle$ as $\langle a_{\xi,i} : \xi, i < \omega_1 \rangle$. We are going to obtain the model in which $j \in a \equiv \exists x\, T_j(x)$.

Let us consider the following notions of forcing Q_i , $i < \omega$: $\langle Q_i, \leqslant \rangle = \langle P_{Y_i}, \leqslant \rangle$ where $Y_i = \{S(a_{\xi,i}) : \xi < \omega_1\}$ (cf. Definition5.3) Put $Q = \prod_{i \in a} Q_i$, and take a Q-generic G_3 over N_3 . As in Jensen and Solovay's paper [5] one can show that in the model $N_4 = N_3[G_3]$ $j \in a \equiv (\exists x)\, T_j(x)$ holds. We think the proof of this fact may be omitted. As in previous sections one can prove that the above equivalence can be expressed in $R_{\omega_1}^{N_4}$. It remains to prove that $a \notin \mathrm{Df}\, R_\beta^{N_4}$ for all $\beta < \omega_1$.

6.2. $a \notin \mathrm{Df}\, R_\beta^{N_4}$. For suppose not, let $B = RO(P)$ be a c.b.a. determined by a notion of forcing P. Let Aut B denote the set of all automorphism of B and $B^* = \{b \in B : (\forall\sigma)[\sigma \in \mathrm{Aut}\, B \rightarrow \sigma b = b]\}$. So B^* is a "rigid" subalgebra of B. (For example, if P is homogeneous then $B^* = \{0,1\}$.) The following lemma is a generalization of Levy's result [6] about the invariance of the class HOD under extension obtained by means of a homogeneous notion of forcing.

Lemma (Vopênka [11]). Let M be a model for $ZF + V = L$. Then $(\mathrm{HOD})^{M[G]} = M[G \cap B^*]$ for any B-generic G.

We shall use this lemma. Recall, we have obtaned a model $N_4 = M[G_1][G_2][a][G_3]$ by a fourfold generic extension. We used iterated notions of forcing. M was a model for $V = L$. In order to use the lemma of Vopenka we have to replace this fourfold iteration by one notion of forcing. The aim of this is to construct a model N_4' such that $R_\beta^{N_4'} = R_\beta^{N_4}$ and $a \notin \mathrm{HOD}^{N_4'}$. This contradiction will finish the proof that $a \in \mathrm{Df}\, R_{\omega_1}^{N_4} \setminus \bigcup_{\beta<\omega_1} \mathrm{Df}\, R_\beta^{N_4}$.

Let us look at this iteration. Note that P_2^* does not depend upon P_1^*, the Cohen notion used to add a does not depend upon the proceeding notions and Q depends upon P_1^* and the Cohen notion. Let us consider the following notion of forcing P:

$$f \in P \equiv f = \langle r;\ \langle s_1,t_1\rangle^{\varepsilon_1}, \langle s_2,t_2\rangle^{\varepsilon_2}, \ldots, \langle s_n,t_n\rangle^{\varepsilon_n}, \ldots : P^*_{\xi,i}\ \xi, i < \omega_1\rangle ,$$

where

1^0 r is a finite function from a subset of ω into $\{0,1\}$ (i.e. r "forces" the set a),

$$\varepsilon_i = \begin{cases} 0 & i \notin \mathrm{dom}\, r \vee r(i) = 0 , \\ 1 & r(i) = 1 \end{cases}$$

$\langle s_i,t_i\rangle^0 = \emptyset$, $\langle s_i,t_i\rangle^1 = \langle s_i,t_i\rangle$. ($\varepsilon_i$ depends on r; we introduce it in order that if $n \notin a$, then Q_n does not occur in the product Q.)

2^0 $P^*_{\xi,i}$ represents the iteration of P_1^* and P_2^* i.e.

$$P^*_{\xi,i} = \langle j_{\xi,i};\ P^0_{\xi,i}, P^1_{\xi,i}, \ldots, P^n_{\xi,i}, \ldots\rangle = \langle j_{\xi,i} ;\ P_{\xi,i}\rangle , \text{ where}$$

$j_{\xi,i}$ is a finite function from a subset of ω into $\{0,1\}$ (i.e. $j_{\xi,i}$ "forces" the set $a_{\xi,i}$), and

$$p^n_{\xi,i} = \begin{cases} \emptyset & n \notin \operatorname{dom} j_{\xi,i} \vee j_{\xi,i}(n) = 0 \ . \\ p^n_{\xi,i} & j_{\xi,i}(n) = 1 \end{cases}$$

We insist that $P^n_{\xi,i} \in P_{f(\xi,i)+1}(n)$ (i.e $p_{\xi,i}$ "forces" $n \in a_{\xi,i}$ iff $\omega_{f(\xi,i)+1}(n)$ has a good subset).

3^0 $\langle s_i, t_i \rangle$ is "to belong" to Q_i ; hence let s_i be a finite function from a subset of ω into $\{0,1\}$. Because in the condition the sets $a_{\xi,i}$ have not yet been defined , t_i cannot ba a collection of sets $S(a_{\xi,i})$. However t_i may contain finite subsets of $S(a_{\xi,i})$. Hence $t_i = \{j'_{\xi_1,i}, \dots, j'_{\xi_k,i}\}$, where $j'_{\xi_l,i}$ is to be a finite subset of $S(a_{\xi,i})$, which need not be defined. Let $j'_{\xi,i} \subseteq \subseteq \{j : s_j$ is an initial segment of $j_{\xi,i}\}$. We insist also that $\operatorname{dom} s_i \subseteq \operatorname{dom} j_{\xi_1,i} \cap \dots \cap \operatorname{dom} j_{\xi_k,i}$. A partial ordering $\leq$ we define as follows : $\langle s_i, t_i \rangle \leq \langle s'_i, t'_i \rangle \equiv 1^0$ $s_i \supseteq s'_i$ and $t_i \supseteq t'_i$ 2^0 $s^*_i \cap A = s'^*_i \cap A$ for $A \in t'_i$, where $s^*_i = \{n : s_i(n) = 1\}$.

We have to show that for some G , P-generic over M , we obtain N_4 . It is obvious that there is G , P-generic over M , such that $\bigcup pr_1 G = a$, $\bigcup pr_1 \circ pr_{\omega+(\xi,i)} G = a_{\xi,i}$, and $\bigcup pr_2 \circ pr_{\omega+(\xi,i)} G = \bigcup pr_{\xi,i} G_2$. (This last condition means that G adds the same good subsets as G_2 to the same cardinals.) The function ε_i guarantees that the nth condition $\langle s_n, t_n \rangle$ appears in any condition from P only if $n \in a$. Let $n \in a$, then $pr_1 \circ pr_{1+n} G$ is a real x_n which has a finite intersection with each set $S(a_{\xi,n})$, $\xi < \omega_1$. (We identify x_n and x^*_n) . Indeed, there is no condition forcing the contrary . Because for any $\langle s_n, t_n \rangle$ there is no stronger condition forcing $|x_n \cap S(a_{\xi,k})| < \omega$ $(k \neq n)$, so $\langle s_n, t_n \rangle$ forces $|x_n \cap S(a_{\xi,k})| = \omega$. Hence there is such a P-generic G for which $M[G] = N_4$.

We shall build the model N'_4 . Suppose that there is a $\beta < \omega_1$ such that $a \in \operatorname{Df} R^{N_4}_\beta$. Establish this β . Note, that for $\xi, i > \beta$ $a_{\xi,i}$ has no definition in $R^{N_4}_\beta$ because $n \in a_{\xi,i}$ iff $\omega_{f(\xi,i)+1}(n)$ has a good subset, and $f_{(\xi,i)+1}(n) > \beta$. (Hence the above equivalence cannot be expressed in $R^{N_4}_\beta$.) In the model N'_4 there will not be any definition of $a_{\xi,i}$ for $\xi, i > \beta$. Because a was definable as a result of the definability of the sequence $\{a_{\xi,i}\}$, a will not be definable in N'_4 .

Consider the following notion of forcing P'. Let $f \in P' \equiv f = \langle r; \langle s_1 t_1\rangle^{\varepsilon_1}, \langle s_2 t_2\rangle^{\varepsilon_2}, \ldots; p^*_{\xi,i} \quad \xi, i < \omega_1\rangle$ and P' satisfy the conditions 1^0, 3^0 and $2^{0'}$: $p^*_{\xi,i} = \langle j_{\xi,i}, \emptyset\rangle$ for $\xi, i > \beta$ (for $\xi, i \leq \beta$ as in 2^0). This means that in $p^*_{\xi,i}$ there is no condition forcing a definition of $a_{\xi,i}$ for $\xi, i > \beta$. $P' \in M$ and $P' \subseteq P$, hence if G is P-generic that $G' = P' \cap G$ is P'-generic (over M). Take $N_4' = M[G']$. The proof of the following fact is easily checked.

Fact: If $Y_1 = \{S(a_\alpha): \alpha < \omega_1\} \cup \{S(b_i): i < \omega\}$ and $Y_2 = \{S(a_\alpha): \alpha < \omega_1\}$ then $RO(P_{Y_1})$ and $RO(P_{Y_2})$ are isomorphic.

Now, we shall prove that for every $f \in P'$ there is an automorphism $\sigma: B' \to B'$ ($B' = RO(P')$) such that $\sigma f \neq f$. By the lemma of Vopěnka this means $a \notin HOD^{N_4'}$. Let f be as above. We define r' as follows:

$$r' = \begin{cases} \{(n_1,0),(n_2,1)\} & \text{for some } (n_1,0)(n_2,1) \in r, \\ \{(n_1,0),(n_2,1)\} & \text{for some } (n_1,0) \in r \text{ and } n_2 \notin \operatorname{dom} r \\ & \text{(if there is no } n_2 \text{ such that } (n_2,1) \in r) \\ \{(n_1,0),(n_2,1)\} & \text{for some } (n_2,1) \in r \text{ and } n_1 \notin \operatorname{dom} r \\ & \text{(if there is no } n_1 \text{ such that } (n_1,0) \in r) \end{cases}$$

Then r' is compatible with r. Let $\bar{r}' = \{(n_1, 1-\delta_1), (n_2, 1-\delta_2)\}$ for $(n_1,\delta_1),(n_2,\delta_2) \in r'$. So $\bar{r}'$ is not compatible with r. Now there is an automorphism σ_0 of the Cohen notion of forcing such that $\sigma_0(r') = \bar{r}'$. Then $\sigma_0(r) \neq r$.

Let $Q_i' = \{\langle s_i, t_i\rangle \in Q_i : t_i$ does not contain any $j_{\xi,i}$ for $\xi, i \leq \beta\}$. Of course $Q_i' = Q_j'$ for all i,j. (Q_i saticfies condition 3^0). By the above Fact we have $RO(Q_i) = RO(Q_j)$ for every i,j. Notice also that $\prod_{\xi>\beta} P_{\xi,i}$ and $\prod_{\xi>\beta} P_{\xi,j}$ are identical and every isomorphism $\sigma_{n_1 n_2}$ of $RO(Q_{n_1})$ and $RO(Q_{n_2})$ determines an isomorphism of $\prod_{\xi>\beta} P_{\xi,n_1}$ and $\prod_{\xi>\beta} P_{\xi,n_2}$ which we denote also by $\sigma_{n_1 n_2}$. Let $C(n_1,n_2)$ be a permutation such that each set of coordinate n_1 (resp. n_2) is mapped to a set of coordinate n_2 (resp. n_1) and such that each set of coordinates ξ, n_1 (resp. ξ, n_2) is mapped to a set of coordinates ξ, n_2 (resp. ξ, n_1) for each $\xi > \beta$. Now we define $\sigma : B' \to B'$ as follows:

$\sigma = \langle \sigma_0; id, \ldots id, \sigma_{n_2 n_1}, id, \ldots id, \sigma_{n_1 n_2}, id, \ldots ; \sigma_{n_1 n_2}, \sigma_{n_2 n_1}, id, \ldots \rangle \circ C(n_1, n_2$

The idea behind is illustrated by the following diagram:

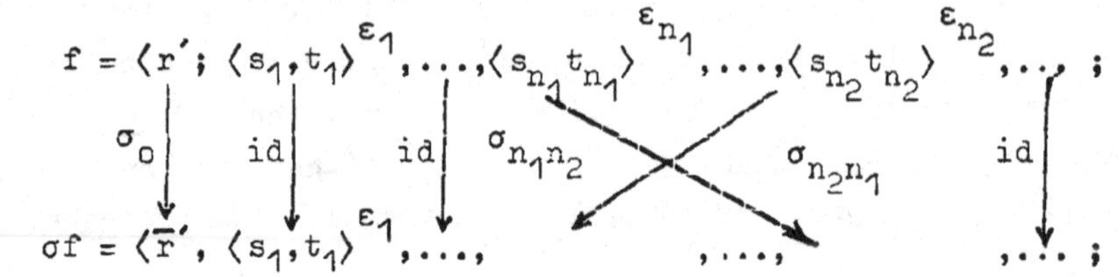

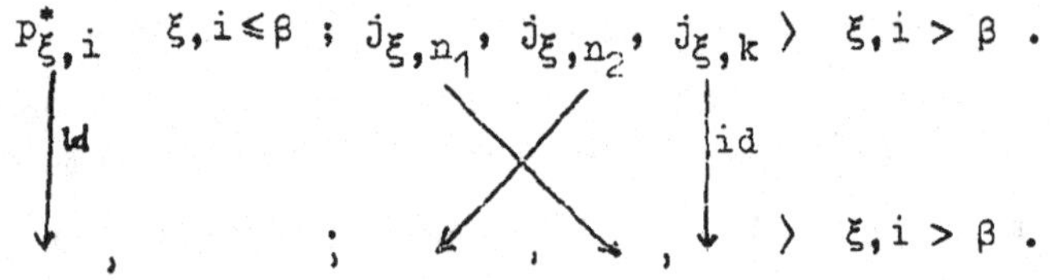

It is easy to show that such a σ (determined by σ_0) is an automorphism of B'. In order to finish the proof it suffices to observe that for every $f \in B'$ of the form *) $f = \langle u_0; u_1^{\varepsilon_1}, \ldots ;$ $u_{\xi,i}$ $\xi, i \leq \beta$; $u_{\xi,i}$ $\xi, i > \beta \rangle$ if $u_0 \neq 0,1$, then there is $\sigma : B' \to B'$ such that $\sigma_0(u_0) \neq u_0$, and hence $\sigma f \neq f$, what is an obvious consequence of the above facts.

So we have proved that any f of the form*) does not belong to B'^{*}, which establishes that $a \notin HOD^{N_4'}$. It remains to show that $R_\beta^{N_4} = R_\beta^{N_4'}$. Note that $N_4' = M[G_1][G_2 \restriction \beta + 1][a][G_3] =$ $= M[G_1][a][G_3][G_2 \restriction \beta + 1]$. Since $P_2^* \restriction \omega_1 \setminus (\beta + 1)$ satisfies $|R_\beta|$-d.c.c. (cf. Jech [4]), the equality $R_\beta^{N_4} = R_\beta^{N_4'}$ holds.

The model N_4 can be extended to a model N for **ZF** + V = HOD. Hence, we have proved the following:

<u>Theorem</u> 6.3. Every countable, transitive model for ZF + V = L can be extended to a model N for ZFC $\pm$ V = HOD such that

$$N \models (\exists a)[a \subseteq \omega \wedge a \in Df\, R_{\omega_1} \setminus \bigcup_{\beta < \omega_1} Df\, R_\beta] .$$

In this model there are ω_1 definable (below ω_1) generic reals.

Now, it is easy to generalize the above result.

Theorem 6.4. Let M be a countable, transitive model for $ZF + V = L$, let $cf^M(\alpha_0) = \varkappa$, where $\varkappa$ is an uncountable cardinal in M. Then there is a model $N \supseteq M$ for $ZFC \pm V = HOD$ such that $N \models (\exists a)[a \subseteq \omega \wedge a \in DfR_{\alpha_0} \setminus \bigcup_{\beta<\alpha_0} DfR_\beta] \pm (\forall b)[b \in DfR_\beta \wedge \beta < \alpha_0 \rightarrow b \in L]$.

Proof. Let $T_j(x) \equiv (\forall \xi)_\varkappa (\forall n)_\omega [|x \cap A_{\xi,n}| < \varkappa \equiv n = j]$. Add a good subset $A_{\xi,n}$ for every $\xi, n < \varkappa$ just using a product of $\varkappa$ copies of the notion of forcing introduced in § 3. Then extend the model obtained by adding a Cohen real a. Let us denote by N_1 the new model. Let Q_j be the following set of conditions : $\langle s,t \rangle \in Q_j$ iff $s \subseteq \varkappa$, $t \subseteq \{A_{\xi,j} : \xi < \varkappa\}$, $|s| < \varkappa$ and $|t| < \varkappa$. (cf. 5.3 and § 6.) We can suppose that $A_{\xi,n}$ for $\xi, n < \varkappa$ are all almost disjoint. We can now extend N_1 to $N_2 = N_1[G_2]$ where G_2 is $\prod_{j \in a} Q_j$-generic over N_1. In this model, by the argument of Jensen and Solovay [5], we have that $j \in a \equiv \exists x\, T_j(x)$. Let $f: \varkappa \nearrow \alpha_0$. It remains to add to N_2 definitions of all $A_{\xi,n}$ simply by extending it to a model N_3 in which $\eta \in A_{\xi,n}$ iff "$R^L_{f(\xi,\eta,n)+1}$ has a good subset". (We use here the fact that $\varkappa \times \varkappa \times \omega \approx \varkappa$ and the method introduced in § 4.) Since $f(\xi,\eta,n)+1 < \alpha_0$ and the notion "be constructible" is expressible in $R^{N_4}_{\alpha_0}$, the above equivalence holds in $R^{N_4}_\beta$. As in the proof of Theorem 6.4. we show that a cannot be definable in $R^{N_4}_\beta$ for all $\beta < \alpha_0$. By the same argument $(\forall b)[b \in Df\, R^{N_4}_\beta \wedge \beta < \alpha_0 \rightarrow b \in L]$. The other case of the negation of this sentence was obtained in the proof of Theorem 6.3 We do not need to make any changes. In $N_4 \models V \neq HOD$, but N_4 can be extended to a model in which $V = HOD$ holds.

§ 7.

Theorem 7.1. All the results about definable generic reals are valid if we replace the symbol "Df" by "Def" in theorems 4.1, 4.3, 6.3, 6.4 .

Proof. By repeating all proofs writing "Def" instead of "Df".

§ 8. Gaps in the GD-universe. In this final section of the paper we shall draw some conclusions about gaps from the theorems proved above. First we define the following different kinds of gaps :

<u>Definition</u> 8.1. α is the beginning of a gap of length β in the sense $*)$, $**)$ respectively if :

$*)$ i) $(\mathrm{Df}\ R_{\alpha+\eta} \setminus \bigcup_{\xi<\eta} \mathrm{Df}\ R_\xi) \cap P(\omega) = \emptyset$ for all $\eta < \beta$.

ii) $(\mathrm{Df}\ R_{\alpha+\beta} \setminus \bigcup_{\xi<\alpha+\beta} \mathrm{Df}\ R_\xi) \cap P(\omega) \neq \emptyset$

iii) $(\exists\alpha')(\forall\gamma)(\exists\gamma'>\gamma)[\alpha' \leq \gamma < \alpha \rightarrow (\mathrm{Df}\ R_{\gamma'} \setminus \bigcup_{\beta<\gamma} \mathrm{Df}\ R_\beta) \cap P(\omega) \neq \emptyset \wedge$

$\wedge \gamma' < \alpha]$

$**)$ i) $(\mathrm{Df}\ R_{\alpha+\eta} \setminus \bigcup_{\xi<\alpha+\eta} \mathrm{Df}\ R_\xi) \cap P(\omega) = \emptyset$ for $0 < \eta < \beta$.

ii) $(\mathrm{Df}\ R_{\alpha+\beta} \setminus \bigcup_{\xi<\alpha+\beta} \mathrm{Df}\ R_\xi) \cap P(\omega) \neq \emptyset$

iii) $(\mathrm{Df}\ R_\alpha \setminus \bigcup_{\xi<\alpha} \mathrm{Df}\ R_\xi) \cap P(\omega) \neq \emptyset$

<u>Definition</u> 8.2. α is the beginning of a gap of length β in the sense $*')$, $**')$ respectively if $*)$, $**)$ hold while "Df" is replaced by "Def" .

<u>Definition</u> 8.3. α is the beginning of a generic gap of length β if :

i) $[(\mathrm{Df}\ R_\alpha \setminus \bigcup_{\xi<\alpha} \mathrm{Df}\ R_\xi) \cap P(\omega)] \setminus L[\bigcup_{\xi<\alpha} \mathrm{Df}\ R_\xi \cap P(\omega)] \neq \emptyset$

ii) $[(\mathrm{Df}\ R_{\alpha+\beta} \setminus \bigcup_{\xi<\alpha+\beta} \mathrm{Df}\ R_\xi) \cap P(\omega)] \setminus L[\bigcup_{\xi<\alpha+\beta} \mathrm{Df}\ R_\xi \cap P(\omega)] \neq \emptyset$

iii) $[(\mathrm{Df}\ R_\eta \setminus \bigcup_{\xi<\eta} \mathrm{Df}\ R_\xi) \cap P(\omega)] \setminus L[\bigcup_{\xi<\eta} \mathrm{Df}\ R_\xi \cap P(\omega)] = \emptyset$,

for all η such that $\alpha < \eta < \beta$.

<u>Definition</u> 8.4. α is the beginning of a gap of length β if i) ii) iii) of 8,3. hold when we replace "Df" by "Def" .

<u>Theorem</u> 8.5.

a) $ZF \vdash \omega_1^L = \omega_1 \rightarrow$ there is no $*)$ gap below ω_1

b) $ZF \vdash$ there are no $**)$ gap at all

c) ZF $\vdash$ If $cf(\alpha) = \varkappa$ and α is the beginning of a *) gap then $|P(\omega) \cap HOD| \geq \varkappa$

d) ZF $\vdash$ If α is the beginning of a *) gap of length $\beta > 0$, then $\alpha \neq \gamma + \eta$ for every $\eta < \omega_1^L$ and every γ.

Let M be a countable, transitive model for ZF + V = L. Then:

e) for any $\beta \geq 2$ such that $\beta \neq \lambda + 1$ and λ is a limit ordinal, there is a model $N \supseteq M$ in which there is a *) gap of length β.

f) every α such that $\alpha \neq \gamma + \eta$ for every $\eta < \omega_1$ (and every γ) can be the beginning of a *) gap.

g) Con ZF $\rightarrow$ Con (ZF + "there is a *) gap beginning at α such that $\omega_1^L < \alpha < \omega_1$")

Proof. a) is just Theorem 2.5. ; b) Suppose conversely. Let $a_1 \in \text{Df } R_\alpha \setminus \bigcup_{\beta<\alpha} \text{Df } R_\beta$ (** iii). Let $\tilde{x} = \text{code}(x_0, x_1, \dots)$ mean that $J(n,m) \in \tilde{x} \equiv m \in x_n$ (J is a pairing function). Let $\underline{a_m} = \text{code}(a_1, \text{Th } R_\alpha, \dots, \text{Th } R_{\alpha+m})$. Then $\underline{a_m} \in \text{Df } R_{\alpha+m+1} \setminus \bigcup_{\xi\leq\alpha+m} \text{Df } R_\xi$. This contradics ** i) ; c) by * iii) ; d) Suppose conversely i.e. $\alpha = \gamma + \eta$ and $\eta < \omega_1^L$. By *iii) we can define in R_α
$\xi_0 = (\mu\xi)(\exists\eta)[\eta < \omega_1^L \wedge \xi + \eta = \alpha \wedge (\exists a_\xi \subseteq \omega)[a_\xi \in \text{Df } R_\xi \setminus \bigcup_{\delta<\xi} \text{Df } R_\delta]]$.
ξ_0 determines η_0 such that $\xi_0 + \eta_0 = \alpha$. Of course $(\xi_0 \eta_0) \in \text{Df } R_\alpha$. Let $f_0 : \omega \xrightarrow[\text{onto}]{1-1} \eta_0$ be the first, in the sense of $<_L$, which has this property. Then $f_0 \in \text{Df } R_\alpha$. Hence, we can find in R_α the real which codes all the sets $\text{Th } R_\beta \times a_\beta$ for all β such that $\xi_0 \leq \beta < \xi_0 + \eta_0$. This real cannot be definable below α. This contradicts *i) $(\eta = 0)$; e) Take $\alpha_0 = \omega_1 + \beta$ and as in one of the proved theorems build a model in which there is a generic real definable in R_{α_0} ; f) Add, as in the proof of Theorem 6.3, a generic real definable in $R_{\gamma+\omega}$ for all $\gamma < \alpha$. Then * iii) holds. It is easy to show that $(\text{Df } R_\alpha \setminus \bigcup_{\beta<\alpha} \text{Df } R_\beta) \cap P(\omega) = \emptyset$. We can now extend the model obtained by adding a generic real definable in $R_{\alpha+\beta}$ for some $\beta \geq 2$, as we did in Theorem 4.1. or 4.3 or 6.3 ; g) Take Dawson's [2] model in which $\omega_1^L < \gamma_0 < \omega_1$ holds. In this model there are countably many definable reals. Hence there is α, $\gamma_0 \leq \alpha < \omega_1$ such that $(\text{Df } R_\alpha \setminus \bigcup_{\eta<\alpha} \text{Df } R_\eta) \cap P(\omega) = \emptyset$. If we extend this model

by adding a generic real definable in R_{ω_1+4} for example, then we obtain a *) gap beginning at α .

Theorem 8.6.

a) $ZF \vdash \omega_1^L = \omega_1 \rightarrow$ there is no *′) gap below ω_1 .

b) $ZF \vdash$ If a *′) gap begins at α and $cf(\alpha) = \varkappa$ then $|P(\omega) \cap HOD| \geq \varkappa$.

c) $ZF \vdash$ there are no **′) gap below ω_1^L .

d) for every α such that $\alpha = \varkappa + n$, $\varkappa \geq \omega_1^L$, $n > 2$ or limit ordinal α there is a model with *′) gap beginning at α and length $\beta \neq \lambda + 1$.

e) there are models with **′) gaps beginning at $\alpha > \omega_1^L$ and of length β , for every $\alpha \neq \lambda_1 + 1$ (for λ_1 a limit ordinal) and for every β such that $\alpha + \beta \neq \lambda_2 + 1$ and λ_2 is a limit ordinal.

Proof. a), b) obvious by Theorem 8.5 a) c) ; c) by a) ; d) $\varkappa \geq \omega_1^L$. Hence we can add a generic real definable in $\varkappa + n - 1$. We do this as in Theorem 4.3. It is easy to check that $(\mathrm{Def}\ R_\alpha \setminus \bigcup_{\beta<\alpha} \mathrm{Def}\ R_\beta) \cap P(\omega) = \emptyset$. If α is limit see the proof of 8.5 f). Now it is easy to extend the model obtained to a model with a *′) gap beginning at α and of length β . We use the methods introduced in § 4, § 6, § 7 ; e) First we add a generic real definable on R_α and then a generic real definable on $R_{\alpha+\beta}$. We use the methods introduced in § 4, § 6, § 7.

Theorem 8.7. For every $\alpha \geq \omega + 1$ ($\alpha \neq \lambda + 1$ and λ is a limit ordinal), for every $\beta \neq \lambda + 1$ there is a model with a generic gap in the sense of 8.3.

Theorem 8.8. For every $\alpha \geq \omega + 1$ ($\alpha \neq \lambda + 1$) , every $\beta \neq \lambda + 1$ there is a model with a generic gap in the sense of 8.4.

For the proof of 8.7. and 8.8. see the proof of 8.6. e).

Remark 8.9. There are models in which $(\exists a \subseteq \omega)[a \in \mathrm{Def}\ R_\alpha \wedge a \notin \bigcup_{\beta\leq\alpha} \mathrm{Df}\ R_\beta]$. For example $a = \mathrm{Th}\ R_\alpha \times a_\alpha$ where

$a_\alpha \in Df\ R_\alpha \setminus \bigcup_{\beta<\alpha} Df\ R_\beta$. A question arises: Are there models in which $(\exists a \subseteq \omega)[a \in Def\ R_\alpha \wedge a \notin L[\bigcup_{\beta\leq\alpha} Df\ R_\beta]]$? We do not have an answer to this question at present.

Similarly, we have no answer to the very natural question whether there are models such that $(\exists a \subseteq \omega)[a \notin L \wedge a \in Df\ R_{\lambda+1}] \wedge \wedge (\forall b)[b \in Df\ R_\alpha \wedge \alpha \leq \lambda \rightarrow b \in L]$ for a limit ordinal $\lambda > \omega$.

INSTITUTE OF MATHEMATICS , TECHNICAL UNIVERSITY , WROCŁAW .

References

[1] J. Addison, Some consequences of the axiom of constructibility, Fund. Math. 46.

[2] J. Dawson, Ordinal definability in the rank hierarchy, Ann. Math. Log. 6 (1973).

[3] F. R. Drake, Set Theory, NHPC 1974.

[4] Th. Jech, Lectures in set theory , Springer-Verlag 1971.

[5] R.B. Jensen, R.M. Solovay, Some applications of almost disjoint sets, Proc. of I. C. Jerusalem 1968, NHPC 1970.

[6] A. Levy, Definability in axiomatic set theory 1, Proc. 1964. I. C. Log. Math. and Phil. of Science, NHPC 1965.

[7] K. Mc Aloon, Consistency results about ordinal definability, Ann Math. Log. 2 (1971).

[8] J. Myhill, D. Scott, Ordinal definability, Proc. of Symp. in Pure Math. 13.I. (1971).

[9] H. Putnam, A note on constructible sets of integers, Notre Dame J. Formal. Log. 4 (1963).

[10] M. Srebrny, β-models and constructible reals, Ph. D. (preprint) Warsaw 1973.

[11] P. Vopěnka, P. Hajék, The theory of semisets, NHPC 1972.

GENERIC EXTENSION OF ADMISSIBLE SETS

by A. Zarach

J. Barwise in Appendix B of his Doctoral Thesis constructs a generic extension of an admissible set $L_\tau[b]$ (where τ is countable). He adds a function $g: \omega \xrightarrow{\text{onto}} b$ such that $L_\tau[g,b]$ is an admissible set. He uses in this construction a method of ramified forcing. For the use of this method the following facts were very material: a notion of forcing is a set in $L_\tau[b]$ and the set of terms of ramified languages which have a rank less than α $(\alpha < \tau)$ is also a set in $L_\tau[b]$.

We shall show that one can omit the second postulate; and furthermore we can replace the set of conditions by a Δ-coherent, continuous notion of forcing, under the assumption that a given admissible set is a Σ-tower.

We use a simple observation of Shoenfield's definition of forcing, namely: in the definition of $p \Vdash^* a \in b$ and $p \Vdash^* a \neq b$ we use only 3-tuples $\langle q,c,d\rangle$ such that q is a condition and $c,d \in TC(\{a,b\})$.

In the first part we prove the following

Lemma 1.9. Let $\varphi(x_1,\dots,x_n)$ be a Δ_0-formula of the language L_{ZF}. Let $\mathcal{P} = \langle P, \leqslant\rangle \in A$ be a notion of forcing, where A is an admissible set. Then $\{\langle p,a_1,\dots,a_n\rangle : p \in P \wedge a_1,\dots,a_n \in A \wedge \wedge p \Vdash \varphi(a_1,\dots,a_n)\}$ is a class Δ-definable over A.

Computing formulae used in proofs we obtain

Theorem 1. Let A be an admissible, countable set, and let $\mathcal{P} = \langle P, \leqslant\rangle \in A$ be a notion of forcing. Let G be $\mathcal{P}$-generic over A. Then

a) $A[G]$ is an admissible set, $A \subseteq A[G]$, $G \in A[G]$ and $A \cap On = A[G] \cap On$.

b) If N is an admissible set such that $A \subseteq N$ and $G \in N$ then $A[G] \subseteq N$

c) If in A the $\Sigma_n(\Pi_n)$-comprehension schema holds, then it holds in $A[G]$ too.
The same applies for the collection schema.

d) If the axiom of power-set holds in A, then it holds in $A[G]$ too.

and

Theorem 2. Let A be a Σ-tower. Let $\mathbf{C} = \langle C, \leq \rangle$ be a Δ-coherent continuous notion of forcing in A. Let C_α be the α-th stage of C and G be C-generic over A. Then $G \cap C_\alpha$ is C_α-generic over A, $A[G] = \bigcup_{\alpha \in On \cap A} A[G_\alpha]$ and $A[G]$ is an admissible set. If N is an admissible set such that $A \subseteq N$ and $G_\alpha \in N$ for $\alpha \in On \cap A$, then $A[G] \subseteq N$.

In the second part we give an example of forcing with a proper class. We add to a trasitive, countable model M for $Th(L_{\aleph^L_\omega})$ generic functions $f_n: \omega \xrightarrow{onto} \aleph^M_n$. This way we obtain a model $M[G]$ for $Z^- + \Sigma$-collection $+ V = HC + (x)(\exists y)(y = TC(x)) + \neg\, \Pi_1$-replacement schema.
$M[G]$ is a β-model, because in $KP + \Sigma_1$-comprehension schema the formula "to be a well-ordering" is a Δ_1-formula. $\mathcal{P}^{M[G]}(\omega)$ is a β-model for $A_2^- + \neg\, \Pi^1_2$ - choice schema. The correspondence $n \to \aleph^L_n$ is Π_1-definable and thus no model for $KP + \Pi_1$ - replacement schema can have height $h(M)$. In particular the model $M[G]$ cannot be extended to a transitive model for ZF^- with the same height. Similarly $\mathcal{P}^{M[G]}(\omega)$ cannot be extended to a β-model for A_2 with the same height.

§ 1. Unramified forcing for admissible sets.

The Σ-collection schema is the following:

$$(x)(\exists y)\, \varphi(x,y) \to (a)(\exists b)\, (x)_a\, (\exists y)_b\, \varphi(x,y)$$

for φ a Σ-formula .
The Π_1 - replacement schema is the following:

$$(x)(\exists ! y)\ \varphi(x,y) \rightarrow (a)(\exists b)\ (y)(y \in b \leftrightarrow (\exists x)_a\ \varphi(x,y))$$

for φ a Π_1 - formula.
Similarly we define collection and replacement schemas for other classes of formulae.
Clearly, from the pairing axiom, the existence of $TC(x)$ and Δ_0 - -collection we obtain Σ-collection.

<u>Definition</u> 1.1. A sentence is fundamental if it is an atomic sentence or the negation of an atomic sentence.

<u>Definition</u> 1.2. Let φ_1 (φ_2) be a fundamental formula with parameters a,b (respectively c,d) . Then $\varphi_1 \prec \varphi_2 \overset{df}{\equiv} a,b \in TC(\{c,d\})$ $\wedge [(\mathrm{rank}(a \cup b) < \mathrm{rank}(c \cup d)) \vee (\mathrm{rank}(a \cup b) = \mathrm{rank}(c \cup d) \wedge$ $\wedge \min(\mathrm{rank}(a), \mathrm{rank}(b)) < \min(\mathrm{rank}(c), \mathrm{rank}(d))) \vee \varphi_2 = \neg \varphi_1]$.

Fact 1.1. The relation $\prec$ is transitive and Δ^{KP} - definable ; every fundamental sentence has a <u>set</u> of predecessors, and $\prec$ is well-founded.

Let A be a fixed admissible set and let $\mathcal{P} = \langle P, \leq \rangle \in A$ be a notion of forcing. We define inductively on $\prec$ the relation $p \Vdash^* \varphi$ as follows:

$$p \Vdash^* a \in b \equiv (\exists c)(\exists q)_{\geqslant p}\ (\langle c,q \rangle \in b \wedge p \Vdash^* a = c)$$

$$p \Vdash^* a \neq b \equiv (\exists c)(\exists q)_{\geqslant p}(\langle c,q \rangle \in b \wedge p \Vdash^* c \notin a) \vee$$
$$\vee (\exists c)(\exists q)_{\geqslant p}(\langle c,q \rangle \in a \wedge p \Vdash^* c \notin b)$$

$$p \Vdash^* \neg \varphi \equiv (q)_{\leqslant p} \sim (q \Vdash^* \varphi)$$

$$p \Vdash^* \varphi \vee \psi \equiv p \Vdash^* \varphi \vee p \Vdash^* \psi \ .$$

From Fact 1.1. we obtain

<u>Lemma</u> 1.1. If $\varphi(x,y)$ is a formula of L_{ZF} without quantifiers, then $\{\langle p,a,b \rangle : a,b \in A \wedge p \Vdash^* \varphi(a,b)\}$ is Δ-definable over A .

Obviously $p \Vdash^* (\exists x)\varphi(x) \overset{\mathrm{df}}{\equiv} (\exists a)\, p \Vdash^* \varphi(a)$.
Clearly Cohen's Lemmas hold. In particular, if G is a $\mathcal{P}$-generic over A , $\varphi(x_1,\dots,x_n)$ is a L_{ZF}-formula $a_1,\dots,a_n \in A$ then $A[G] \models \varphi\,[K_G(a_1),\dots,K_G(a_n)]$ iff $(\exists p)_G\, p \Vdash^* \varphi(a_1,\dots,a_n)$, where $K_G(a) \overset{\mathrm{df}}{\equiv} \{K_G(b): (\exists p)_G(\langle b,p\rangle \in a)\}$, $A[G] = \{K_G(a) : a \in A\}$.

Definition 1.3. $\vdash_G \varphi(a_1,\dots,a_n) \overset{\mathrm{df}}{\equiv} A[G] \models \varphi[K_G(a_1),\dots,K_G(a_n)]$
$p \Vdash \varphi(a_1,\dots,a_n) \overset{\mathrm{df}}{\equiv} (G)(p \in G \wedge G$ $\mathcal{P}$-generic over $A \to \vdash_G \varphi(a_1,\dots,a_n))$

The following lemma holds:

Lemma 1.2. $p \Vdash \varphi(a_1,\dots,a_n) \Longleftrightarrow p \Vdash^* \neg\neg\,\varphi(a_1,\dots,a_n)$.

Now we remind the reader of the definition of a coherent notion of forcing which is given in [1] (or [2]). Let $C'(\alpha,x)$ be a formula L_{ZF} with parameters from A such that $C_\alpha = \{x \in A: A \models C'(\alpha,x)\} \in A$ and $C_\alpha \subseteq C_\beta$ for $\alpha < \beta$. Let $C = \bigcup_{\alpha \in On} C_\alpha$. Let $\leq$ be a partial ordering of C defined over A . 1_C is the weakest element of C . Assume for any two compatible conditions $p,q \in C$ there is the weakest extension which we denote by $p \wedge q$. Additionaly there is a formula $C''(\alpha,x)$ such that $\wedge:\ C_\alpha \times C^\alpha \to C$ is an order isomorphism, where $C''_\alpha = \{x \in C : A \models C''(\alpha,x)\}$. Then the pair $\mathbf{C} = \langle C,\leq\rangle$ is called a coherent notion of forcing. If $p \in C$ then $p_{(\alpha)}$ and $p^{(\alpha)}$ are the unique r,s such that $r \in C_\alpha$, $s \in C^\alpha$ and $r \wedge s = p$.

Definition 1.4. $st'(p) = \min_\alpha (p \in C_\alpha)$ for $p \in C$

Remark 1. From now we assume that if $a \in A$, then $C \cap a$, $\leq \cap\, a$ and $\{\langle p,\alpha\rangle: \alpha = st'(p) \wedge p \in a\}$ are also elements of A .

Definition 1.5. $st(a) = \bigcup\{\max(st(b), st'(p)): \langle b,p\rangle \in a\}$
$st(a,b,p) = \max\,(st(a), st(b), st'(p))$.

Lemma 1.3. $(x)_A(\{\langle y, st(y)\rangle : y \in x\} \in A)$

Proof. $\langle y,p\rangle \in x \to y,p \in TC(x)$. But $C \cap TC(x) \in A$ and $\{\langle p, st'(p)\rangle : p \in C \cap TC(x)\} \in A$.

Definition 1.6. $a^{(\alpha)} = \{\langle b^{(\alpha)},p\rangle : \langle b,p\rangle \in a \wedge p \in C_\alpha\}$

Obviously $a^{(\alpha)} \in A$ for any $a \in A$ and $\alpha \in On$.

Remark 2. If G is $\mathcal{P} = \langle P, \leq \rangle$ generic over A and $D \subseteq P$ is a dense class definable over A , then $G \cap D \neq \emptyset$. Clearly D need not be from A if even $\mathcal{P} \in A$.

In a standard way we prove the following

Lemma 1.4. If $\mathbf{C} = \langle C, \leq \rangle$ is a coherent notion of forcing in A and G is $\mathbf{C}$-generic over A , then $G_\alpha = G \cap C_\alpha = \{p_{(\alpha)}: p \in G\}$ is $\mathbf{C}_\alpha$-generic over A . Moreover $A[G] = \bigcup_{\alpha \in On \cap A} A[G_\alpha]$, $G_\alpha \in A[G_\alpha]$ and $K_{G_\alpha}(a) = K_{G_\beta}(a) = K_G(a)$ for $\alpha, \beta \geq st(a)$.

It follows from Lemma 1.3. that the definition and the essential facts about the relation $\Vdash^*$ for $\mathbf{C}$ can be formulated as in [2, pp. 9,10].

Definition 1.7. Let $\Vdash^*_\alpha$ denote the forcing relation for C_α . Then

$$p \Vdash^* a \in b \overset{df}{\equiv} p \Vdash^*_\beta a \in b$$

$$p \Vdash^* a \neq b \overset{df}{\equiv} p \Vdash^*_\beta a \neq b , \qquad \text{where } \beta = st(a,b,p)$$

$$p \Vdash^* \neg \varphi \overset{df}{\equiv} (q)_{\leq p} \neg (q \Vdash^* \varphi)$$

$$p \Vdash^* \varphi \vee \psi \overset{df}{\equiv} p \Vdash^* \varphi \vee p \Vdash^* \psi$$

$$p \Vdash^* (\exists x)\, \varphi(x) \overset{df}{\equiv} (\exists a)\, p \Vdash^* \varphi(a)$$

The formal definition of $p \Vdash \varphi$ is the same as in Definition 1.3.

Lemma 1.5. If $st(a,b,p) \leq \alpha, \beta$ and $\varphi(a.b)$ is a fundamental sentence, then $p \Vdash^* \varphi \iff p \Vdash^*_\alpha \varphi \iff p \Vdash^*_\beta \varphi$.

Main Lemma: The three Cohen Lemmas hold for $\Vdash^*$.

$$p \Vdash \varphi \quad \text{iff} \quad p \Vdash^* \neg\neg \varphi$$

and for every formula $\varphi(x_1,\dots,x_n) \in L_{ZF}$ there is a formula $Forc_\varphi\ (x_0,y_1,\dots,x_n) \in L_{ZF}(A)$ such that for every $a_1,\dots,a_n \in A$, $p \in C$

$$p \Vdash \varphi(a_1,\dots,a_n) \quad \text{iff} \quad A \models \mathrm{Forc}_\varphi\,[p,a_1,\dots,a_n]$$

and

$$\vdash_G \varphi(a_1,\dots,a_n) \quad \text{iff} \quad (\exists p)_G\, p \Vdash \varphi(a_1,\dots,a_n)$$

In the case of ZF^- theory the only important thing was the existence of a formula Forc_φ, but in the case of KP (Kripke-Platek theory) it is very important to know the rank of the formula Forc_φ in the Levy hierarchy.

Now we are going to estimate the rank of the formula Forc_φ in terms of the rank of φ.

<u>Lemma</u> 1.6.

$$p \Vdash^* a \in b \leftrightarrow (\exists c)(\exists q)_{\geqslant p}\, (\langle c,q\rangle \in b \wedge p \Vdash^* a = b)$$

$$p \Vdash^* a \neq b \leftrightarrow (\exists c)(\exists q)_{\geqslant p}\, (\langle c,q\rangle \in b \wedge p \Vdash^* c \notin a) \vee$$

$$\vee (\exists c)(\exists q)_{\geqslant p}\, (\langle c,q\rangle \in a \wedge p \Vdash^* c \notin b)$$

The proof we obtain from Definition 1.7 and Lemma 1.5. Notice that in the case $C \in A$ Lemma 1.6. is a consequence of the definition of $\Vdash^*$.

<u>Lemma</u> 1.7. Let $\beta = \max(st(a), st(b))$. Then

$$p \Vdash^* a \in b \quad \text{iff} \quad p_{(\beta)} \Vdash^* a \in b \quad ;$$

$$p \Vdash^* a \neq b \quad \text{iff} \quad p_{(\beta)} \Vdash^* a \neq b .$$

<u>Proof</u>. We use induction w.r.t. $\prec$ and Lemma 1.6. Notice that $\langle c,q\rangle \in b \wedge p \leqslant q \rightarrow (st'(q) \leqslant \beta \wedge q_{(\beta)} = q \wedge p_{(\beta)} \leqslant q)$.

<u>Definition</u> 1.8. We say that $\varphi(x_1,\dots,x_n) \in L_{ZF}$ has the <u>restriction property</u> if for every $a_1,\dots,a_n \in A$ and $\alpha = \max(st(a_1),\dots,st(a_n))$ $p \Vdash \varphi(a_1,\dots,a_n)$ iff $p_{(\alpha)} \Vdash \varphi(a_1,\dots,a_n)$

<u>Lemma</u> 1.8. The class of formulas with the restriction property is closed under propositional connectives and bounded quantifiers.

Proof. (i) Negation: suppose $\varphi(x_1,\dots,x_n)$ has the restriction property, $a_1,\dots,a_n \in A$ and $\alpha = \max(st(a_1),\dots,st(a_n))$.
If $p_{(\alpha)} \Vdash \neg\varphi(a_1,\dots,a_n)$ then $p \Vdash \neg\varphi(a_1,\dots,a_n)$. If $\neg(p_{(\alpha)} \Vdash \neg\varphi(a_1,\dots,a_n))$ then there is some $q \leq p_{(\alpha)}$ such that $q \Vdash \varphi(a_1,\dots,a_n)$. But then $q_{(\alpha)} \Vdash \varphi(a_1,\dots,a_n)$.
$s = q_{(\alpha)} \wedge p^{(\alpha)} \leq p$ and $s \Vdash \varphi(a_1,\dots,a_n)$. Thus
$\neg(p \Vdash \neg\varphi(a_1,\dots,a_n))$.

(ii) Disjunction - obvious

(iii) Bounded quantification; Let $\phi(x_1,\dots,x_n) \equiv$
$\equiv (\exists x_0)_{x_1} \varphi(x_0,x_1,\dots,x_n)$, where φ has the restriction property.
From Lemma 1.7. the formula $x \in y$ has the restriction property
From Lemma 1.6. $(\exists c)(s \Vdash^* c \in a_1 \wedge s \Vdash \varphi(c,a_1,\dots,a_n)) \equiv$
$\equiv (\exists d)(\exists r)_{\geq s} (\langle d,r\rangle \in a_1 \wedge s \Vdash \varphi(d,a_1,\dots,a_n))$. Clearly
$\langle d,r\rangle \in a_1 \rightarrow (st(d) \leq \alpha \wedge r_{(\alpha)} = r)$. By assumption
$s \Vdash \varphi(d,a_1,\dots,a_n) \equiv s_{(\alpha)} \Vdash \varphi(d,a_1,\dots,a_n)$. Thus
$p \Vdash (\exists x_0)(x_0 \in a_1 \wedge \varphi(x_0,a_1,\dots,a_n)) \equiv$
$\equiv p_{(\alpha)} \Vdash (\exists x_0)(x_0 \in a_1 \wedge \varphi(x_0,a_1,\dots,a_n))$.
From Lemma 1.7. and Lemma 1.8. we obtain

Restriction Lemma: Every Δ_0-formula had the restriction property.

From Lemma 1.1. and Lemma 1.8. we deduce

Lemma 1.9. Let $\varphi(x_1,\dots,x_n)$ be a Δ_0-formula of the language L_{ZF} . Let $\mathcal{P} = \langle P, \leq\rangle \in A$ be a notion of forcing, where A is an admissible set. Then
$\{\langle p,a_1,\dots,a_n\rangle : p \in P \wedge a_1,\dots,a_n \in A \wedge p \Vdash \varphi(a_1,\dots,a_n)\}$ is a class Δ-definable over A .

From the Restriction Lemma, the Truth Lemma and absoluteness of Δ_0-formulae we have

Lemma 1.10. If $\varphi(x_1,\dots,x_n)$ is a Δ_0-formula, and $\alpha = \max(st(a_1),\dots,st(a_n))$ then $p \Vdash \varphi(a_1,\dots,a_n) \equiv$
$\equiv p_{(\alpha)} \Vdash_{\alpha} \varphi(a_1,\dots,a_n)$

Note: This Lemma may also be deduced from Lemma 1.6. and the Restriction Lemma.

Lemma 1.11. If $\mathcal{P} = \langle P, \leqslant \rangle \in A$ is a notion of forcing, and G is $\mathcal{P}$-generic over A , then Δ_0-comprehension and the axioms of pairing and of union hold in $A[G]$. If $a \in A[G]$ then $TC(a) \in A[G]$.

Proof. If $p \in G$ then $\{K_G(a), K_G(b)\} = K_G(\{\langle a,p\rangle, \langle b,p\rangle\})$. Δ_0-comprehension we can deduce from Lemma 1.9. The remaining part of the Lemma follows from the fact that $TC(K_G(a)) \subseteq K_G(TC(a))$.

From Lemma 1.4. and Lemma 1.11. we get:

Lemma 1.12. If $\mathbb{C}$ is a coherent notion of forcing in A and G is $\mathcal{P}$-generic over A then $A[G]$ is closed under the pair, union and TC operations, and Δ_0-comprehension holds in $A[G]$.

In the case when the notion of forcing is a set the rank of $Forc_\varphi$ depends only on the rank of φ . The situation is different when the notion of forcing is a proper class. We restrict our considerations to Δ-coherent notions of forcing.

Definition 1.9. A coherent notion if forcing is a Δ-coherent notion of forcing if $\leqslant$ is Δ-definable, C has a Π-definition and $x = C_\alpha$ is a Σ-formula with free variables x and α .

Lemma 1.13. If $\mathbb{C}$ is a Δ-coherent notion of forcing, then the following formulae are Δ over A :

(1) $x = C_\alpha$, (2) $x \in C$, (3) $x = \{\langle \beta, C_\beta \rangle : \beta < \alpha\}$,

(4) $\beta = st'(p)$, (5) $p_{(\alpha)} = x$, (6) $\alpha = st(a)$.

Moreover $C \cap a$, $\leqslant \cap a$, $st' \upharpoonright a = \{\langle \alpha, p\rangle : \alpha = st'(p) \wedge p \in a\}$ are elements of A for every $a \in A$.
If φ is a Δ_0-formula, then $Forc_\varphi$ is a Δ-formula.

Proof. It's clear that (1)-(4) hold. (5) follows from the fact that $x = p_{(\alpha)} \equiv [x \in C_\alpha \wedge p \leqslant x \wedge (y)_{C_\alpha} (p \leqslant y \rightarrow x \leqslant y)]$. $C \cap a$ and $\leqslant \cap a$ are elements of A because Δ_0-comprehension holds in A. For the last statement compare Lemma 1.9. and Lemma 1.10.

Lemma 1.14. If φ is a Σ_n-formula (Π_n-formula) for $n > 0$ then $p \Vdash^* \varphi$ is a Σ_n-formula (Π_n-formula) .

Proof. If φ is a Δ_0-formula beginning with the negation symbol i.e. φ is of the form $(x)_a \varphi_1$, $\varphi_1 \wedge \varphi_2$ or $\neg \varphi_1$, then $p \Vdash^* \varphi \equiv p \Vdash \varphi$ and thus $p \Vdash^* \varphi$ is a Δ-formula. We observe that $\Vdash^*$ preserves disjunction ; $p \Vdash^* a \in b$ and $p \Vdash^* a \neq b$ are Δ-formulae and $p \Vdash^* (x)_a \varphi(x) \equiv (c)\, p \Vdash c \in a \wedge \varphi(c)$. Hence $p \Vdash^* \psi$ is a Σ-formula if ψ is a Δ_0-formula. But in KP one can prove that every Σ-formula is a Σ_1-formula, thus $p \Vdash^* \psi$ is a Σ_1-formula. It follows that $p \Vdash^* \varphi$ is a Σ_1-formula, provided that φ is Σ_1. From Lemma 1.13. we deduce that $p \Vdash^* \varphi$ is a Π_1-formula, if φ is Π_1. Obviously the Lemma for Π_n-formula implies the Lemma for Σ_{n+1}-formulae. Since $p \Vdash^* (x)(\exists y)\, \varphi(x,y) \equiv (a)(q)_{\leq p}(\exists s)_{\leq q}(\exists b)\, p \Vdash^* \varphi(a,b)$, hence the Lemma for Π_n implies the Lemma for Π_{n+2}. To complete the proof notice that the Lemma holds for Σ_1, Π_1 and for Π_2-formulae, because $(\exists s)_{\leq q}(\exists b)\, p \Vdash^* \varphi(a.b)$ is Σ_1-formula, if φ is Δ_0-formula.

Now we can prove

THEOREM 1. Let A be an countable admissible set, and let $\mathcal{P} = \langle P, \leq \rangle \in A$ be a notion of forcing. Let G be $\mathcal{P}$-generic over A. Then

a) $A[G]$ is an admissible set, $A \subseteq A[G]$, $G \in A[G]$ and $A \cap On = A[G] \cap On$.

b) If N is an admissible set such that $A \subseteq N$ and $G \in N$, then $A[G] \subseteq N$

c) If in A the Σ_n (Π_n) - comprehension schema holds, then it holds in $A[G]$, too.
The same applies for the collection schema.

d) If holds the power set axiom in A, then it holds in $A[G]$ too.

Proof. a) From Lemma 1.14. it follows that Σ-collection holds in $A[G]$. The rest was shown earlier.

b) In the theory KP the functional $K_S(x) = \{K_S(y) : (\exists p)_S(\langle y,p\rangle \in x)\}$ is Δ-definable.

c) This follows from Lemma 1.14.

d) $p \Vdash a \subseteq b$ is a Δ-formula and we can repeat the usual proof for ZF.

Definition 1.10. The admissible set A is called a Σ-tower if $\omega \in A$ and there is a sequence $\langle A_\alpha : \alpha \in On \cap A \rangle$ such that

(1) $A_\alpha \in A$ and A_α is transitive ,

(2) $A_\alpha \subseteq A_\beta$ for $\alpha < \beta$

(3) $A_\lambda = \bigcup_{\alpha<\lambda} A_\alpha$ for limit λ ,

(4) $A = \bigcup_{\alpha \in On\, A} A_\alpha$ and $x = A_\alpha$ is a Σ-formula over A with free variables x and α .

Remark 3. If $A \models V = L[a]$ for some $a \in A$, then A is a Σ-tower.

Fact 1.2. If A is a Σ-tower, then $x = \{\langle \alpha, A_\alpha \rangle : \alpha < \beta\}$ is a Δ-formula over A .

Definition 1.11. A Δ-coherent notion of forcing $\mathbb{C}$ is continuous if $C_\lambda = \bigcup_{\alpha<\lambda} C_\alpha$ for every limit $\lambda > 0$,

Lemma 1.15. If A is a Σ-tower, $\mathbb{C}$ is a Δ -coherent, continuous notion of forcing, and G is $\mathbb{C}$-generic over A , then $A[G] \models \Delta_0$-collection.

Proof. Assume $\varphi(x,y)$ is a Δ_0-formula, $p_0 \in G$ and $p_0 \Vdash (x)(\exists y)\, \varphi(x,y)$. Let a be a name for some element of $A[G]$. Let α_0 be an ordinal such that $\alpha_0 \geq st(a)$, $\alpha_0 \geq st'(p_0)$ and $\alpha_0 \geq st(u)$, where u is a parameter from φ . Clearly $(c)_{TC(a)}\, (q)_{\leq p_0}\, (\exists d)(\exists s)_{\leq q}\;\; s \Vdash \varphi(c,d)$.

Hence $(c)_{TC(a)}(\alpha)[(q)_{C_\alpha}(q \leq p_0 \rightarrow (\exists\beta)_{\geq\alpha_0}(\exists d)_{A_\beta}(\exists s)_{C_\beta}(s \leq q \wedge st(d) \leq \beta \wedge \wedge\, s \Vdash \varphi(c,d)))]$. Let $\psi(c,\alpha,\beta) \equiv (q)_{C_\alpha}(q \leq p_0 \rightarrow (\exists d)_{A_\beta}(\exists s)_{C_\beta}(s \leq q \wedge \wedge\, st(d) \leq \beta \wedge s \Vdash \varphi(c,d)))$ and $\nu(\alpha,\beta) \equiv \beta \geq \alpha_0 \wedge (c)_{TC(a)}\psi(c,\alpha,\beta) \wedge \wedge\, (\gamma)_{<\beta}(\gamma \geq \alpha_0 \rightarrow \neg\, (c)_{TC(a)}\psi(c,\alpha,\gamma))$. Then ψ and ν are Δ-formulae.

By Lemma 1.13, Fact 1.2. and Σ-collection (in A) we obtain $(c)_{TC(a)}(\alpha)(\exists\beta)_{\geq\alpha_0}\psi(c,\alpha,\beta)$ and $(\alpha)(\exists!\beta)\nu(\alpha,\beta)$. Clearly $\psi(c,\alpha,\beta) \rightarrow \alpha \leq \beta$ and $\nu(\alpha,\beta) \rightarrow \alpha \leq \beta$. Recall that $\omega \in A$; thus the sequence $\langle \beta_n : n < \omega \rangle$ such that $\beta_0 = \alpha_0$ and $\nu(\beta_n, \beta_{n+1})$ for $n \in \omega$ is an element of A . Let $\lambda_0 = \bigcup_{n\in\omega} \beta_0$. Then $\lambda_0 \in A$, $\lambda_0 \geq \alpha_0$

and

$(c)_{TC(a)}(q)_{C_{\lambda_0}}(q \leq p_0 \to (\exists d)_{A_{\lambda_0}} (\exists s)_{C_{\lambda_0}} (s \leq q \wedge st(d) \leq \lambda_0 \wedge s \Vdash \varphi(c,d)))$.

By Lemma 1.10 we may replace $s \Vdash \varphi(c,d)$ by $s \Vdash_{\lambda_0} \varphi(c,d)$. Thus

$(c)_{TC(a)}\ p_0 \Vdash_{\lambda_0} (\exists y)\ \varphi(c,y)$. Obviously $p_0 \in G_{\lambda_0}$ and

$K_G(x) \in K_G(a) \to (\exists c)_{TC(a)}(K_G(c) = K_G(x))$. Hence $\vdash_{G_{\lambda_0}} (x)_a (\exists y)\varphi(x,y)$.

By THEOREM 1 $A[G_{\lambda_0}]$ is an admissible set i.e. for some $b \in A$

$\vdash_{G_{\lambda_0}} (x)_a (\exists x)_b\ \varphi(x,y)$ and we can assume $st(b) \leq \lambda_0$. Thus

$\vdash_G (x)_a (\exists y)_b\ \varphi(x,y)$.

Because Δ_0-collection implies Σ-collection we have proved the following

THEOREM 2. Let A be a Σ-tower. Let $\mathbf{C} = \langle C, \leq \rangle$ ba a Δ-coherent continuous notion of forcing in A . Let C_α be the α-th stage of C and G be $\mathbf{C}$-generic over A . Then $G \cap C_\alpha$ is C_α-generic over A , $A[G] = \bigcup_{\alpha \in On \cap A} A[G_\alpha]$ and $A[G]$ is an admissible set.
If N is an admissible set such that $A \subseteq N$ and $G_\alpha \in N$ for $\alpha \in On \cap A$, then $A[G] \subseteq N$.

§ 2 . Application to second order arithmetic.

From Levy's theorem about Σ-formulae and Gödel's condensation Lemma it follows that $L_{\aleph_\omega^L} \models Z + \Sigma$-collection. Moreover

(1) $x \in L_{\aleph_\omega^L} \to TC(x) \in L_{\aleph_\omega^L}$

(2) $L_{\aleph_\omega^L} \models Card(x) \leftrightarrow (\exists n)_\omega\ (x = \aleph_n^L)$

(3) $\{\langle k, \aleph_k^L \rangle : k < n\} \in L_{\aleph_\omega^L}$ for all $n \in \omega$

(4) $L_{\aleph_\omega^L} \models (\alpha)(\exists n)_\omega\ (\alpha < \aleph_n^L)$

(5) $L_{\aleph_\omega^L}$ is a Σ-tower

Let M be a countable standard model for $Th(L_{\aleph_\omega^L})$. Let C be the class of all finite functions $f \in M$ such that $dom(f) \subseteq On \times \omega \wedge \wedge rg(f) \subseteq On \wedge (\langle\langle\alpha,n\rangle,\beta\rangle \in f \rightarrow \beta < \alpha)$. C is ordered by inverse inclusion and $p \wedge q \overset{df}{=} p \cup q$. $p_{(\gamma)} = \{\langle\langle\alpha,n\rangle,\beta\rangle \in p : \alpha < \gamma\}$, $p^{(\gamma)} = p \setminus p_{(\gamma)}$. Then $\langle C, \leq\rangle$ is a Δ-coherent continuous notion of forcing in M and M is a Σ-tower. If G is $\mathbb{C}$-generic over M then $M[G]$ is an admissible set. Moreover $G_{\aleph_{n+1}^M} \in M[G]$,

$f_n = \{\langle m,\beta\rangle : \langle\langle \aleph_n^M, m\rangle, \beta\rangle \in \bigcup G_{\aleph_{n+1}^M}\} \in M[G]$ and $f_n : \omega \xrightarrow{onto} \aleph_n^M$.

Thus $M[G] \models V = HC$.

Now we show

<u>Lemma</u> 2.1. If $\mathbb{C}$ is a notion of forcing as defined above and $\varphi(x_1,\dots,x_n)$ is a formula of L_{ZF} then φ has the restriction property.

<u>Proof</u>. Let $p_0, p_1 \in C$ and $\{\langle\alpha_i,n_i\rangle : i < k\}$ be the set of all elements from $dom(p_0) \cap dom(p_1)$ such that $p_0(\alpha_i,n_i) \neq p_1(\alpha_i,n_i)$ for $i < k$. Let $\psi_{p_0,p_1} : \bigcup C \rightarrow \bigcup C$ be a mapping such that

$\psi_{p_0,p_1}(\langle\langle\alpha_i,n_i\rangle, p_j(\alpha_i,n_i)\rangle) = \langle\langle\alpha_i,n_i\rangle, p_{1-j}(\alpha_i,n_i)\rangle$ for $i < k$, $j = 0,1$ and on other elements ψ_{p_0,p_1} is the identity mapping. For $s \in C$ let $\varphi_{p_0,p_1}(s) = \psi''_{p_0,p_1} s$. Then φ_{p_0,p_1} is an automorphism of C such that $\varphi_{p_0,p_1}(p_0) \cup p_1 \in C$ and $x = \varphi_{p_0,p_1}(s)$ is a Δ-formula over M. Let $\varphi^*_{p_0,p_1}(a) = a \setminus (V \times C) \cup$

$\cup \{\langle\varphi^*_{p_0,p_1}(b), \varphi_{p_0,p_1}(q)\rangle : \langle b,q\rangle \in a\}$. Then $x = \varphi^*_{p_0,p_1}(y)$ is a Δ-formula over M. Now we are able to prove the permutation lemma. Moreover, if $\alpha < \alpha_i$ for $i < k$ then $\varphi_{p_0,p_1} \restriction C_\alpha = id \restriction C_\alpha$ and $\varphi^*_{p_0,p_1}(a) = a$ if $st(a) \leq \alpha$.

From the above Lemma we obtain

THEOREM 3. The admissible set $M[G]$ is a β-model for $Z^- + \neg\Pi_1$-replacement schema. Moreover no standard model for $KP + \Pi_1$-replacement can have the same height as M.

$\mathcal{P}^{M[G]}(\omega)$ is a β-model for $A_2^- + \neg\Pi_2^1$-choice schema and $\mathcal{P}^{M[G]}(\omega)$ cannot be extended to any β-model for A_2 with the same height.

<u>Note:</u> Z^- is Zermelo theory without the power set axiom A_2^- is A_2 without the Choice Schema.

<u>Proof</u>. If $\varkappa$ is a cardinal in L, then $L_\varkappa \models KP + V = L$ and $L_\varkappa \models \mathrm{Card}(\alpha) \Longleftrightarrow \mathrm{Card}^L(\alpha)$. $x = L_\beta$ is a Δ^{KP}-formula and $\mathrm{Card}^L(\varkappa)$ is a Π_1-formula, so "x is the n-th constructible aleph" is a Π_1-formula. From Lemma 2.1 $M[G] \models Z^-$. This proves the first part of the theorem. Let $\varphi(n,X)$ be a formula of the language of A_2 such that $\varphi(n,X) \equiv (i)_{<n}\, \mathrm{Bord}(X^{(i)}) \wedge (i)_{<n}(Y)(k)[\mathrm{Func}(Y) \wedge \mathrm{Constr}(Y) \to$
$\to Y''X^{(i)} \restriction k \prec X^{(i)}) \wedge (i)_{<n} \neg \mathrm{Fin}(X^{(i)}) \wedge (i+1)_{<n}(X^{(i)} \prec X^{(i+1)})$.

$\varphi(n,X)$ is a Π_2^1-formula and $\mathcal{P}^{M[G]}(\omega)$ is a continuum of the model in which "to be a well-ordering" is a Δ_1-formula. So $\mathcal{P}^{M[G]}(\omega)$ is a β-model for A_2^-, because $M[G] \models Z^-$. If $g_n\colon \omega \xrightarrow[1-1]{\text{onto}} \aleph_n^L$ then $\{2^l 3^k\colon g(l) \in g(k)\}$ is a well-ordering in $\mathcal{P}^{M[G]}(\omega)$ which has type $\aleph_n^L$. Thus $(n)(\exists X)\,\varphi(n,X)$ but obviously in $\mathcal{P}^{M[G]}(\omega)$ the sentence $(\exists X)(n)\,\varphi(n,X^{(n)})$ does not hold. Moreover the same situation obtains in every β-extension of $\mathcal{P}^{M[G]}(\omega)$ which has the same height as $^{M[G]}(\omega)$.

References.

[1] R. Chuaqui, Forcing for the impredicative theory of classes, the Journal of Symbolic Logic 37(1972) pp. 1-18.

[2] A. Zarach, Forcing with proper classes, Fundamenta Mathematicae LXXXI (1973), pp. 1-27.

TECHNICAL UNIVERSITY OF WROCŁAW,
MATHEMATICAL INSTITUTE.

SOME REMARKS ON OBSERVATIONAL MODEL-THEORETIC LANGUAGES

Petr Hájek
Mathematical Institute
Czechoslovak Academy of Sciences
115 67 Prague, Czechoslovakia

We use the term "model-theoretic languages" in the sense of Feferman [6]; a <u>model-theoretic language</u> L is given by a non-empty set Typ_L of admitted <u>types</u> and for each $\tau \in Typ_L$ by a set $Stc_L(\tau)$ of abstract objects called <u>sentences</u> of the type τ, by a non-empty set $S_L(\tau)$ of admitted <u>structures</u> of the type τ and a relation $\models_{L,\tau}$ which is a subset of $Stc_L(\tau) \times S_L(\tau)$ and called the <u>truth relation</u> for the type τ. Equivalently, L may be given by the relation Rel_L consisting of all quadruples $\langle \tau, \varphi, \underset{\sim}{M}, \varepsilon \rangle$ where $\tau \in Typ_L$, $\varphi \in Stc_L(\tau)$, $\underset{\sim}{M} \in S_L(\tau)$ and $\varepsilon = 1$ iff $\underset{\sim}{M} \models_{L,\tau} \varphi$ and $\varepsilon = 0$ otherwise. As usual examples serve languages L_A for an admissible set A, languages with generalized quantifiers, ω-logic, second order logic etc. Feferman [7] and Barwise [3] formulate some good (desired) properties of model-theoretic languages, which may serve as tentative axioms for "abstract model theory".

In the present note we are going to consider some unusual examples of model-theoretic languages; unusual since our sets S_L of admitted structures will contain only <u>finite</u> structures (i.e. structures with finite domains). The impuls for the study of such languages comes from investigations of inductive inference in connection with mechanized hypothesis formation, thus from Artificial Intelligence (cf. [9]). Carnap distinguishes between the observational and the theoretical level of a scientific language; a typical feature of an observational language is that the truth value of a sentence is effectively computable as a function of two arguments: sentences and finite models (data). This leads to the notion of an <u>observational model-theoretic language</u> as a model-theoretic language L such that Rel_L is a recursive relation. (Imagine elements of S_L as structures whose domains are finite subsets of the set ω of natural numbers; the type of each structure is assumed to be finite. Such structures can be obviously coded by natural numbers.)

Some results on observational model-theoretic languages are contained in [8] where observational generalized quantifiers are considered. (Cf. also [12].) In the present paper we shall study projective classes (in the sense of [6]) both in languages with one-sorted structures and in languages with many-sorted structures; we shall also describe some observational languages satisfying a weak form of the Souslin-Kleene interpolation theorem.

We shall mention relations of notions concerning observational languages to problems of the theory of complexity of computations. The proof of the main result of § 1 uses the ultraproduct as a model of the theory of semisets; but no knowledge of the theory of semisets is assumed.

§ 1. Projective classes in one-sorted languages.

1.1 A one-sorted observational structure has the form

$$\underset{\sim}{M} = \langle M, R_1, \ldots, R_p, a_1, \ldots, a_q \rangle$$

where M is a finite non-empty set of natural numbers, each R_i is a n_i-ary relation and each a_i is an element of M. The type of $\underset{\sim}{M}$ is $\langle\langle n_1, \ldots, n_p \rangle, \underbrace{\langle 0, \ldots, 0 \rangle}_{q \text{ times}}\rangle$. The language $L^{obs}_{\omega\omega}$ has all types of the form just described for its types, $Stc_L(\tau)$ is the set of all formulas of $L_{\omega\omega}$ of the type τ (i.e. finite formulas with classical quantifiers and connectives), $S_L(\tau)$ consists of all observational structures of the type τ and satisfaction is defined following Tarski. Obviously, $L^{obs}_{\omega\omega}$ is an observational language.

1.2 A class $K \subseteq S_L(\tau)$ is elementary (w.r.t. L) if there is a $\varphi \in Stc_L(\tau)$ such that $K = \{\underset{\sim}{M} \in S_L(\tau);\ \underset{\sim}{M} \models_{L,\tau} \varphi\}$. The notion "$\underset{\sim}{M}'$ is an expansion of $\underset{\sim}{M}$" is defined as usual (cf. [6] or [15]). A class $K \subseteq S_L(\tau)$ is projective (PC) if there is a $\tau' \supseteq \tau$ and a $\varphi \in Stc_L(\tau')$ such that, for each $\underset{\sim}{M} \in S_L(\tau)$, $\underset{\sim}{M} \in K$ iff $\underset{\sim}{M}$ can be expanded to a model of φ (cf. [6]). It is easy to show that the Souslin-Kleene interpolation theorem $(I)^{\Delta}_{L}$ (saying that if both K and the complement of K are projective then K is elementary) does not hold for $L^{obs}_{\omega\omega}$ (take e.g. $K = \{\langle M \rangle;\ card(M) \text{ is odd}\}$).

1.3 Call K doubly projective if both K and the complement of K are projective. We have the following very natural question: is each projective class doubly projective (w.r.t. L^{obs})? This question includes the question whether each spectrum (in the sense of Scholz) is a double spectrum (cf. [2]). It turns out that the former question is very difficult and is equivalent to the question whether languages (= sets of words) recognizable by non-deterministic Turing machines operating in polynomial time (so-called NP-langauges) are closed under complementation, which is one of the famous open questions of the theory of complexity of computations. In particular, assumed an appropriate coding of observational structures, a class $K \in S_L(\tau)$ which is closed under isomorphisms is projective iff the set Cod(K) of codes of elements of K is NP. (Jones and Selman; see [4] and [13]. In [9] I referred to a work of Johnsson et al. in ignorance of [4]. For NP--classes see [1] Chapter 10.) Thus we have a good "computational" characterization of projective classes of L^{obs}; but nothing is known concerning the above questions.

1.4 There is a modification of the notion of projective classes which has been considered in the literature. A class $K \subseteq S_L(\tau)$ is called 1-projective if there is a type $\tau' \supseteq \tau$ containing besides elements of τ only unary predicates and a $\varphi \in Stc_L(\tau')$ such that $K = \{\underset{\sim}{M} \in S_L(\tau); \underset{\sim}{M}$ can be expanded to a model of $\varphi\}$ Fagin proved in [5] that the class of all connected graphs is not 1-projective (whereas the class of all disconnected graphs is evidently 1-projective), showing in this way that 1-projective classes are not closed under complementation. I reproved later the same fact in [9] in ignorance of Fagin's work and by other methods. (Fagin used Fraissé's games and I used semisets.) Unfortunately the fact that 1-projective classes are not closed under complementation does not say anything to the above question since one easily sees that the classof connected graphs is doubly projective. In [9] I stated without proof the fact that the class of all planar graphs is not 1-projective. This fact, not covered by [5], will be proved in the rest of the present section. We shall use the language of semisets, but no knowledge of the theory of semisets is assumed (the reader may consult [16] and [11] if desired); our present application of semisets is rather modest and the word "semiset" can be viewed as a mere "façon de parler" in the present context.

1.5 Instead of describing a system of the theory of semisets we shall simply describe and use a particular model of semisets, namely the ultrapower of the universe. The ultrapower of the universe $(V, \in)$

as a model of the theory of semisets is studied in details in [16]; the short description in [10] 1(b) is sufficient for our purpose. Let z be a non-principal ultrafilter on ω; the ultrapower $(V, \in)^{\omega}/z$ yields a non-standard extension $(V_1, \in_1)$ of $(V, \in)$ and we have the elementary embedding $\iota:(V, \in) \to (V_1, \in_1)$. For each $x \in V_1$, let $\tilde{x} = \{y \in V_1;\ y \in_1 x\}$; then $\tilde{x} \subseteq V_1$, $\tilde{x}$ is a set and if we put $V_2 = \{\tilde{x};\ x \in V_1\}$, $\tilde{x} \in_2 \tilde{y}$ iff $x \in_1 y$ then $(V_2, \in_2)$ is obviously isomorphic to $(V_1, \in_1)$. There are subsets of V_1 not having the form $\tilde{x}$ for any $x \in V_1$, e.g. the set $\iota''\omega$, which is a proper subset of $\widetilde{\iota(\omega)}$. Call elements of V_2 sets in the new sense and subsets of V_1 semisets. (Arbitrary subclasses of V_1 may be called classes in the new sense.) In this way we obtain a model of semisets (semisets are precisely all subclases of sets). In the theory of semisets, a natural number n is called <u>absolute</u> (or standard) if there is no semiset one-one mapping of n onto its successor $n + 1$. In the model, the class An of all absolute natural numbers coincides with $\iota''\omega$ and is a proper subsemiset of $\omega^{x} = \widetilde{\iota(\omega)}$. Using the elementary embedding we obtain immediately the following

1.6 <u>Metatheorem.</u> Let $\varphi(x)$ be a ZF-formula. If $(\forall n \in An)\varphi(n)$ is provable in the theory of semisets then $(\forall n \in \omega)\varphi(n)$ is provable in ZF.

This is proved by observing that if $(\forall n \in An)\varphi(n)$ holds in the model then $(\forall n \in \omega)\varphi(n)$ holds in $(V, \in)$; the last fact is the only fact we actually need below since instead of proving in the theory of semisets the reader may verify everything in the particular model described. (Alternatively, one could use the language of non-standard analysis.)

1.7 From now on, we proceed in the conservative extension of ZF admitting semisets and modelled by the above model. (Cf. [9] Part III.) Assume that formulas are coded by natural numbers in the usual way. Call a class K of structures <u>absolutely 1-projective</u> if there is a $\varphi \in An$ 1-projectively defining K (i.e. K is 1-projectively defined by a short formula). By 1.6, it suffices to prove that the class of planar graphs is not absolutely 1-projective. Say that a semiset σ <u>respects</u> a set x if σ maps x onto a set (in general the image could be a proper semiset). Notice that validity of short (absolute) formulas is preserved even by semiset isomorphisms. Hence the following suffices for a variety K not to be absolutely 1-projective:

There is a structure $\underset{\sim}{M} \in K$ such that, for each short (absolutely long) tuple $x_1, \ldots, x_k$ of subsets of M, there is a σ mapping $\underset{\sim}{M}$ isomorphically onto a (set) structure $\sigma(M) \notin K$ and respecting each $x_1,\ldots,x_k$. Such an M will be called <u>critical</u> for K.

1.8 Remark. Pudlák's results [14] imply that in the theory of semisets with the axiom (EEI) one can prove the following: A class K is absolutely 1-projective <u>if and only if</u> there is no structure critical for K.

1.9 We are going to construct a critical graph for the class of planar graphs. Remember that a graph containing a subgraph of the form

(*)

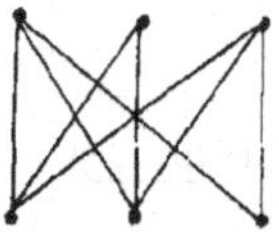

("three factories, three utilities") is not planar. For each n, let Z_n be the graph

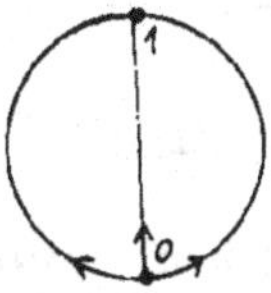

where the vertices 0, 1 are connected by three paths of the length n: the left, middle and right one. Let K_n be the direct sum of n disjoint copies of Z_n. Evidently, K_n is planar.

<u>Claim:</u> If $n \notin An$ then K_n is critical for planar graphs.

Proof. Let $x_1,\ldots,x_k$ be few (absolutely many) subsets of the field of K_n. Associate with each vertex a the k-tuple $u(a) = \langle u_1,\ldots,u_k\rangle$ where $u_i = 1$ iff $a \in x_i$, and $u_i = 0$ otherwise. Let Z be a copy of Z_n in K_n containing the vertex a and let R, L be the right and left path of Z respectively. If $a \neq 0, 1$ then the successor a^+ of a is uniquely determined; if a is <u>far</u> from 1, i.e. the distance of a from 1 is non-absolute then let $a^{An} = \underbrace{\langle a^+, a^{++},\ldots\rangle}_{An\ \text{times}}$ and let $\tau(a)$ (the <u>type</u> of a) be the sequence $\underbrace{\langle u(a^+), u(a^{++}),\ldots\rangle}_{An\ \text{times}}$ ($\tau(a)$ is a proper semiset). If $h \in An$ then the h-type $\tau_h(a)$ of a is the initial segment of $\tau(a)$ of the length h ($\tau_h(a)$ is a set). Evidently, for each absolute h there are three distinct copies Z_n^1, Z_n^2, Z_n^3 of Z_n in K_n and elements $a_i \in L(Z_n^i)$, $b_i \in R(Z_n^i)$ such that $\tau_h(a_1) = \tau_h(a_2) = \tau_h(a_3)$ and and $\tau_h(b_1) = \tau_h(b_2) = \tau_h(b_3)$ (Schubladenprinzip). Since An has no largest element, the following must hold: There are three distinct copies Z_n^1, Z_n^2, Z_n^3 of Z_n in K_n and elements $a_i \in L(Z_i)$, $b_i \in R(Z_i)$ such that $\tau(a_1) = \tau(a_2) = \tau(a_3)$ and $\tau(b_1) = \tau(b_2) = \tau(b_3)$.

Remove edges $a_i \longrightarrow a_i^+$ and $b_i \longrightarrow b_i^+$; instead, add edges $a_1 \longrightarrow a_3^+$, $b_1 \longrightarrow b_2^+$, $a_2 \longrightarrow a_1^+$, $b_2 \longrightarrow b_3^+$, $a_3 \longrightarrow a_2^+$, $b_3 \longrightarrow b_1^+$. Then the modified graph K_n' contains a subgraph of the form (*). Define a mapping σ of K_n onto K_n' as follows: a_1^{An} is mapped pointwise onto a_3^{An} (a_1^+ onto a_3^+ etc.), b_1^{An} onto b_2^{An}, a_2^{An} onto b_3^{An}, a_3^{An} onto a_2^{An}, and b_3^{An} onto b_1^{An}; σ is identical otherwise. Then σ is a semiset isomorphism of K_n onto K_n' mapping x_1 onto x_1, ... , x_k onto x_k, hence respecting $x_1,\dots,x_k$. Thus K_n is critical.

1.10 Corollary. The class of planar graphs is not 1-projective.

§ 2. Projective classes in many-sorted languages.

2.1 A many-sorted observational structure has the form

$$\underset{\sim}{M} = \langle M_1,\dots,M_k, R_1,\dots,R_p, a_1,\dots,a_q \rangle$$

where each M_j is a finite non-empty set of natural numbers, each R_i is a n_i-ary relation on $\bigcup_{1\leq j\leq k} M_j$ and $a_i \in M_{j_i}$; its type is k, $n_1,\dots,n_p$, $j_1,\dots,j_q$. An expansion of M has the form

$$\underset{\sim}{M}' = \langle M_1,\dots,M_k,\dots M_{\bar{k}}, R_1,\dots,R_p,\dots,R_{\bar{p}}, a_1,\dots,a_q,\dots,a_{\bar{q}} \rangle,$$

i.e. $\underset{\sim}{M}'$ has some additional universes, relations and constants. $\underset{\sim}{M}'$ is a strict expansion if $\bar{k} = k$, i.e. there are no new universes. (Cf.[7].) The many-sorted $L^{obs}_{\omega\omega}$ (briefly, m-$L^{obs}_{\omega\omega}$) is the obvious generalization of L^{obs} and has the same formulas as the many-sorted $L_{\omega\omega}$. A class K of structures of a type τ is projective (w.r.t. m-$L^{obs}_{\omega\omega}$) if there is a sentence φ of a richer type such that

$$K = \{\underset{\sim}{M} \text{ of type } \tau;\ (\exists\, \underset{\sim}{M}' \text{ expansion of } \underset{\sim}{M})(\underset{\sim}{M}' \models \varphi)\ ;$$

K is strictly projective if there is a φ of a richer type but with no variables of sorts not in τ such that

$$K = \{\underset{\sim}{M} \text{ of type } \tau;\ (\exists\, M' \text{ strict expansion of } \underset{\sim}{M})(\underset{\sim}{M}' \models \varphi).$$

K is doubly strictly projective if both K and the complement of K are strictly projective.

2.2 It is easy to see that the computational characterization of projective classes in 1.3 applies to strictly projective classes in the present situation: a class $K \subseteq S_L(\tau)$ which is closed under isomorphisms is strictly projective iff the set Cod(K) of codes of elements of K is NP. Here we give a computational characterization of projective classes in the many-sorted $L^{obs}_{\omega\omega}$.

2.3 Theorem. A class K of observational many-sorted structures of a type τ is projective iff the set Cod(K) of codes of elements of K is recursively enumerable and K is closed under isomorphisms.

Proof. First note that here we need not be so much careful with coding as if the computational complexity is concerned; any natural Gödel numbering is sufficient.

The implication $\Rightarrow$ is obvious:

$$\underset{\sim}{M} \in K \text{ iff } (\exists \underset{\sim}{M}')(\underset{\sim}{M}' \text{ expansion of } \underset{\sim}{M} \text{ and } \underset{\sim}{M}' \models \varphi)$$

- hence the definition is an existential quantification of a recursive condition.

We sketch the proof of $\Leftarrow$. Let T be a Turing machine such that, for each $\underset{\sim}{M}$, $\underset{\sim}{M} \in K$ iff T halts when computing with the code of $\underset{\sim}{M}$ as input. With each $\underset{\sim}{M}$ and each computation of T with the code of $\underset{\sim}{M}$ as input associate an expansion $\underset{\sim}{M}'$ of $\underset{\sim}{M}$ dascribed as follows: $\underset{\sim}{M}'$ has one new domain M' viewed as a rectangular matrix; use a relation $<_{hor}$ saying that an element of M' is in the same row but more at left than another element and similarly for $<_{ver}$. Introduce unary predicates corresponding to the alphabet and internal states of T and interpret them in such a way that M' is made to a sequence of instantaneous descriptions; an extra predicate denotes the position of the head in each row and the current internal state is ascribed as a property (say) of the first element of each row. Some finitely many additional predicates interrelating M and M' guarrantee that the initial instantaneous description is the code of $\underset{\sim}{M}$. Define a formula φ_0 describing all of these; then φ says "φ_0 and the computation is halting". For more details use e.g. [4] Theorem 6.

2.4 Corollary. (1) Projective classes in m-$L^{obs}_{\omega\omega}$ are not closed under complementation (since there are recursively enumerable non-recursive sets). (2) There is a doubly projective class K which is not strictly projective and thus not elementary (since there is a recursive set non recognizable in polynomial time; to name a prominent example, take the set of all tautologies of Presburger arithmetic).

2.5 Note that each strictly projective class is doubly projective since it is NP and hence recursive. (Of course, we do not claim that it is doubly strictly projective.)

2.6 Concerning arbitrary observational many sorted model-theoretic languages, observe the following: For each such L, if K is elementary w.r.t. L (i.e. there is a $\varphi \in Stc_L(\tau)$ such that $K = \{M;\ M \models_{L,\tau} \varphi\}$) then K is recursive (since L is observational) and therefore K is a doubly projective class w.r.t. $m\text{-}L^{obs}_{\omega\omega}$.

Conversely, each class K doubly projective w.r.t. $m\text{-}L^{obs}_{\omega\omega}$ is elementary in an observational language, since K may be used to define an observational generalized quantifier. This leads us to the question whether the minimal extension of $m\text{-}L^{obs}_{\omega\omega}$ in which all doubly projective classes of $m\text{-}L^{obs}_{\omega\omega}$ are elementary can be represented as an observational language. We shal show that this is not the case; but if we restrict ourselves to doubly strictly projective classes of $m\text{-}L^{obs}_{\omega\omega}$ then the answer is positive.

2.7 Theorem. There is no observational many-sorted model-theoretic language L such that each doubly projective class of $m\text{-}L^{obs}_{\omega\omega}$ is elementary in L.

Proof. Such an L would correspond to a universal recursive relation: fix a type τ, enumerate effectively all pairwise non-isomorphic observational structures of the type τ and all sentences of the type τ Then define $M_n \in K_o$ iff $M_n \models_{L,\tau} \neg\varphi_n$ and let K be the set of all observational structures isomorphic to an element of K_o. Then K is recursive, hence doubly projective w.r.t. $m\text{-}L^{obs}_{\omega\omega}$ and thus elementarily defined by a $\varphi_{n_o} \in Stc_L$ - a diagonal contradiction.

2.9 Theorem. There is an observational many-sorted model-theoretic language L such that each doubly strictly projective class of $m\text{-}L^{obs}_{\omega\omega}$ is elementary in L and conversely, each class elementary in L is doubly strictly projective in $m\text{-}L^{obs}_{\omega\omega}$. L satisfies the following form of the Souslin-Kleene interpolation theorem: Each doubly strictly projective class of L is elementary in L.

Proof. The language L will have the usual connectives and infinitely many generalized quantifiers $Q_{t\varphi\psi}$ where t is a type, φ is a classical sentence of a type t' richer than t but with no new sorts and the same for ψ. A model $\underset{\sim}{M}$ of the type t belongs to the class $K(Q_{t\varphi\psi})$ determining the semantics of $Q_{t\varphi\psi}$ iff the following holds:

(i) $\underset{\sim}{M}$ can be strictly expanded to a model of φ and

(ii) for each $\underset{\sim}{N}$ of the type t such that card(N) $\leq$ card(M), $\underset{\sim}{N}$ can be strictly expanded to a model of φ iff N cannot be strictly expanded to a model of ψ.

Obviously, our language is observational, since the satisfaction relation is recursive. Evidently, if a class K of the type t is doubly strictly projective w.r.t. $m\text{-}L_{\omega\omega}^{obs}$, K being strictly projectively defined by φ and the complement of K by ψ, then $K = K(Q_{t\varphi\psi})$ and hence K is elementary w.r.t. L.

Conversely, we prove that if K is elementary in L then K is doubly strictly projective w.r.t. $m\text{-}L^{obs}$. To see this, first observe that each $K(Q_{t\varphi\psi})$ is doubly strictly projective. This is clear if φ strictly projectively defines $K(Q_{t\varphi\psi})$ and ψ strictly projectively defines the complement of $K(Q_{t\varphi\psi})$; but otherwise $K(Q_{t\varphi\psi})$ contains only finitely many non-isomorphic elements and hence is elementary in $m\text{-}L_{\omega\omega}^{obs}$. Then one proves the desired assertion by induction on the (definitional) complexity of sentences of L. The only non-trivial step concerns quantifiers. For example, consider the sentence $(Qx_0x_1x_2)(\alpha(x_0), \beta(x_1,x_2))$ and denote it by Φ. By the induction assumption, the classes defined elementarily by the sentences $\alpha(c_0)$, $\beta(c_1,c_2)$ are doubly strictly projective w.r.t. $m\text{-}L_{\omega\omega}^{obs}$; furthermore, the class K(Q) is doubly strictly projective w.r.t. $m\text{-}L^{obs}$. Thus let

$$\alpha(c_0) \Leftrightarrow (\exists \underline{X})\, \alpha^+(c_0,\underline{X}) \Leftrightarrow \neg(\exists \underline{X})\, \alpha^-(c_0,\underline{X}),$$

$$\beta(c_1,c_2) \Leftrightarrow (\exists \underline{Y})\, \beta^+(c_1,c_2,\underline{Y}) \Leftrightarrow \neg(\exists \underline{Y})\, \beta^-(c_1,c_2,\underline{Y}),$$

$$(Qx_0x_1x_2)(P(x_0), R(x_1,x_2)) \Leftrightarrow (\exists \underline{Z})\, \varkappa^+(P,R,\underline{Z}) \Leftrightarrow \neg(\exists \underline{Z})\, \varkappa^-(P,R,\underline{Z}),$$

where α^+, α^-, β^+, β^-, $\varkappa^+$, $\varkappa^-$ are formulas of the classical second order logic with no quantified predicates and $\Leftrightarrow$ means logical equivalence on finite structures. Then

$$\Phi \Leftrightarrow (\exists P,R)((\forall x_0)(\alpha(x_0) \equiv P(x_0)) \,\&\, (\forall x_1x_2)(\beta(x_1,x_2) \equiv R(x_1,x_2)) \,\&\, \& (Qx_0x_1x_2)(P(x_0),R(x_1x_2)).$$

Observe that

$$(\forall x_0)(\alpha(x_0) \equiv P(x_0)) \Leftrightarrow$$
$$\Leftrightarrow (\forall x_0)\left[(P(x_0) \,\&\, (\exists \underline{X})\, \alpha^+(x_0,\underline{X})) \vee (\neg P(x_0) \,\&\, (\exists \underline{X})\, \alpha^-(x_0,\underline{X}))\right] \Leftrightarrow$$
$$\Leftrightarrow (\exists \underline{X}')(\forall x_0)\left[(P(x_0) \,\&\, \alpha^+(x_0,\underline{X}'_{x_0})) \vee (\neg P(x_0) \,\&\, \alpha^-(x_0,\underline{X}'_{x_0}))\right] \Leftrightarrow$$
$$\Leftrightarrow (\exists \underline{X}')A(P,\underline{X}'),$$

where $\alpha^+(x_0,\underline{X}'_{x_0})$ results from $\alpha^+(x_0,\underline{X})$ by replacing each subformula $X_j(\underline{z})$ by the subformula $X'_j(x_0,\underline{z})$ etc; the formula $A(P,\underline{X}')$ is classical. Similarly one finds a classical formula $B(R,\underline{Y}')$ such that $(\forall x_1x_2)(\beta(x_1,x_2) \equiv R(x_1,x_2)) \Leftrightarrow (\exists \underline{Y}')B(R,\underline{Y}')$. Consequently,

$$\Phi \Leftrightarrow (\exists P,R,\underline{X}',\underline{Y}',\underline{Z})(A(P,\underline{X}') \,\&\, B(R,\underline{Y}') \,\&\, \varkappa^+(P,R,\underline{Z}));$$

the formula $A(P,X') \,\&\, B(R,Y') \,\&\, \varkappa^+(P,R,Z)$ is classical and strictly projectively defines the class elementarily defined by Φ. Similarly, $A(P,\underline{X}') \& B(R,\underline{Y}') \& \varkappa^{-}(P,R,Z)$ strictly projectively defines the class elementarily defined by $\neg\Phi$. We have proved that the class elementarily defined by Φ is doubly strictly projective in m-$L^{obs}_{\omega\omega}$.

It remains to prove that the Souslin-Kleene interpolation theorem for doubly strictly projective classes holds. But observe that if a class K is strictly projective in L then K is strictly projective in m-$L^{obs}_{\omega\omega}$; the result then follows immediately.

References

1. Aho A.V., Hopcroft J.E. and Ullman J.D.: The design and analysis of computer algorithms, Addison-Wesley 1974
2. Asser G: Das Repräsentantenproblem im Prädikatenkalkül der ersten Stufe mit Identität, Zeitschr. f. Math. Logik 1 (1955) 252-263
3. Barwise K.J.: Axioms for abstract model theory, Annals of Math. Logic 7 (1974) 221-266
4. Fagin R.: Generalized first order spectra and polynomial-time recognizable sets, in: Complexity of Computations, SIAM-AMS Proc. vol. 7, ed. R. Karp 1974, 43-73
5. Fagin R.: Monadic generalized spectra, Zeitschrift f. Math. Logik 21 (1975) 89-96
6. Feferman S: Applications of many-sorted interpolation theorems, in: Proceedings of the Tarski symposium (has appeared)
7. Feferman S.: Two notes on abstract model thoery, Fundamenta Math. 82 (1974) 153-165 and 89 (1975) 111-180
8. Hájek P.: Generalized quantifiers and finite sets, in: Proc. of the autumn school in set theory and hierarchy theory Karpacz 1974 (to appear)
9. Hájek P.:On logics of discovery, in: Mathematical foundations of computer science 1975, Lecture Notes in Computer Science vol 32 ed. J. Bečvář, Springer-Verlag 1975, p. 15-30
10. Hájek P.: On semisets, in: Logic Colloquium '69, ed. R.O.Gandy and C.M.E.Yates, Studies in Logic vol. 61, North-Holland 1971, 67-
11. Hájek P.: Why semisets? Comment. Math. Univ. Carolinae 14 (1973), 397-420
12. Havránek T.: Statistical quantifiers in observational calculi, Theory and Decision 6 (1975) 213-230
13. Pudlák P.: The observational predicate calculus and complexity of computations, Comment. Math. Univ. Carolinae 16 (1975) 395-398

14. Pudlák P.: Generalized quantifiers and semisets, Proceedings of the autumn school in set theory and hierarchy theory Karpacz 1974 (to appear)

15. Shoenfield J.R.: Mathematical Logic, Addison-Wesley 1967

16. Vopěnka P. and Hájek P.: The theory of semisets, Studies in Logic vol. 70, North-Holland 1972

Vol. 399: Functional Analysis and its Applications. Proceedings 1973. Edited by H. G. Garnir, K. R. Unni and J. H. Williamson. II, 584 pages. 1974.

Vol. 400: A Crash Course on Kleinian Groups. Proceedings 1974. Edited by L. Bers and I. Kra. VII, 130 pages. 1974.

Vol. 401: M. F. Atiyah, Elliptic Operators and Compact Groups. V, 93 pages. 1974.

Vol. 402: M. Waldschmidt, Nombres Transcendants. VIII, 277 pages. 1974.

Vol. 403: Combinatorial Mathematics. Proceedings 1972. Edited by D. A. Holton. VIII, 148 pages. 1974.

Vol. 404: Théorie du Potentiel et Analyse Harmonique. Edité par J. Faraut. V, 245 pages. 1974.

Vol. 405: K. J. Devlin and H. Johnsbråten, The Souslin Problem. VIII, 132 pages. 1974.

Vol. 406: Graphs and Combinatorics. Proceedings 1973. Edited by R. A. Bari and F. Harary. VIII, 355 pages. 1974.

Vol. 407: P. Berthelot, Cohomologie Cristalline des Schémas de Caracteristique p > o. II, 604 pages. 1974.

Vol. 408: J. Wermer, Potential Theory. VIII, 146 pages. 1974.

Vol. 409: Fonctions de Plusieurs Variables Complexes, Séminaire François Norguet 1970–1973. XIII, 612 pages. 1974.

Vol. 410: Séminaire Pierre Lelong (Analyse) Année 1972–1973. VI, 181 pages. 1974.

Vol. 411: Hypergraph Seminar. Ohio State University, 1972. Edited by C. Berge and D. Ray-Chaudhuri. IX, 287 pages. 1974.

Vol. 412: Classification of Algebraic Varieties and Compact Complex Manifolds. Proceedings 1974. Edited by H. Popp. V, 333 pages. 1974.

Vol. 413: M. Bruneau, Variation Totale d'une Fonction. XIV, 332 pages. 1974.

Vol. 414: T. Kambayashi, M. Miyanishi and M. Takeuchi, Unipotent Algebraic Groups. VI, 165 pages. 1974.

Vol. 415: Ordinary and Partial Differential Equations. Proceedings 1974. XVII, 447 pages. 1974.

Vol. 416: M. E. Taylor, Pseudo Differential Operators. IV, 155 pages. 1974.

Vol. 417: H. H. Keller, Differential Calculus in Locally Convex Spaces. XVI, 131 pages. 1974.

Vol. 418: Localization in Group Theory and Homotopy Theory and Related Topics. Battelle Seattle 1974 Seminar. Edited by P. J. Hilton. VI, 172 pages. 1974.

Vol. 419: Topics in Analysis. Proceedings 1970. Edited by O. E. Lehto, I. S. Louhivaara, and R. H. Nevanlinna. XIII, 392 pages. 1974.

Vol. 420: Category Seminar. Proceedings 1972/73. Edited by G. M. Kelly. VI, 375 pages. 1974.

Vol. 421: V. Poénaru, Groupes Discrets. VI, 216 pages. 1974.

Vol. 422: J.-M. Lemaire, Algèbres Connexes et Homologie des Espaces de Lacets. XIV, 133 pages. 1974.

Vol. 423: S. S. Abhyankar and A. M. Sathaye, Geometric Theory of Algebraic Space Curves. XIV, 302 pages. 1974.

Vol. 424: L. Weiss and J. Wolfowitz, Maximum Probability Estimators and Related Topics. V, 106 pages. 1974.

Vol. 425: P. R. Chernoff and J. E. Marsden, Properties of Infinite Dimensional Hamiltonian Systems. IV, 160 pages. 1974.

Vol. 426: M. L. Silverstein, Symmetric Markov Processes. X, 287 pages. 1974.

Vol. 427: H. Omori, Infinite Dimensional Lie Transformation Groups. XII, 149 pages. 1974.

Vol. 428: Algebraic and Geometrical Methods in Topology. Proceedings 1973. Edited by L. F. McAuley. XI, 280 pages. 1974.

Vol. 429: L. Cohn, Analytic Theory of the Harish-Chandra C-Function. III, 154 pages. 1974.

Vol. 430: Constructive and Computational Methods for Differential and Integral Equations. Proceedings 1974. Edited by D. L. Colton and R. P. Gilbert. VII, 476 pages. 1974.

Vol. 431: Séminaire Bourbaki – vol. 1973/74. Exposés 436–452. IV, 347 pages. 1975.

Vol. 432: R. P. Pflug, Holomorphiegebiete, pseudokonvexe Gebiete und das Levi-Problem. VI, 210 Seiten. 1975.

Vol. 433: W. G. Faris, Self-Adjoint Operators. VII, 115 pages. 1975.

Vol. 434: P. Brenner, V. Thomée, and L. B. Wahlbin, Besov Spaces and Applications to Difference Methods for Initial Value Problems. II, 154 pages. 1975.

Vol. 435: C. F. Dunkl and D. E. Ramirez, Representations of Commutative Semitopological Semigroups. VI, 181 pages. 1975.

Vol. 436: L. Auslander and R. Tolimieri, Abelian Harmonic Analysis, Theta Functions and Function Algebras on a Nilmanifold. V, 99 pages. 1975.

Vol. 437: D. W. Masser, Elliptic Functions and Transcendence. XIV, 143 pages. 1975.

Vol. 438: Geometric Topology. Proceedings 1974. Edited by L. C. Glaser and T. B. Rushing. X, 459 pages. 1975.

Vol. 439: K. Ueno, Classification Theory of Algebraic Varieties and Compact Complex Spaces. XIX, 278 pages. 1975.

Vol. 440: R. K. Getoor, Markov Processes: Ray Processes and Right Processes. V, 118 pages. 1975.

Vol. 441: N. Jacobson, PI-Algebras. An Introduction. V, 115 pages. 1975.

Vol. 442: C. H. Wilcox, Scattering Theory for the d'Alembert Equation in Exterior Domains. III, 184 pages. 1975.

Vol. 443: M. Lazard, Commutative Formal Groups. II, 236 pages. 1975.

Vol. 444: F. van Oystaeyen, Prime Spectra in Non-Commutative Algebra. V, 128 pages. 1975.

Vol. 445: Model Theory and Topoi. Edited by F. W. Lawvere, C. Maurer, and G. C. Wraith. III, 354 pages. 1975.

Vol. 446: Partial Differential Equations and Related Topics. Proceedings 1974. Edited by J. A. Goldstein. IV, 389 pages. 1975.

Vol. 447: S. Toledo, Tableau Systems for First Order Number Theory and Certain Higher Order Theories. III, 339 pages. 1975.

Vol. 448: Spectral Theory and Differential Equations. Proceedings 1974. Edited by W. N. Everitt. XII, 321 pages. 1975.

Vol. 449: Hyperfunctions and Theoretical Physics. Proceedings 1973. Edited by F. Pham. IV, 218 pages. 1975.

Vol. 450: Algebra and Logic. Proceedings 1974. Edited by J. N. Crossley. VIII, 307 pages. 1975.

Vol. 451: Probabilistic Methods in Differential Equations. Proceedings 1974. Edited by M. A. Pinsky. VII, 162 pages. 1975.

Vol. 452: Combinatorial Mathematics III. Proceedings 1974. Edited by Anne Penfold Street and W. D. Wallis. IX, 233 pages. 1975.

Vol. 453: Logic Colloquium. Symposium on Logic Held at Boston, 1972–73. Edited by R. Parikh. IV, 251 pages. 1975.

Vol. 454: J. Hirschfeld and W. H. Wheeler, Forcing, Arithmetic, Division Rings. VII, 266 pages. 1975.

Vol. 455: H. Kraft, Kommutative algebraische Gruppen und Ringe. III, 163 Seiten. 1975.

Vol. 456: R. M. Fossum, P. A. Griffith, and I. Reiten, Trivial Extensions of Abelian Categories. Homological Algebra of Trivial Extensions of Abelian Categories with Applications to Ring Theory. XI, 122 pages. 1975.

Vol. 457: Fractional Calculus and Its Applications. Proceedings 1974. Edited by B. Ross. VI, 381 pages. 1975.

Vol. 458: P. Walters, Ergodic Theory – Introductory Lectures. VI, 198 pages. 1975.

Vol. 459: Fourier Integral Operators and Partial Differential Equations. Proceedings 1974. Edited by J. Chazarain. VI, 372 pages. 1975.

Vol. 460: O. Loos, Jordan Pairs. XVI, 218 pages. 1975.

Vol. 461: Computational Mechanics. Proceedings 1974. Edited by J. T. Oden. VII, 328 pages. 1975.

Vol. 462: P. Gérardin, Construction de Séries Discrètes p-adiques. »Sur les séries discrètes non ramifiées des groupes réductifs déployés p-adiques«. III, 180 pages. 1975.

Vol. 463: H.-H. Kuo, Gaussian Measures in Banach Spaces. VI, 224 pages. 1975.

Vol. 464: C. Rockland, Hypoellipticity and Eigenvalue Asymptotics. III, 171 pages. 1975.

Vol. 465: Séminaire de Probabilités IX. Proceedings 1973/74. Edité par P. A. Meyer. IV, 589 pages. 1975.

Vol. 466: Non-Commutative Harmonic Analysis. Proceedings 1974. Edited by J. Carmona, J. Dixmier and M. Vergne. VI, 231 pages. 1975.

Vol. 467: M. R. Essén, The Cos $\pi\lambda$ Theorem. With a paper by Christer Borell. VII, 112 pages. 1975.

Vol. 468: Dynamical Systems – Warwick 1974. Proceedings 1973/74. Edited by A. Manning. X, 405 pages. 1975.

Vol. 469: E. Binz, Continuous Convergence on C(X). IX, 140 pages. 1975.

Vol. 470: R. Bowen, Equilibrium States and the Ergodic Theory of Anosov Diffeomorphisms. III, 108 pages. 1975.

Vol. 471: R. S. Hamilton, Harmonic Maps of Manifolds with Boundary. III, 168 pages. 1975.

Vol. 472: Probability-Winter School. Proceedings 1975. Edited by Z. Ciesielski, K. Urbanik, and W. A. Woyczyński. VI, 283 pages. 1975.

Vol. 473: D. Burghelea, R. Lashof, and M. Rothenberg, Groups of Automorphisms of Manifolds. (with an appendix by E. Pedersen) VII, 156 pages. 1975.

Vol. 474: Séminaire Pierre Lelong (Analyse) Année 1973/74. Edité par P. Lelong. VI, 182 pages. 1975.

Vol. 475: Répartition Modulo 1. Actes du Colloque de Marseille-Luminy, 4 au 7 Juin 1974. Edité par G. Rauzy. V, 258 pages. 1975.

Vol. 476: Modular Functions of One Variable IV. Proceedings 1972. Edited by B. J. Birch and W. Kuyk. V, 151 pages. 1975.

Vol. 477: Optimization and Optimal Control. Proceedings 1974. Edited by R. Bulirsch, W. Oettli, and J. Stoer. VII, 294 pages. 1975.

Vol. 478: G. Schober, Univalent Functions – Selected Topics. V, 200 pages. 1975.

Vol. 479: S. D. Fisher and J. W. Jerome, Minimum Norm Extremals in Function Spaces. With Applications to Classical and Modern Analysis. VIII, 209 pages. 1975.

Vol. 480: X. M. Fernique, J. P. Conze et J. Gani, Ecole d'Eté de Probabilités de Saint-Flour IV-1974. Edité par P.-L. Hennequin. XI, 293 pages. 1975.

Vol. 481: M. de Guzmán, Differentiation of Integrals in R^n. XII, 226 pages. 1975.

Vol. 482: Fonctions de Plusieurs Variables Complexes II. Séminaire François Norguet 1974–1975. IX, 367 pages. 1975.

Vol. 483: R. D. M. Accola, Riemann Surfaces, Theta Functions, and Abelian Automorphisms Groups. III, 105 pages. 1975.

Vol. 484: Differential Topology and Geometry. Proceedings 1974. Edited by G. P. Joubert, R. P. Moussu, and R. H. Roussarie. IX, 287 pages. 1975.

Vol. 485: J. Diestel, Geometry of Banach Spaces – Selected Topics. XI, 282 pages. 1975.

Vol. 486: S. Stratila and D. Voiculescu, Representations of AF-Algebras and of the Group $U(\infty)$. IX, 169 pages. 1975.

Vol. 487: H. M. Reimann und T. Rychener, Funktionen beschränkter mittlerer Oszillation. VI, 141 Seiten. 1975.

Vol. 488: Representations of Algebras, Ottawa 1974. Proceedings 1974. Edited by V. Dlab and P. Gabriel. XII, 378 pages. 1975.

Vol. 489: J. Bair and R. Fourneau, Etude Géométrique des Espaces Vectoriels. Une Introduction. VII, 185 pages. 1975.

Vol. 490: The Geometry of Metric and Linear Spaces. Proceedings 1974. Edited by L. M. Kelly. X, 244 pages. 1975.

Vol. 491: K. A. Broughan, Invariants for Real-Generated Uniform Topological and Algebraic Categories. X, 197 pages. 1975.

Vol. 492: Infinitary Logic: In Memoriam Carol Karp. Edited by D. W. Kueker. VI, 206 pages. 1975.

Vol. 493: F. W. Kamber and P. Tondeur, Foliated Bundles and Characteristic Classes. XIII, 208 pages. 1975.

Vol. 494: A Cornea and G. Licea. Order and Potential Resolvent Families of Kernels. IV, 154 pages. 1975.

Vol. 495: A. Kerber, Representations of Permutation Groups II. V, 175 pages. 1975.

Vol. 496: L. H. Hodgkin and V. P. Snaith, Topics in K-Theory. Two Independent Contributions. III, 294 pages. 1975.

Vol. 497: Analyse Harmonique sur les Groupes de Lie. Proceedings 1973–75. Edité par P. Eymard et al. VI, 710 pages. 1975.

Vol. 498: Model Theory and Algebra. A Memorial Tribute to Abraham Robinson. Edited by D. H. Saracino and V. B. Weispfenning. X, 463 pages. 1975.

Vol. 499: Logic Conference, Kiel 1974. Proceedings. Edited by G. H. Müller, A. Oberschelp, and K. Potthoff. V, 651 pages 1975.

Vol. 500: Proof Theory Symposion, Kiel 1974. Proceedings. Edited by J. Diller and G. H. Müller. VIII, 383 pages. 1975.

Vol. 501: Spline Functions, Karlsruhe 1975. Proceedings. Edited by K. Böhmer, G. Meinardus, and W. Schempp. VI, 421 pages. 1976.

Vol. 502: János Galambos, Representations of Real Numbers by Infinite Series. VI, 146 pages. 1976.

Vol. 503: Applications of Methods of Functional Analysis to Problems in Mechanics. Proceedings 1975. Edited by P. Germain and B. Nayroles. XIX, 531 pages. 1976.

Vol. 504: S. Lang and H. F. Trotter, Frobenius Distributions in GL_2-Extensions. III, 274 pages. 1976.

Vol. 505: Advances in Complex Function Theory. Proceedings 1973/74. Edited by W. E. Kirwan and L. Zalcman. VIII, 203 pages. 1976.

Vol. 506: Numerical Analysis, Dundee 1975. Proceedings. Edited by G. A. Watson. X, 201 pages. 1976.

Vol. 507: M. C. Reed, Abstract Non Linear Wave Equations. VI, 128 pages. 1976.

Vol. 508: E. Seneta, Regularly Varying Functions. V, 112 pages. 1976.

Vol. 509: D. E. Blair, Contact Manifolds in Riemannian Geometry. VI, 146 pages. 1976.

Vol. 510: V. Poénaru, Singularités C^∞ en Présence de Symétrie. V, 174 pages. 1976.

Vol. 511: Séminaire de Probabilités X. Proceedings 1974/75. Edité par P. A. Meyer. VI, 593 pages. 1976.

Vol. 512: Spaces of Analytic Functions, Kristiansand, Norway 1975. Proceedings. Edited by O. B. Bekken, B. K. Øksendal, and A. Stray. VIII, 204 pages. 1976.

Vol. 513: R. B. Warfield, Jr. Nilpotent Groups. VIII, 115 pages. 1976.

Vol. 514: Séminaire Bourbaki vol. 1974/75. Exposés 453 – 470. IV, 276 pages. 1976.

Vol. 515: Bäcklund Transformations. Nashville, Tennessee 1974. Proceedings. Edited by R. M. Miura. VIII, 295 pages. 1976.